W9-CPO-107

Statistics for Engineering
and Information Science

Series Editors
M. Jordan, S.L. Lauritzen, J.F. Lawless, V. Nair

Springer
New York
Berlin
Heidelberg
Barcelona
Hong Kong
London
Milan
Paris
Singapore
Tokyo

Statistics for Engineering and Information Science

Akaike and Kitagawa: The Practice of Time Series Analysis.
Cowell, Dawid, Lauritzen, and Spiegelhalter: Probabilistic Networks and Expert Systems.
Fine: Feedforward Neural Network Methodology.
Hawkins and Olwell: Cumulative Sum Charts and Charting for Quality Improvement.

Robert G. Cowell
A. Philip Dawid
Steffen L. Lauritzen
David J. Spiegelhalter

Probabilistic Networks and Expert Systems

With 45 Illustrations

Springer

Robert G. Cowell
School of Mathematics, Actuarial Science, and Statistics
City University, London
London EC1V 0HB
United Kingdom
r.g.cowell@city.ac.uk

A. Philip Dawid
Department of Statistical Science
University College London
London WC1E 6BT
United Kingdom
dawid@stats.ucl.ac.uk

Steffen L. Lauritzen
Department of Mathematical Sciences
Aalborg University
DK-9220 Aalborg
Denmark
steffen@math.auc.dk

David J. Spiegelhalter
MRC Biostatistics Unit
Institute of Public Health
Cambridge CB2 2SR
United Kingdom
david.spiegelhalter@mrc-bsu.cam.ac

Series Editors

Michael Jordan
Department of Computer Science
University of California, Berkeley
Berkeley, CA 94720
USA

Steffen L. Lauritzen
Department of Mathematical Sciences
Aalborg University
DK-9220 Aalborg
Denmark

Jerald F. Lawless
Department of Statistics
University of Waterloo
Waterloo, Ontario N2L 3G1
Canada

Vijay Nair
Department of Statistics
University of Michigan
Ann Arbor, MI 48109
USA

Library of Congress Cataloging-in-Publication Data
Probabilistic networks and expert systems/Robert G. Cowell . . . [et al.].
p. cm. — (Statistics for engineering and information science)
Includes bibliographical references and index.
ISBN 0-387-98767-3 (hardcover: alk. paper)
1. Expert systems (Computer science) 2. Probabilities.
I. Cowell, Robert G. II. Series.
QA76.76.E95P755 1999
006.3′3—dc21 99-13242

Printed on acid-free paper.

Production managed by Frank McGuckin; manufacturing supervised by Jacqui Ashri.
Camera-ready copy prepared from the authors' LaTeX files.
Printed and bound by R.R. Donnelley and Sons, Harrisonburg, VA.
Printed in the United States of America.

9 8 7 6 5 4 3 2 1

ISBN 0-387-98767-3 Springer-Verlag New York Berlin Heidelberg SPIN 10712463

Preface

This book arises out of a long-standing collaboration between the authors, which began in 1985 when a subset of us discussed the potential connection between graphical modelling in contingency tables and the type of diagrams being used to represent qualitative knowledge in expert systems. Our various enthusiasms for multivariate analysis, Bayesian statistics, computer algorithms, conditional independence, graph theory, decision-support systems, and so on, have since found a common area of application in probabilistic networks for expert systems, and we have been fortunate to enjoy a long and fruitful period of joint work which has been little hindered by the intervening North Sea.

Over this time we have benefited greatly from interactions with a range of researchers who are too numerous to list individually, although for both their scholarly insights and personal good company we must mention David Heckerman, Judea Pearl, Glenn Shafer, Prakash Shenoy, Jim Smith, Joe Whittaker, and the Danes in the ODIN group, particularly Stig Andersen, Finn Jensen, Frank Jensen, Uffe Kjærulff, and Kristian Olesen.

Financial support has been received at various times from the European Community SCIENCE fund, the UK Engineering and Physical Science Research Council, Glaxo Pharmaceuticals, the Danish Research Councils through the PIFT programme, and all our home institutions.

Finally, we would like to thank our families for putting up (usually in good humour) with the numerous meetings, visits, and late nights that this collaboration has entailed.

RGC *City University, London, UK*
APD *University College London, UK*
SLL *Aalborg University, Denmark*
DJS *MRC Biostatistics Unit, Cambridge, UK*

Contents

1
Introduction

1.1 What is this book about?

The term *expert system* is perhaps an anachronism in 1999, but is a convenient label for a computer program intended to provide reasoned guidance on some complex but fairly tightly delineated task. The appropriate way of dealing with the many sources of uncertainty in such problems has long been a matter of dispute, but since the late 1980s, particularly following the publication of Judea Pearl's classic text (Pearl 1988), a fully probabilistic approach has steadily gained acceptance.

The crucial realization of Pearl and others was that 'brute force' probabilistic manipulations in high-dimensional problems could never become either technically feasible or substantively acceptable, and that the path ahead was to find some way of introducing 'modularity', so enabling a large and complex model, and its associated calculations, to be split up into small manageable pieces. The best way to do this turns out to be through the imposition of meaningful simplifying conditional independence assumptions. These, in turn, can be expressed by means of a powerful and appealing graphical representation, and the resulting networks are often termed *Bayesian networks*, although in this book we prefer the term *probabilistic networks*, reflecting an increased generality in the representations we consider. Such graphs not only provide an attractive means for modelling and communicating complex structures, but also form the basis for efficient algorithms, both for propagating evidence and for learning about parameters.

Over the last ten years there has been a steady shift in the focus of attention from algorithms for propagating evidence towards methods for learning parameters and structure from data. This has been accompanied by a broadening of scope, into the general area of *graphical modelling*: a term with its roots in Statistics, but which also incorporates neural networks, hidden Markov models, and many other techniques that exploit conditional independence properties for modelling, display, and computation. The range of researchers now involved in this field, from Computer Science, Engineering, Statistics and the Social Sciences, amongst others, has ensured an exciting and diverse research environment. Nevertheless, we believe that there is still a particular contribution to be made from the statistical perspective emphasizing the generality of the concepts and attempting to place them on a rigorous footing.

1.2 What is in this book?

The content of this book arose, to a considerable extent, from the collaborative research work of the authors, here brought together in a coherent and consistent format. The book divides into three broad sections. Chapters 2 to 5 cover the basic ideas, describing the problems that can be tackled by this approach, the kinds of structures that can be handled, what the algorithms do and how they work, and the mathematical definitions and background required. In Chapters 6 to 8, which form the heart of the book, the main evidence propagation algorithms and their properties are rigorously described and justified. Chapters 9 to 11 deal with various issues around learning and model criticism in the light of observed data, and include some previously unpublished results.

Chapter 2, after briefly describing the history of uncertainty management in expert systems, presents simple examples to illustrate the ability of probabilistic networks to provide an intuitive yet rigorous framework for revising uncertainty in the light of new evidence. Chapter 3 discusses the determination of the qualitative and quantitative ingredients of such networks in real problems and illustrates how the graphical structure itself provides the basis for the evidence propagation algorithm. The illustrations make use of various concepts and results of graph theory and conditional independence, which are fully detailed in Chapters 4 and 5 respectively. These chapters provide a central reference for definitions and properties relating to directed and undirected graphs, and to their hybrid, chain graphs; and describe how any of these may be manipulated into a decomposable undirected graph, the fundamental graphical object on which the computational algorithms operate.

Chapter 6 deals with propagation methods for networks of discrete variables and presents the central unifying algorithm, operating by means of flows of information between adjacent sets of variables arranged in a structure known as a *junction tree*. The generality of the approach is illustrated by its application to a variety of tasks beyond the simple calculation of marginal and conditional probabilities, such as finding the most probable configuration of all the variables. Detailed numerical examples are given. Chapter 7 extends this theory to networks in which some of the variables are continuous, and Chapter 8 illustrates how it can be applied to decision problems involving chain graphs.

Up to this point in the book it has been assumed that the network is precisely specified, with no revision to its qualitative or probabilistic structure after observing data. It is, however, of practical importance to devise methods for adapting these aspects of the model in the light of further information. Chapter 9 presents some exact and approximate methods for learning about unknown probabilities in a network, for both complete and incomplete data. Chapter 10 confronts the important issue of criticizing a network in the light of observed data by establishing systems of monitors to identify failings of various aspects of the model structure. Chapter 11 deals with some aspects of the currently active topic of learning about the graphical structure of a problem from data. This chapter differs from the others in that few technical results are established; instead, it is hoped that the structured outline of and guide to the current literature will be of help to understanding this rapidly developing field.

Each chapter contains a guide to further reading. The book concludes with three appendices: on Bayesian conjugate analysis of discrete data, on stochastic simulation and Gibbs sampling, and on information and software available on the World Wide Web.

1.3 What is not in this book?

The area of study which this book treats is a highly active one, and we can not even attempt to do justice to all the extensive and fast-moving new developments. We have deliberately restricted ourselves to what we consider the central core of the subject, the theory and applications of exact methods in probabilistic networks, although at a number of points we do introduce both analytic and simulation-based approximations. This book therefore makes no claim to be a comprehensive guide to the whole area of graphical modelling. In particular, it does not deal in detail with such topics as maximum likelihood estimation in general graphical models, Monte Carlo methods, neural network learning, pattern recognition applications, or mean-field approximations. For this broader perspective, see, for example, Lauritzen (1996), Gilks et al. (1996), Neal (1996), Frey (1998), and Jordan (1998). The last is a particularly good guide to the wide variety of models that now shelter under the graphical umbrella.

1.4 How should this be book be used?

This book is not intended to be read linearly. It may profitably be read in various ways by different audiences. Researchers already knowledgeable in probabilistic expert systems may want to concentrate on the technical results contained in Chapters 6, 7, 8, 9, and 10, while newcomers to the area, or those seeking an overview of developments, could focus on the more descriptive material in Chapters 2, 3, parts of 9 and 10, and 11. Those interested in expert systems but unconcerned with learning algorithms could read Chapters 2, 3, 6, 7, 8. Chapters 4 and 5 can be used as references for important definitions and results or for individual study in their own right.

2
Logic, Uncertainty, and Probability

In this chapter we discuss characteristics of expert systems that relate to their ability to deal with the all-pervasive problem of uncertainty. We begin by describing one of the earliest approaches to the computer-based representation of expert knowledge, so called *rule-based* systems. The limitations of such systems when faced with uncertainty, and some of the alternatives that have been proposed in the literature, are highlighted. We then focus on the probabilistic representation of uncertainty, emphasizing both its strong theoretical basis and its possibility of a subjective interpretation. Bayes' theorem then forms the fundamental tool for belief revision, and 'Bayesian networks' can be formed by superimposing a probability model on a graph representing qualitative conditional independence assumptions. The resulting structure is capable of representing a wide range of complex domains.

Here we can only give a brief informal overview of the background to probabilistic expert systems; further reading material is indicated at the end of the chapter. The World Wide Web is a major resource in such a rapidly changing area; addresses of specific sites for information and software are given in Appendix C.

2.1 What is an expert system?

The *Concise Oxford English Dictionary* defines *expert* as "person having special skill or knowledge." Informally, an expert is someone you turn to when you are faced with a problem that is too difficult for you to solve on your own or that is outside your own particular areas of specialized knowledge, and whom you trust to reach a better solution to your problem than you could by yourself. Expert systems are attempts to crystallize and codify the knowledge and skills of one or more experts into a tool that can be used by non-specialists. Usually this will be some form of computer program, but this need not be the case.

An *expert system* consists of two parts, summed up in the equation:

$$\textbf{Expert System} = \textbf{Knowledge Base} + \textbf{Inference Engine.}$$

The *knowledge base* contains the domain-specific knowledge of a problem, encoded in some manner. The *inference engine* consists of one or more algorithms for processing the encoded knowledge of the knowledge base together with any further specific information at hand for a given application. Both parts are important for an expert system. Modern expert systems strive for the ideal of a clean separation of both components. This allows the knowledge base to be improved in the light of further information, and facilitates learning from the experience of making mistakes.

The knowledge base is the core of an expert system; no matter how sophisticated the inference procedures are for manipulating the knowledge in a knowledge base, if the content of the knowledge base is poor then the inferences will be correspondingly poor. Nevertheless it is vital to have a good inference engine to take full advantage of the knowledge base.

2.2 Diagnostic decision trees

A *diagnostic decision tree* (also known as a *classification tree*, *flowchart*, or *algorithm*) is a structured sequence of questions in which each response determines the next question to ask. The inference process involves simply walking through the algorithm, selecting the appropriate path from the answers to the questions contained in the nodes. The system encodes the expert knowledge in the order and form in which the questions are structured. At a certain stage a diagnosis or conclusion is reached.

An example of part of a diagnostic decision tree is shown in Figure 2.1. The background is as follows. The Great Ormond Street Hospital for Sick Children in London (here abbreviated to GOS) acts as a referral centre for newborn babies with congenital heart disease. Early appropriate treatment is essential, and a preliminary diagnosis must be reached using information reported over the telephone. This may concern clinical signs, blood gases, ECG, and X-ray. A decision tree, intended to help the junior doctors in GOS, was constructed from expert judgement. It contained 66 nodes, and discriminated 27 diagnostic categories in neonates, including lung disease masquerading as heart disease. It was developed and evaluated on 400 cases (Franklin et al. 1991).

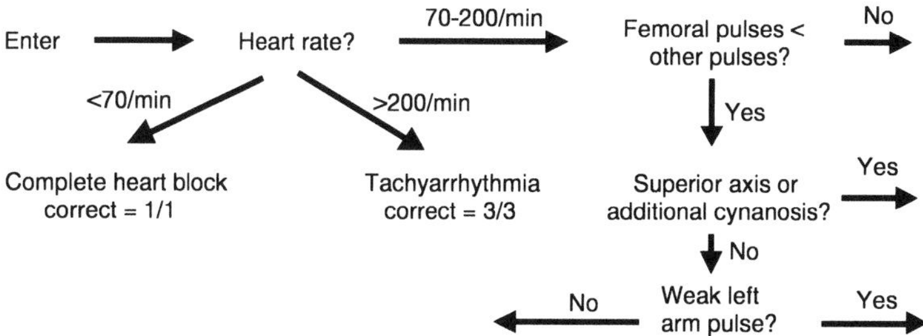

FIGURE 2.1. Initial part of Great Ormond Street diagnosis decision tree for diagnosing problems in newborn babies. The first question is *Heart rate?*, and, depending on the answer, one of three paths is chosen. For example, if the heart rate is greater than 200 beats a minute, an immediate diagnosis of *Tachyarrhythmia* is made. The *correct* = *3/3* in the figure indicates that in the available database there were 3 cases that went down this path, all of which actually had a correct final diagnosis of Tachyarrhythmia.

A classification tree does not necessarily require a computer for implementation and is generally easy to explain and use. If it performs badly for a particular case, it is usually possible to pinpoint where the wrong branch was taken. However, despite their appealing nature, classification trees suffer from some drawbacks. An incorrect conclusion can be reached after a *single* unexpected response, due for example to observer error. They are inflexible with respect to missing information. Typically default responses are assumed when a question cannot be answered; for example, in the GOS algorithm the default is to assume a negative response where data are missing. Such systems usually provide little opportunity for adaptation as data become available. We might interpret their faults as stemming from the lack of separation of the knowledge base and the inference engine, leading to a rigid non-modular system.

2.3 Production systems

A more flexible type of expert system is the *production system*, also called *rule-based system*. Such a system has its origin in the attempt to perform *symbolic reasoning* using logical rules. Generally, in a rule-based system, domain knowledge is encapsulated by a collection of implications, called *production rules*, having the form: IF (A_1 & A_2 & ... & A_k) THEN B; where $\{A_i\}$ are assertions and B may be an assertion or action. The following are examples of production rules (taken from Winston (1984)).

- IF the animal has hair THEN it is a mammal.
- IF the animal gives milk THEN it is a mammal.
- IF the animal has feathers THEN it is a bird.
- IF the animal flies & it lays eggs THEN it is a bird.

Since an assertion A_i may itself be a consequence of these modular rules, chains of reasoning are established. A trace through such a chain provides a degree of explanation for a particular case under consideration.

The collection of rules forms a modular knowledge base in that it is possible easily to add further rules if desired. Although there is a reliance on logical reasoning, the questions or rules do not need to be applied in a predetermined and inflexible manner, as in classification trees. Computer programs can be written, for example in languages such as LISP or Prolog, which manipulate such symbolic production rules and logic (see Lucas and van der Gaag (1991) for examples). The inference engine is embodied as a control mechanism in the program, which can select rules relevant to the particular case under consideration and suggest additional assertions that, if true, could be useful. It can also make valid deductions from a given set of assertions, a process called *forward chaining*, and perform the reverse operation to determine whether assertions exist that can validate a conjectured property (*backward chaining*).

However, there are problems with production systems. They focus on specific *assertions*, rather than *questions* with a choice of answer. They do not automatically distinguish "found to be false" and "not found to be true" (for example, the question may not be asked). The application of the laws of logic seems somewhat incomplete, particularly with respect to negation: for example, they can go from A with $A \rightarrow B$ to B, but the reverse, $\bar{B}$ with $A \rightarrow B$ may not necessarily lead to $\bar{A}$. The number of rules can grow enormously, and it is necessary to confirm consistency and eliminate redundancy. Exceptions to the rules have to be dealt with (for example penguins are birds that do not fly). Finally, the actual chains of reasoning may become too complex to comprehend.

2.4 Coping with uncertainty

Originally, production systems involved only *logical* deductions. Although this can be adequate for representing complex but determinate structures such as legislation, some problems of a general nature arise. In particular, the data available on an individual of interest may be inadequate or insufficiently reliable to enable a conclusion to be reached, or the production rules themselves may not be logically certain.

To deal with such situations, we need to quantify *uncertainty* in the conclusions. An early attempt by the artificial intelligence (AI) community concentrated on the logical certainty of the production rules, attaching a numerical value called a *certainty factor* (CF) to each production rule. For example, a system for medical diagnosis might have productions of the form:

- IF headache & fever THEN influenza (certainty 0.7)
- IF influenza THEN sneezing (certainty 0.9)
- IF influenza THEN weakness (certainty 0.6)

An early example of a backward chaining system with certainty factors is the MYCIN program (Shortliffe and Buchanan 1975), designed to assist doctors in prescribing treatment for bacteriological blood disorders. It employs about 300 productions and was the first system to separate its knowledge base from its inference engine.

It is still possible that one production can trigger others in a chain. However, with the additional numerical structure this requires that the certainty factors associated with such a chain be combined in some manner. It may also happen that two or more different productions yield the identical assertion or action, and then the various certainty factors again have to be combined in some manner. Thus, there arises the need for an *algebra* or *calculus* of certainty factors, as illustrated in Figure 2.2. To postulate or develop a plausible calculus requires some interpretation to be given to the meaning of the numbers.

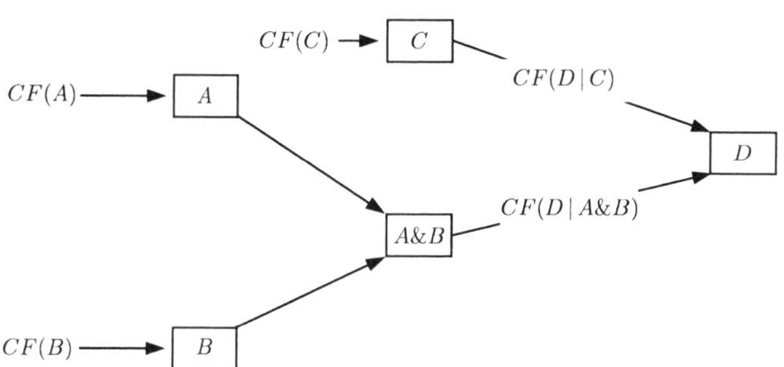

FIGURE 2.2. Combining certainty factors: How do $CF(A)$, $CF(B)$, $CF(C)$, $CF(D \mid C)$, and $CF(D \mid A \,\&\, B)$ combine to yield $CF(D \mid A \,\&\, B \,\&\, C)$?

Certainty factors can be, but have typically not been, regarded as statements of conditional probability. Although this may seem an appealing interpretation, there can be major consistency problems with this interpretation. This is because an arbitrary set of such production rules might not be compatible with any overall probability distribution, and if it is, that distribution might not be unique. Also, while 'IF A THEN B' is logically equivalent to 'IF $\bar{B}$ THEN $\bar{A}$', it is generally false that '$P(B \mid A) = q$' implies '$P(\bar{A} \mid \bar{B}) = q$'. We thus see that the desire for large modular systems made up of many smaller components or productions, together with local combination of certainty factors, appears to argue against a probabilistic interpretation of the certainty factors. This impression led to probability theory being abandoned by most of the AI community. Instead other ad hoc rules for manipulating and combining certainty factors were developed, or alternative measures of uncertainty were developed or applied, that allowed modularity to be retained, for example *fuzzy logic* (Zadeh 1983) and *belief functions* (Dempster 1967; Shafer 1976).

However, in a detailed examination of the MYCIN system, Heckerman (1986) showed that the "original definition of certainty factors is inconsistent with the functions used in MYCIN to combine the quantities." By redefining the interpretation of certainty factors he established a connection with probability theory, specifically that certainty factors can be interpreted as monotone functions of likelihood ratios. Furthermore, he showed that consistency can only be maintained or satisfied in tree-like structures.

Other reasons why probability theory has been proclaimed useless for expert systems are: first, that it is irrelevant because the uncertainty in the knowledge that is being represented does not match that of a conceptual chance mechanism underlying an observable event; secondly, that if a system is to be judged by rank order of hypotheses then a non-probabilistic calculus may be adequate; and thirdly, that a full probability distribution over many quantities would require assessment of too many numbers. We now give a short overview of how these perceived barriers were overcome.

2.5 The naïve probabilistic approach

In probabilistic terms the basic problem and solution can be stated as follows. We have a collection of unknown quantities $(A, B, \dots)$, we observe the true values for a subset of these, and we wish to derive appropriate expressions of uncertainty about the others. The solution, in principle, is quite simple. We need a *joint distribution* over all the unknown quantities, in which a probability is assigned to each possible combination of values. Then we must use the laws of probability to *condition* on the discovered facts, and hence obtain the appropriate conditional probability of those quantities that are still unknown. However, this simple solution has the following snags:

- Generally many probability assignments will be required to form the joint distribution.
- There is no modularity — just one huge computation.

- The approach is non-intuitive and lacks explanatory power.

Up to the early 1980s these problems made the application of probability theory appear infeasible in expert systems, but subsequent theoretical developments have managed to overcome or mitigate these problems and to address other conceptual concerns of the early workers in the AI community over the use of probability theory. This book deals with developments in probabilistic networks that address these and other points. The first of these is the introduction of the subjectivist Bayesian interpretation of probability into the AI context.

2.6 Interpretations of probability

The interpretation of probability continues to be a subject of intense debate, with important implications for the practice of probability modelling and statistical inference, both in general and in expert systems applications. One major division (Gillies 1994) is between *objective* and *epistemological* understandings of $P(A)$, the probability of an event A — or, more generally, of $P(A \mid B)$, the probability of A conditional on the happening of another event B.

Objective theories regard such probabilities as real-world attributes of the events they refer to, unrelated to and unaffected by the extent of our knowledge. Popper's *propensity theory* (Popper 1959), for example, in which probability measures an innate disposition of an event to occur in identified circumstances, is one such objective theory. The most influential objective interpretation has been the *frequentist* interpretation (Venn 1866; von Mises 1939), in which probabilities of events are defined as limiting proportions in an infinite ensemble or sequence of experiments. This has been the dominant interpretation of probability for most of this century and forms the basis of the influential frequentist approach to statistical inference, as developed by Neyman and Pearson (1967). However, because it only allows probabilities to be meaningfully assigned to outcomes of strictly repeatable situations and takes an uncompromising physical view of their nature, its scope is severely limited. It was the adoption of this specific interpretation by the early AI pioneers that led to the perception that there were fundamental conceptual obstacles to the incorporation of probability theory into expert systems.

Epistemological theories eschew possibly problematic 'true probabilities', instead regarding $P(A \mid B)$ as describing a state of mental uncertainty about A in the knowledge of B, where now A and B can be singular propositions, as opposed to repeatable events. From this viewpoint the probability calculus can be considered a generalization of Boolean logic (which historically came later), allowing numerical quantification of uncertainty about propositions, and describing how such numerical uncertainties should combine and change in the light of new information. Epistemological theories further divide into *logical* theories (Keynes 1921; Jeffreys 1939; Carnap 1950) and *subjectivist* theories. Logical theories posit the existence of a unique rational numerical degree of uncertainty about a proposition A in the light of information B. However, attractive though this viewpoint may be, no satisfactory theory or method for the evaluation of logical probabilities has yet been devised. In recent years the subjectivist interpretation has become popular. This does not impose any particular numerical evaluation of probabilities, but merely requires that all the probability assessments an individual makes should 'cohere' appropriately. In such a theory, $P(A)$ is a numerical measure of a particular person's subjective degree of belief in A, with probability 1 representing certain belief in the truth of A, and probability 0 expressing certainty that A is false. Thus, it is more appropriate to think of $P(A)$ as representing the probability *for* A — a characteristic of both A and the person whose probability it is. We can measure a person's subjective probabilities $P(A)$ or $P(A \mid B)$ either directly by offering bets at various odds, or indirectly by observing the subject's behaviour in situations whose uncertain consequences depend on the actions he or she takes.

From a subjectivist standpoint, it is possible to assign probabilities to individual propositions, or to treat unknown constants or parameters as random variables, even though there may be no physical stochastic mechanism at work. For example, a person could assert "My probability that the Suez canal is longer than the Panama canal is 0.2." Clearly, such a subjective probability must be relative to that person's degree of background knowledge or information. Direct frequentist interpretation of such a probability statement is not possible: there is no stochastic element or ensemble of relevant repetitions, and so no basis for assigning a frequentist probability (other than 0 or 1) to the content of the statement. However, when many subjective probability assessments are made, they do have implications for the behaviour of certain real-world frequencies, which can be investigated empirically (see Dawid (1986)).

Many authors have sought to justify a subjectivist interpretation and calculus of probability from more basic axiomatic foundations. The influential approaches of Ramsey (1926), de Finetti (1937), and Savage (1954) (see also de Finetti (1975), Savage (1971), and Lindley (1982)) are based on a decision-theoretic interpretation of subjective probability as a determinant of action, and a principle of *coherence*, which requires that an individual should not make a collection of probability assessments that could put him in the position of suffering a sure loss, no matter how the relevant uncertain events turn out. It can then be shown that coherence is attained if and only if the probability assessments satisfy the standard probability axioms (see Section 2.7).

Artificial Intelligence researchers often refer to Cox (1946), which was based on a logical interpretation of probability, but applies equally to a subjectivist one. Cox asked: if one agrees that it is possible to assign numerical values to represent degrees of rational belief in a set of propositions, how should such values combine? He assumed, for instance, the existence of some function F such that $P(C \cap B \mid A) = F\{P(C \mid A \cap B), P(B \mid A)\}$ for any three propositions A, B, and C. He then showed that there must exist a transformation of the initial numerical belief values to values in the real interval [0,1], such that the transformed values combine according to the rules of the probability calculus. Cox's paper required certain differentiability assumptions, but these have been relaxed to assumptions of continuity only (Aczél 1966); more recently Aleliunas (1990) has produced a stronger result in a discrete setting. Interestingly, the introduction of certainty factors into MYCIN was just such an attempt to use numerical values to represent uncertainty, and, as mentioned in Section 2.4, Heckerman (1986) showed that for a consistent interpretation of the way that certainty factors combine there has to be a monotonic mapping of their values to ratios of conditional probabilities.

2.7 Axioms

We assume that the reader has had some contact with probability theory, and we shall use standard results and definitions as required. However, it is useful to review the basic probability axioms. We can regard these as applying either to propositions, combining under the logical operations of the propositional calculus, or to events (subsets of a suitable sample space), combining according to the operations of set theory. Although the former interpretation is more intuitive, we shall generally use the more standard notation and terminology of the latter. For the logical conjunction $A\&B$ of events A and B we may use, interchangeably, $A \cap B$ or (A, B).

Axiom 1: $0 \leq P(A) \leq 1$, with $P(A) = 1$ if A is certain.

Axiom 2: If events (A_i) $(i = 1, 2, \ldots)$ are pairwise incompatible, then $P(\bigcup_i A_i) = \sum_i P(A_i)$.

Axiom 3: $P(A \cap B) = P(B \mid A)P(A)$.

These axioms are not quite in the form given in the standard account by Kolmogorov (1950). Our Axiom 3 relates unconditional and conditional probabilities, regarded as having independent existence on an equal footing (indeed, from our viewpoint any 'unconditional' probability is only really so by appearance, the background information behind its assessment having been implicitly assumed and omitted from the notation). Kolmogorov's approach takes unconditional probability as the primitive concept, and would therefore treat our Axiom 3 as *defining* conditional probability.

There is continuing discussion over whether the union in Axiom 2 should be restricted to finite, rather than countably infinite, collections of events (de Finetti 1975). For our purposes this makes little difference, and for convenience we shall assume full countable additivity.

Many other properties of probabilities may be deduced from the axioms, including *Bayes' Theorem* (see Section 2.8), which shows how to interchange the outcome and the conditioning events in a conditional probability.

In an epistemological approach, all quantities, be they observables or parameters, are jointly modelled as random variables with a known joint distribution. *Statistical inference* then consists simply in calculating the conditional distribution of still unknown quantities, given data. Since Bayes' theorem is the principal (though not the only) tool for performing such calculations, this approach to statistics has come to be called 'Bayesian'. This in turn has logical and subjectivist branches, although there are difficulties in constructing a fully consistent logical account (Dawid 1983). A good account of modern Bayesian statistics may be found in Bernardo and Smith (1994).

In this book we mostly adopt both the subjectivist interpretation of probability and the subjectivist Bayesian approach to statistical inference. However, in later chapters we shall also make use of non-Bayesian statistical techniques when dealing with model construction and criticism.

2.8 Bayes' theorem

Bayes' theorem is the basic tool for making inferences in probabilistic expert systems. From Axiom 3 and the fact that $P(A \cap B) = P(B \cap A)$, we immediately have

$$P(A \cap B) = P(A \mid B)P(B) = P(B \mid A)P(A). \tag{2.1}$$

By rearrangement we obtain *Bayes' theorem*:

$$P(A \mid B) = \frac{P(B \mid A)P(A)}{P(B)}. \tag{2.2}$$

This can be interpreted as follows. Suppose we are interested in A and we begin with a *prior probability* $P(A)$, representing our belief about A before observing any relevant evidence. Suppose we then observe B. By (2.2), our revised belief for A, the *posterior probability* $P(A \mid B)$, is obtained by multiplying the prior probability $P(A)$ by the ratio $P(B \mid A)/P(B)$.

We now extend attention beyond simple events to *random variables*. Informally, a random variable is an unknown quantity that can take on one of a set of mutually exclusive and exhaustive outcomes. Such a variable, say M with values $m \in \mathcal{M}$, will have a distribution, its *prior distribution*

$P(M)$, specifying the probabilities $P(m) = P(M = m)$, $m \in \mathcal{M}$. Then for any value d of another variable D, the expression $P(d \mid M)$, with values $P(d \mid m) = P(D = d \mid M = m)$, considered as a function of m, is called the *likelihood function* for M on data d. The *posterior distribution* for M given the data, $P(M \mid d)$, can then be expressed, using (2.2), by the relationship:

$$P(M \mid d) \quad \propto \quad P(d \mid M) \quad \times \quad P(M), \tag{2.3}$$

that is,

$$\textbf{Posterior} \quad \propto \quad \textbf{Likelihood} \quad \times \quad \textbf{Prior},$$

where the proportionality arises since the denominator $P(d)$ in (2.2) is the same for all values of M, and can thus be reconstructed as the normalizing constant needed to scale the right-hand side to sum to 1 over all outcomes of M.

The above discussion assumes that the random variables involved are discrete, but the identical formula (2.3) continues to hold in the case of continuous variables (or a mixture of discrete and continuous variables), so long as, when M (for example) is continuous, we interpret $P(m)$ as the *probability density* of M at m. In that case, the normalization constant $P(d)$ is given by the integral of the right-hand side.

In Figure 2.3 we display this 'prior-to-posterior inference' process pictorially. Both of the diagrams represent the structure of the joint distribution $P(M, D)$. Diagram (a) decomposes $P(M, D)$ in terms of its 'prior' components $P(M)$ and $P(D \mid M)$: often, we will think of M as a possible 'cause' of the 'effect' D, and the downward arrow represents such a causal interpretation. Diagram (b) decomposes $P(M, D)$ in terms of its 'posterior' components $P(M \mid D)$ and $P(D)$: the 'inferential' upward arrow then represents an 'argument against the causal flow', from the observed effect to the inferred cause.

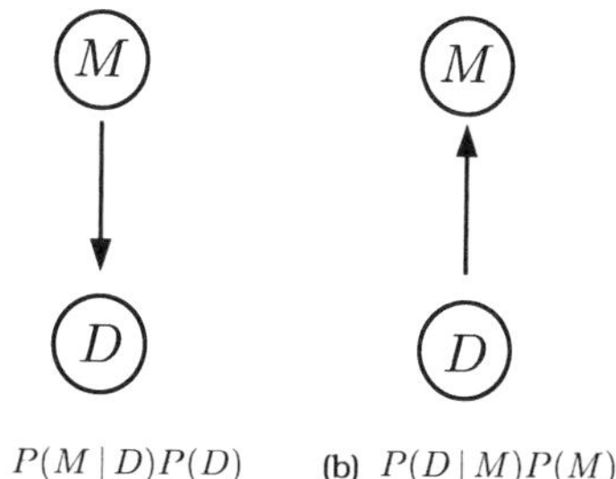

(a) $P(M \mid D)P(D)$ (b) $P(D \mid M)P(M)$

FIGURE 2.3. Bayesian inference as reversing arrows.

A generalization of Figure 2.3 is illustrated in Figure 2.4. Here the variable D represents some unknown member of a set of alternative diseases, and influences the chance of the occurrence of each of a set of potential symptoms or features (F_i). We shall see later that this graph encodes an assumption that the features (F_i) are *conditionally independent* given the disease D, and hence that the joint distribution of all variables satisfies

$$P(D, F_1, \ldots, F_K) = \left(\prod_{i=1}^{K} P(F_i \mid D)\right) P(D). \tag{2.4}$$

This requires as numerical inputs only the distribution for the disease and the conditional distribution of each of the features in each of the disease categories. These can be readily estimated if we have a random sample of 'training data', in which for each case we observe the disease and some or all of the features. Calculating the posterior probability of each disease on the basis of observed findings is extremely straightforward: this simple model has been termed *naïve* — or even *idiot's* — Bayes (Titterington et al. 1981).

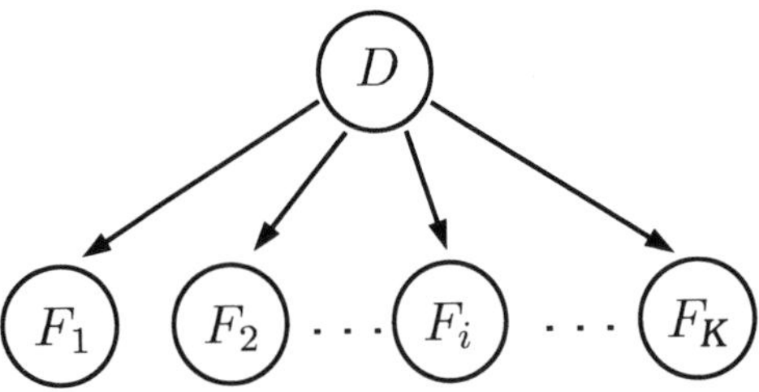

FIGURE 2.4. Directed graphical model representing conditional independence of feature variables within each disease class — the naïve Bayes model.

The naïve Bayes model was first used by Warner et al. (1961) for the diagnosis of congenital heart disease. Later applications of this model are too numerous to list, but a notable example is the acute abdominal pain system (de Dombal et al. 1972), which has been implemented in a number of hospitals and remote sites such as submarines, and has been claimed to have a significant impact on care and resources (Adams et al. 1986).

It has been argued that in most applications the assumptions underlying such a model are blatantly inappropriate. For (2.4) implies that once the disease class is known information about some feature variables is of no further relevance to predicting the values of any others. This property of the model of Figure 2.4 — the independence of features conditional on knowing the state of D — can be verified by performing the necessary calculations on the joint probability distribution. However, it can also be deduced simply from the figure, without knowing explicitly any numerical values attached to the probabilities of the model. This follows from the theory of *Markov distributions on graphs*, to be discussed in Chapter 5, which in turn relies on some aspects of graph theory described in Chapter 4.

The model of Figure 2.4 allows a simple use of Bayes' theorem, since the conditional independence assumptions mean that each item of evidence can be considered in turn, with the posterior probability distribution for the disease after observing each item becoming the prior probability distribution for the next. Thus, the sparseness of the graph leads directly to a modular form for the inference.

2.9 Bayesian reasoning in expert systems

Pearl (1982) realized that this modular approach could be generalized to more complex graphical structures and presented some elegant techniques for exploiting this vital idea of 'local computation' in graphs that are more complex than Figure 2.4 but still have a tree structure, so that removing any edge disconnects the graph. A simple example illustrates a number of points.

Suppose we wish to reason about possible personal computer failure. Let C be the variable *Computer failure?*, allowing answers "yes" and "no." The possible causes, with their assumed probabilities, are E: *Electricity failure?*, with $P(E = \text{yes}) = 0.1$, and M: *Malfunction?*, with $P(M = \text{yes}) = 0.2$. We assume that these possible precipitating events are independent, in that we have no reason to believe that the occurrence of one should influence the occurrence of the other. We also adopt the following conditional probabilities for failure:

$$\begin{aligned}
P(C = \text{yes} \mid E = \text{no}, M = \text{no}) &= 0\\
P(C = \text{yes} \mid E = \text{no}, M = \text{yes}) &= 0.5\\
P(C = \text{yes} \mid E = \text{yes}, M = \text{no}) &= 1\\
P(C = \text{yes} \mid E = \text{yes}, M = \text{yes}) &= 1
\end{aligned}$$

The left-hand diagram in Figure 2.5 shows a directed graphical model of this system, with each variable labelled by its current probability of taking the value "yes" (the value $P(C = \text{yes}) = 0.19$ is calculated below).

Suppose you turn your computer on and nothing happens. Then the event "$C = \text{yes}$" has occurred, and you wish to find the conditional probabilities for E and M, given this computer failure. By Bayes' theorem,

$$P(E, M \mid C = \text{yes}) = \frac{P(C = \text{yes} \mid E, M)\ P(E, M)}{P(C = \text{yes})}.$$

The necessary calculations are laid out in Table 2.1. Note that, owing to the assumed independence, $P(E, M) = P(E)P(M)$. Also $P(C = \text{yes}, E, M) = P(C = \text{yes} \mid E, M)P(E, M)$, and when summed this provides $P(C = \text{yes}) = 0.19$.

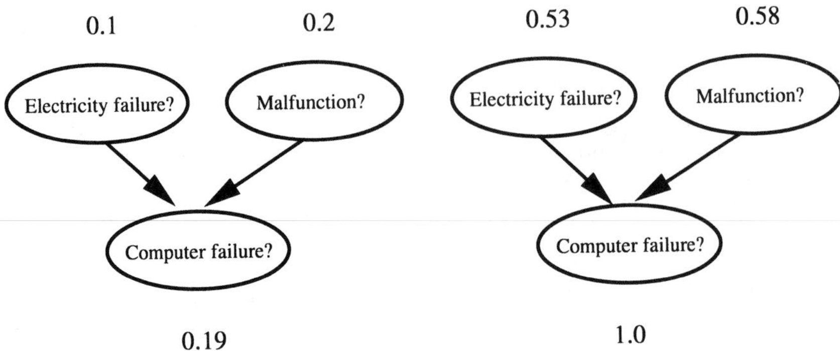

FIGURE 2.5. Directed graphical model representing two independent potential causes of computer failure, with probabilities of a 'yes' response before and after observing computer failure.

By summing over the relevant entries in the joint posterior distribution of E and M we thus obtain $P(E = \text{yes} \mid C = \text{yes}) = 0.42 + 0.11 = 0.53$ and $P(M = \text{yes} \mid C = \text{yes}) = 0.47 + 0.11 = 0.58$. These values are displayed in the right-hand diagram of Figure 2.5. Note that the observed failure has induced a strong dependency between the originally independent possible causes; for example, if one cause could be ruled, out the other *must* have occurred.

TABLE 2.1.

E $[P(E)]$	no [0.9]		yes [0.1]		
M $[P(M)]$	no [0.8]	yes [0.2]	no [0.8]	yes [0.2]	
$P(E, M)$	0.72	0.18	0.08	0.02	1
$P(C = \text{yes} \mid E, M)$	0	0.50	1	1	
$P(C = \text{yes}, E, M)$	0	0.09	0.08	0.02	0.19
$P(E, M \mid C = \text{yes})$	0	0.47	0.42	0.11	1

We now extend the system to include the possible failure, denoted by L, of the light in the room, assuming that such a failure depends only on the electricity supply, and that

$$
\begin{aligned}
P(L = \text{yes} \mid E = \text{yes}) &= 1 \\
P(L = \text{yes} \mid E = \text{no}) &= 0.2
\end{aligned}
$$

so that $P(L = \text{yes}) = P(L = \text{yes} \mid E = \text{yes}) P(E = \text{yes}) + P(L = \text{yes} \mid E = \text{no}) P(E = \text{no}) = 1 \times 0.1 + 0.2 \times 0.9 = 0.28$. The extended graph is shown in Figure 2.6.

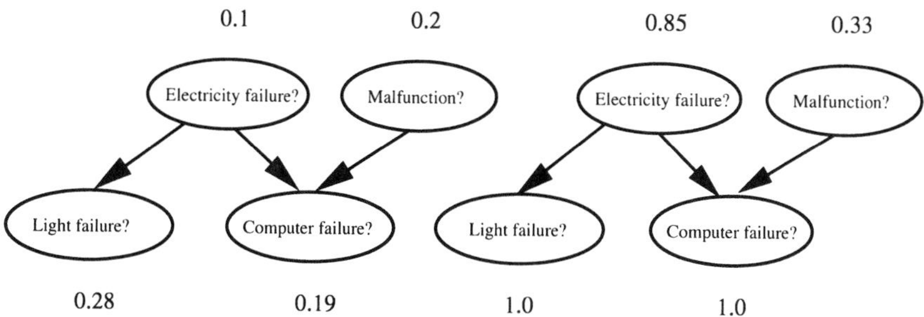

FIGURE 2.6. Introducing the roomlight into the system, before and after observing that neither the light nor the computer work.

Suppose we now find the light does not work (L = yes). Our previous posterior distribution $P(E, M \mid C = \text{yes})$ now becomes the prior distribution for an application of Bayes' theorem based on observing that the light has failed. Note that $P(L = \text{yes} \mid E, M, C) = P(L = \text{yes} \mid E)$, since only the electricity supply directly affects the light.

TABLE 2.2.

E	no		yes		
M	no	yes	no	yes	
$P(E, M \mid C = \text{yes})$	0	0.47	0.42	0.11	
$P(L = \text{yes} \mid E, M, C = \text{yes})$	0.2	0.2	1	1	
$P(L = \text{yes}, E, M \mid C = \text{yes})$	0	0.094	0.42	0.11	0.624
$P(E, M \mid C = \text{yes}, L = \text{yes})$	0	0.15	0.67	0.18	1

The calculations are displayed in Table 2.2. We obtain $P(E = \text{yes} \mid C = \text{yes}, L = \text{yes}) = 0.85$, $P(M = \text{yes} \mid C = \text{yes}, L = \text{yes}) = 0.33$. Thus, observing "light off" has *increased* the chance of "electricity failure," and *decreased* the chance of "malfunction": the original computer fault has been *explained away*. This ability to withdraw a tentative conclusion on the basis of further information is extremely difficult to implement within a system based on logic, even with the addition of measures of uncertainty. In contrast, it is both computationally and conceptually straightforward within a fully probabilistic system built upon a conditional independence structure.

The above example has heuristically argued for the explanatory power of probabilistic models based on Bayesian reasoning, following closely the insights of Pearl (1986b) and Pearl (1988), which largely provided the foundation for probabilistic evidence propagation in complex systems. We have not directly illustrated the specific techniques developed in these references for updating belief on any part of a tree-structured graph given evidence on any other part, techniques which can be used to organize and streamline our brute force calculations above. The approach developed in this book is more general, dealing with graphs with a more complex structure. However, many parallels with Pearl's work may be drawn.

The following fictitious example, ASIA , due to Lauritzen and Spiegelhalter (1988), illustrates the nature of the more complex graphical structures we shall be analysing in this book.

> Shortness–of–breath (dyspnoea) may be due to tuberculosis, lung cancer or bronchitis, or none of them, or more than one of them. A recent visit to Asia increases the chances of tuberculosis, while smoking is known to be a risk factor for both lung cancer and bronchitis. The results of a single chest X-ray do not discriminate between lung cancer and tuberculosis, as neither does the presence or absence of dyspnoea.

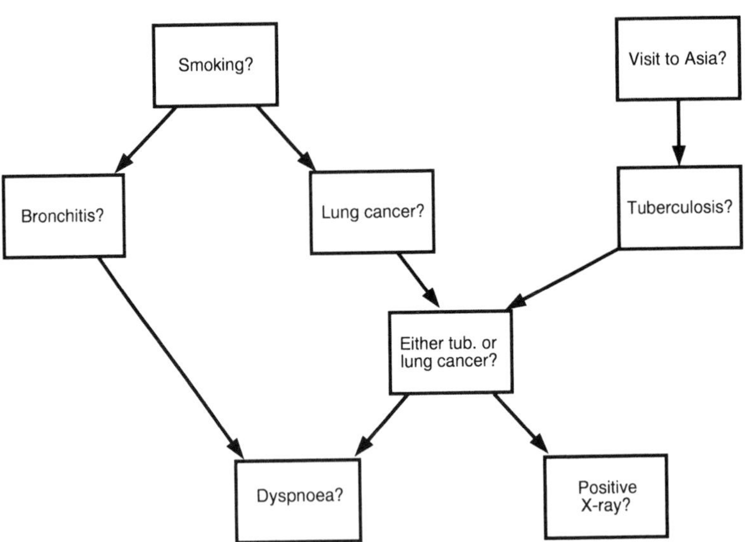

FIGURE 2.7. The ASIA network.

The qualitative structure of this example is given in Figure 2.7. Note that, as opposed to the previous example, *Smoking?* is connected to *Dyspnoea?* via two alternative routes. The quantitative specification is given in Table 2.3. Here we use B (for example) to denote the variable *Bronchitis?*, and $b, \bar{b}$ respectively for "*Bronchitis?* = yes," "*Bronchitis?* = no."

The model might be applied to the following hypothetical situation. A patient presents at a chest clinic with dyspnoea, and has recently visited Asia. Smoking history and chest X-ray are not yet available. The doctor

TABLE 2.3. Conditional probability specifications for the ASIA example.

A:	$p(a)$	$=$	0.01	L:	$p(l \mid s)$	$=$	0.1
					$p(l \mid \bar{s})$	$=$	0.01
B:	$p(b \mid s)$	$=$	0.6	S:	$p(s)$	$=$	0.5
	$p(b \mid \bar{s})$	$=$	0.3				
D:	$p(d \mid b, e)$	$=$	0.9	T:	$p(t \mid a)$	$=$	0.05
	$p(d \mid \bar{b}, e)$	$=$	0.7		$p(t \mid \bar{a})$	$=$	0.01
	$p(d \mid b, \bar{e})$	$=$	0.8				
	$p(d \mid \bar{b}, \bar{e})$	$=$	0.1				
E:	$p(e \mid l, t)$	$=$	1	X:	$p(x \mid e)$	$=$	0.98
	$p(e \mid \bar{l}, t)$	$=$	1		$p(x \mid \bar{e})$	$=$	0.05
	$p(e \mid l, \bar{t})$	$=$	1				
	$p(e \mid \bar{l}, \bar{t})$	$=$	0				

would like to know the chance that each of these diseases is present, and if tuberculosis were ruled out by another test, how would that change the belief in lung cancer? Also, would knowing smoking history or getting an X-ray contribute more information about cancer, given that smoking may 'explain away' the dyspnoea since bronchitis is considered a possibility? Finally, when all information is in, can we identify which was the most influential in forming our judgement?

2.10 A broader context for probabilistic expert systems

We have informally introduced the idea of representing qualitative relationships between variables by graphs and superimposing a joint probability model on the unknown quantities. When the graph is directed and does not contain any (directed) cycles, the resulting system is often called a *Bayesian network*, although later we shall see how broader classes of graphs may be used. Using the terms introduced earlier, we may think of this network and its numerical inputs as forming the knowledge base, while efficient methods of implementing Bayes' theorem form the inference engine used to draw conclusions on the basis of possibly fragmentary evidence.

While Bayesian networks have now become a standard tool in artificial intelligence, it is important to place them in a wider context of what might be called *highly structured stochastic systems* (HSSS). This broad term attempts to bring together areas in which complex interrelationships can be expressed by local dependencies, and hence a graphical representation can be exploited both to help communication and as a basis for computational algorithms. We are led naturally to a unifying system of Bayesian reasoning on graphical structures: by embedding apparently unrelated topics in this common framework, strong similarities are revealed which can lead to valuable cross-fertilization of ideas. Further information can be found on the HSSS Web page (see Appendix C).

A natural example area is genetics, in which familial relationships form the basis for the graph and Mendelian laws of inheritance and relationships between genotype and phenotype provide the elements of the probability distribution. The 'peeling' algorithm for pedigree analysis derived by Cannings et al. (1978) was shown by Spiegelhalter (1990) to be very similar to the local computation algorithm of Lauritzen and Spiegelhalter (1988). Similarly, much of image analysis is dominated by Markov field models which are defined in terms of local dependencies and can be described graphically (Besag and Green 1993), although simulation methods are generally required for inference. Such spatial models are also used in geographical epidemiology (Bernardinelli et al. 1997) and agricultural field trials (Besag et al. 1995). Within the artificial intelligence community, neural networks are natural candidates for interpretation as probabilistic graphical models, and are increasingly being analysed within a Bayesian statistical framework (see, for example, Neal (1996)). Hidden Markov models, which form the basis for work in such diverse areas as speech recognition (Rabiner and Juang 1993) and gene sequencing (Durbin et al. 1998), can likewise be considered as special cases of Bayesian networks (Smyth et al. 1997).

Further reading

Recent years have seen an explosion of interest in graphical models as a basis for probabilistic expert systems, with a number of dedicated books and a wide range of theoretical and practical publications. Textbooks on probabilistic expert systems include the classic Pearl (1988). Neapolitan (1990) explains the basic propagation algorithms, and these are studied in detail by Shafer (1996). Jensen (1996) is a very good tutorial introduction, while Castillo et al. (1997) provides another sound introduction with many worked examples.

Perhaps the best guide to current research is to be found in the *Proceedings* of the annual meeting of the Association for Uncertainty in Artificial

Intelligence, which hosts an excellent Web page providing many relevant links, and provides a forum for discussion of a wide range of issues concerning uncertainty in expert systems (although the arguments between the advocates of probabilistic and non-probabilistic approaches appear to have died down as each group tries to identify the most appropriate domains for its work). Other electronic sources of information include the Bayesian network Web page of the US Air Force Institute of Technology Artificial Intelligence Laboratory, which in particular features informal comments of people who work in industry, and the Web page of the Microsoft Decision Theory and Adaptive Systems group. See Appendix C for addresses and more details. Appendix C also details some free and commercial software available over the World Wide Web.

Other good sources of tutorial material are the special issues of *AI Magazine* (Charniak 1991; Henrion et al. 1991) and of the *Communications of the ACM* (Heckerman and Wellman 1995).

See also Section 3.5 for some pointers to various implementations and applications of probabilistic expert systems.

3

Building and Using Probabilistic Networks

In this chapter we illustrate many of the features to be discussed in detail later on in this book, focusing attention on a specific example. At a first reading the forward references to as yet undefined concepts should be ignored or taken on trust; subsequently, it may be helpful to refer back to this chapter to clarify the more general treatment presented later on.

The development of a probabilistic network model can be broadly split into three phases:

- Define the model, i.e.,
  - Specify relevant variables.
  - Specify structural dependence between variables.
  - Assign component probabilities to the model.
- Construct the inference engine, i.e.,
  - Moralize the graphical model.
  - Triangulate the moral graph.
  - Find the cliques of the moral graph.
  - Make a junction tree from the cliques.
- Use the inference engine for individual case analysis, i.e.,
  - Initialize potentials on the junction tree.
  - Perform local propagation taking account of evidence.

– Find node marginal distributions.

The process of constructing the inference engine from the model specification is sometimes called *compiling* the model. We shall illustrate these phases by describing a probabilistic network model, CHILD, of the 'blue baby' diagnosis problem introduced in Chapter 2. This is a domain appropriate for analysis using Bayesian expert systems because there is a good clinical understanding of the disease process and basic physiological mechanisms are well understood. Also it can exploit the available accumulated data on nearly 600 babies. Such a probabilistic system needs to be based on consideration of the true disease and possible clinical findings as random variables, and requires the specification of a full joint distribution over these variables. It is tasks of this nature that have in the past been considered so off-putting by constructors of expert systems.

3.1 Graphical modelling of the domain

We can divide the construction of a model into three distinct components. The first *qualitative* stage considers only general relationships between the variables of interest, in terms of the *relevance* of one variable to another under specified circumstances. This naturally leads to a graphical representation of *conditional independence*, but one that is not restricted to a probabilistic interpretation. The next *probabilistic* stage introduces the idea of a joint distribution defined on the variables in the model and relates the form of this joint distribution to the structure of the graph. The final *quantitative* step requires the numerical specification of the necessary conditional probability distributions.

As described in Chapter 2, the Great Ormond Street Hospital for Sick Children in London (GOS) acts as a major referral centre for newborn babies with congenital heart disease for South-East England and handles about 200 such babies a year. Congenital heart disease is generally suspected when cyanosis (blue baby) or heart failure (breathlessness) occurs immediately after birth, and it is vital that the child be transported to a specialist referral centre. There can be rapid deterioration and so appropriate treatment should start before transportation, which requires that a diagnosis be reached during the telephone conversation between the referring and the specialist hospital. Diagnosis is not straightforward, and is often made by junior doctors at any time of day or night. The decision is based on the clinical description provided by the referring paediatrician, as well as his reading of a chest X-ray, ECG, and blood gases.

There is considerable interest in improving diagnostic performance and providing teaching aids for doctors training in paediatric cardiology. This interest has led to the development of the diagnosis decision-tree algorithm shown in part in Figure 2.1. This algorithm has been shown to be highly effective (Franklin et al. 1991). However, there are still errors made through the use of this algorithm, owing to missing data, atypical presentations, or observer error in interpreting clinical signs, X-ray, or ECG.

3.1.1 Qualitative modelling

The idea of a *graphical model* is to allow experts to concentrate on building up the qualitative structure of a problem before even beginning to address issues of quantitative specification. As emphasized by Pearl (1988), such models are intended to encode natural judgements of relevance and irrelevance, and can be formed prior to any probabilistic considerations. Vertices in the graph represent variables, while *missing* edges in the graph encode the irrelevance properties.

The graph may be *directed* or *undirected*, or may be a *chain graph*, having both directed and undirected edges. Directed edges represent probabilistic *influence* or *causal* mechanisms, or perhaps more weakly direct *relevance*. Undirected edges represent mere *associations* or *correlations*. To see the difference between these two types of edges, consider building a graphical model in which there are two variables, *Smoking?* and *Lung cancer?*, which are to be connected by an edge. A graph that puts a directed edge from *Smoking?* to *Lung cancer?* models the assumption that smoking has an effect on the incidence of lung cancer in smokers. In contrast, a graph with an undirected link would model an association between the two variables; perhaps this might represent an unknown gene at work — not explicit in the model — which when present in some people predisposes them to smoke and also predisposes them to develop lung cancer. This example shows that even at the qualitative level the details of model structure can reflect subtle assumptions, perhaps having economic, political, or other social consequences.

In Figure 3.1 we show the CHILD probabilistic network model (with a restriction of the *Disease?* node to six possible states), elicited after consultation with an expert paediatrician; this is a purely directed acyclic graph. The graph incorporates, for example, the assumption that the level of oxygen in the lower body (vertex 16) is directly related to the underlying level when breathing oxygen (vertex 11) and whether the hypoxia is equally distributed around the body (vertex 10). In turn, the level when breathing oxygen depends on the degree of mixing of the blood in the heart (vertex 6) and the state of the blood vessels (parenchyma) in the lungs (vertex 7). It is these intermediate variables that are directly influenced by the underlying disease (vertex 2). When we reach the stage of assessing the detailed probabilistic structure associated with a graphical model, such qualitative judgements translate into algebraic properties of the overall joint distribution. (See Section 3.5 for examples of other complex problems that have been modelled as directed acyclic graphs.)

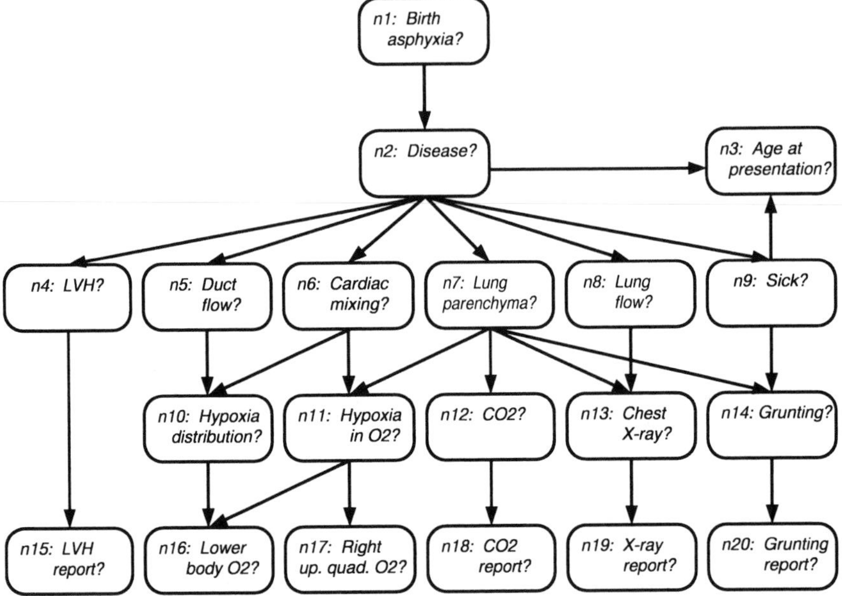

FIGURE 3.1. The CHILD network: Directed acyclic graph representing possible diseases that could lead to a blue baby.

3.1.2 Probabilistic modelling

The quantitative inputs required will differ, depending on the type of graphical model considered. In the case of an undirected graph $\mathcal{G}$ with family $\mathcal{C}$ of cliques, we need to specify, for each $C \in \mathcal{C}$, a potential function a_C such that the overall joint density $p(x)$ has the graphical factorization

$$p(x) = \prod_{C \in \mathcal{C}} a_C(x_C). \tag{3.1}$$

Such a specification may be difficult to think about directly, since the probabilistic interpretation of the potential functions is somewhat complex. Specification is much simpler when we can structure the problem in terms of a directed acyclic graph (DAG) $\mathcal{D}$ with node set V. In this case, for each $v \in V$, we have to specify the conditional distributions of X_v given its 'parents' $X_{\mathrm{pa}(v)}$. Suppose that this density is $p(x_v \,|\, x_{\mathrm{pa}(v)})$. Then the overall joint density is simply

$$p(x) = \prod_{v \in V} p(x_v \,|\, x_{\mathrm{pa}(v)}). \tag{3.2}$$

A problem structured as a chain graph $\mathcal{K}$, with family K of chain components, will require a hybrid specification of the form

$$p(x) = \prod_{k \in K} p(x_k \mid x_{\mathrm{pa}(k)}), \tag{3.3}$$

where each factor admits an additional factorization as in (5.11).

We may use the results of Chapter 5 to examine the precise conditional independence properties (i.e., properties of relevance and irrelevance) that such a factorization of the joint density implies. The definitions introduced there may now be illustrated for this example, in which we look beyond the immediate parents of a vertex to the more distant ancestor and descendant relations — the use of such language reflecting that pedigree analysis and genetic counselling provide examples of natural conditional independence graphs. As an example of the directed Markov property, the factorization (3.2) of the CHILD network implies that, given values for *n7: Lung parenchyma?* and *n8: Lung flow?*, the vertex *n13: Chest X-ray?* is independent of all other vertices except *n19: X-ray report?*.

3.1.3 Quantitative modelling

For our CHILD example we have the simplest type of model to specify, a DAG with probability density factorizing as in (3.2). In addition, all of the random variables of the model are discrete. The joint distribution thus requires as numerical inputs the conditional distribution of each node given any configuration of its parents.

The expert must provide sufficient conditional probabilities to specify fully the joint distribution. For example, Table 3.1 gives assessments for the links *Disease?* → *LVH?* and *LVH?* → *LVH report?*; the elicitation of these assessments is discussed below. These judgements express the view that LVH is essentially a feature of PAIVS, although on fairly rare occasions it can appear with other conditions, and that there is an estimated 5 percent false positive and 10 percent false negative rate in reporting. Criticism and revision of these quantitative assumptions is dealt with in Chapter 9 and 10. In total, for the 20 variables with, on average, 3 states each, 114 distributions were assessed, requiring 230 independent numerical assessments.

3.1.4 Further background to the elicitation process

The network in Figure 3.1 is adapted from one shown in Spiegelhalter and Cowell (1992), which was elicited using the technique of *similarity graphs* described by Heckerman (1990): the expert first identifies sets of diseases that typically present a diagnostic problem (i.e., are similar in their presentation), and the graphical structure appropriate to that differential diagnosis is then elicited. The resulting graphs are then superimposed.

TABLE 3.1. Conditional probability tables subjectively assessed by expert for edges *n2* → *n4* and *n4* → *n15*. Diseases are persistent foetal circulation (PFC), transposition of the great arteries (TGA), tetralogy of Fallot, pulmonary atresia with intact ventricular septum (PAIVS), obstructed total anomalous pulmonary venous connection (TAPVD), and lung disease.

	n4: LVH?	
n2: Disease?	Yes	No
PFC	0.10	0.90
TGA	0.10	0.90
Fallot	0.10	0.90
PAIVS	0.90	0.10
TAPVD	0.05	0.95
Lung	0.10	0.90

	n15:LVH report?	
n4: LVH?	Yes	No
Yes	0.90	0.10
No	0.05	0.95

A series of changes was made in response to analysis of fictitious cases and scrutiny of the model. First, in the original network a link existed between *n10* and *n17*. Not only is this unrealistic, since hypoxia distribution should not influence upper body oxygen, but it was found that the assessed conditional probabilities did not actually show a quantitative dependence. Second, *Age at presentation?* was originally a parent of *Disease?*, since it was felt fairly natural to think of different incidence rates for different ages at presentation. After discussion, this link was reversed and a dependence on *Sick?* introduced as an attempt to model the referral behaviour of distant paediatricians. This issue of modelling the selection process deserves deeper analysis — ideally the network should reflect the structure of an unselected case, and a new vertex *Case selected?* be introduced, as a child of those variables that would influence the decision to refer. Conditioning on the case being selected would then give the appropriate joint distribution over the remaining vertices. This approach was used in Cowell et al. (1993a).

Finally, some conditional probability tables were adjusted in order to give the system reasonable behaviour when 'archetypal' cases were put through the system. Of the 114 distributions, 13 (11 percent) were altered at this stage.

We note that, apart from the somewhat anomalous *Age at presentation?*, the graph naturally falls into a set of five blocks, representing, respectively, risk factors, underlying physiological anomaly (disease), physiological disturbances caused by the disease, clinical signs as elicited by specialist staff in GOS, and reported signs as obtained over the telephone from a paediatrician who is not a specialist in cardiology. Many other clinical problems appear to follow a similar structure, although sometimes the direction of an arrow is not clear. For example, birth asphyxia is a possible *cause* of PFC (persistent foetal circulation), but a possible *effect* of the other diseases.

3.2 From specification to inference engine

3.2.1 Moralization

Whether our initial graph $\mathcal{K}$ is an undirected graph, a DAG, or a chain graph, the first stage in creating the inference engine is to recast the problem in terms of an undirected graph. The basic idea is that the initial probability distribution satisfies the conditional independence restrictions of $\mathcal{K}$ determined by the associated Markov property. Therefore, it also satisfies these properties on the so-called moral graph $\mathcal{K}^m$. (See Lemma 5.9 and Corollary 5.19.) In the case of a DAG the moral graph is obtained by 'marrying parents', i.e., introducing additional undirected edges between any two nodes with a common child and subsequently ignoring directions.

The CHILD graph is a DAG, with its moral graph shown in Figure 3.2. A distribution for the CHILD problem that factorizes as in (3.2) with respect to the graph in Figure 3.1 must be Markov with respect to its moral graph.

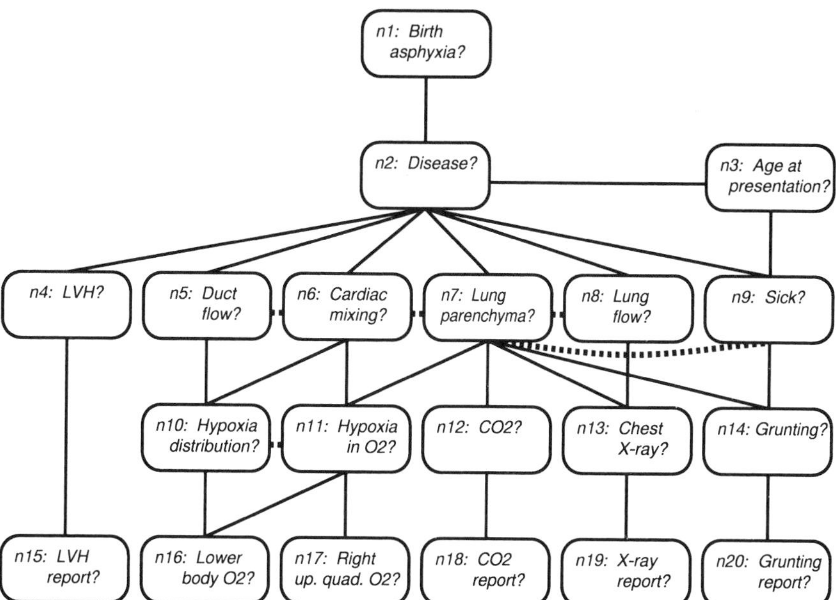

FIGURE 3.2. Moral graph formed from the CHILD network. The joint distribution of the variables is Markov with respect to this graph.

For a somewhat contrived example, we might ask whether, if we were to know the true disease (*n2*), and have measured the hypoxia when breathing oxygen (*n11*), then knowing the CO_2 report (*n18*) would tell us anything additional about the distribution of hypoxia (*n10*)?. This question can be answered by investigating the separation properties in the moral graph of the ancestral set of the variables of interest, displayed in Figure 3.3.

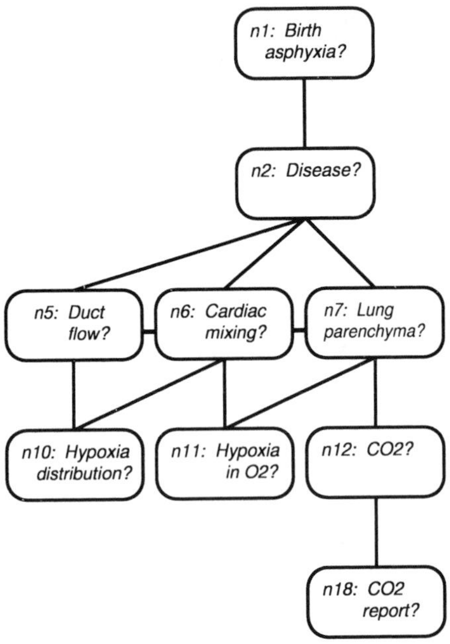

FIGURE 3.3. Moral graph formed from ancestral set of vertices {*n2, n10, n11, n18*}. From the Markov property, *n18* and *n10* are not conditionally independent given {*n2, n11*}, since there is a path between *n18* and *n10* that is not blocked by {*n2, n11*}.

It is clear that a path exists between *n18* and *n10* that bypasses *n2* and *n11*. Hence, *n18* would be informative, essentially since it would tell us more about *n7: Lung parenchyma?*, and hence whether the observed hypoxia level (*n11*) is explained by the state of the lungs. If not, this changes our belief in *n6: Cardiac mixing?*, which finally feeds down to *n10*. It is a crucial feature of these directed structures that they allow reasoning about joint causes through effects being 'explained away', as shown in the example in Section 2.9. As discussed later in Chapter 5 this technique of forming the moral graph of ancestral sets will reveal *all* the conditional independence properties logically implied by the factorization of the joint distribution. Pearl (1986b) presented an alternative way of finding these properties for a DAG in terms of a concept called *d-separation* (see Section 5.3). However, under the re-representation of P as factorizing over $\mathcal{D}^m$, some of the conditional independence properties displayed in the original DAG $\mathcal{D}$, such as the one between *n5* and *n6* given *n2* in Figure 3.1, may lose their representation in graphical form. They still hold, but are now effectively buried in the quantitative component of the model. Only those conditional independence properties that retain a representation in the graph $\mathcal{D}^m$ are exploited in the further analysis.

3.2.2 From moral graph to junction tree

The passage from moral graph to junction tree proceeds in the same way, regardless of the type of the initial graph. The first stage is to add sufficient edges to the moral graph, to make the resulting graph $(\mathcal{K}^m)'$ *triangulated* or *chordal*. Algorithms to do this are discussed in Section 4.4 below. The result is shown in Figure 3.4. The joint distribution, which factorizes on the

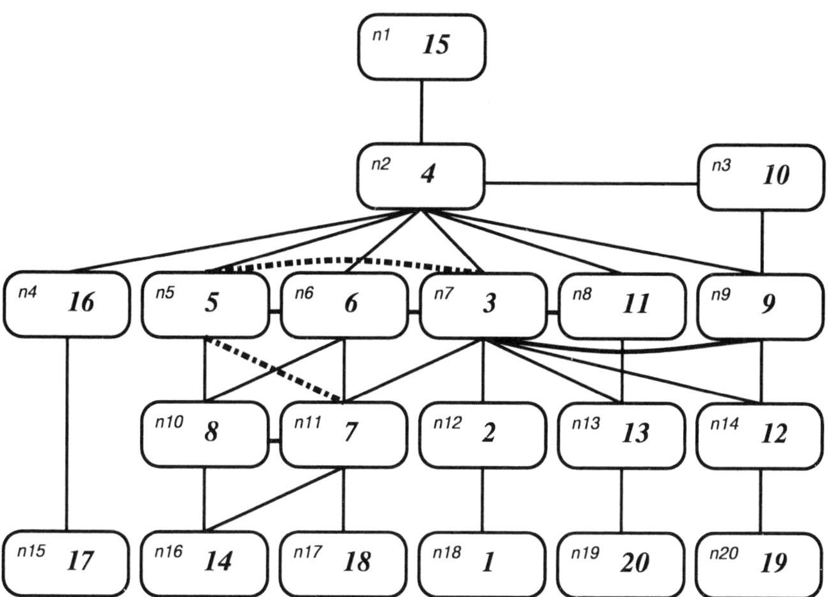

FIGURE 3.4. A triangulation of the moral graph of Figure 3.2, showing a perfect ordering of the nodes arising from maximum cardinality search (Algorithm 4.9).

moral graph $\mathcal{K}^m$, will also factorize on the larger triangulated graph $(\mathcal{K}^m)'$, since each clique in the moral graph will be either a clique or a subset of a clique in the triangulated graph. Hence, by a simple rearrangement of terms, we can transform the joint density into a product of factors on the cliques of the triangulated graph:

$$p(x) = \prod_{C \in \mathcal{C}} a_C(x_C), \tag{3.4}$$

which defines the numerical specification of p in terms of the functions $\{a_C\}$ on the cliques of $(\mathcal{K}^m)'$.

There are many ways of choosing functions $\{a_C\}$ to satisfy (3.4), and indeed this very freedom forms the basis of the computational algorithms to be described. They may be initialized by a two stage process of (1) setting all functions $a_C \equiv 1$; and then (2) taking each factor of the probability density of p on the original probabilistic network and multiplying it into the function of any *one* clique that contains all of the variables of that factor as a subset — by the moralization process there is always at least one such clique — thus redefining that clique function.

It is useful to generalize (3.4) to allow still more freedom, as follows. Let the cliques of the triangulated graph be arranged to form a junction tree (see Section 4.4), and let C_i and C_j be adjacent cliques of the junction tree. We can associate with the edge joining them their *separator*, the set $S = C_i \cap C_j$ of nodes of $\mathcal{G}$ (see Section 4.3). We denote by $\mathcal{S}$ the family of all separators, including any repetitions. As well as having a function a_C for each clique $C \in \mathcal{C}$, we suppose that we have a function b_S for each separator $S \in \mathcal{S}$, such that we can express

$$p(x_V) = \frac{\prod_{C \in \mathcal{C}} a_C(x_C)}{\prod_{S \in \mathcal{S}} b_S(x_S)}. \tag{3.5}$$

The right-hand side of (3.5) is interpreted as 0 whenever any denominator term is 0. The individual a and b functions are called *potential functions*, and when equation (3.5) holds, the collection of functions $(\{a_C, C \in \mathcal{C}\}, \{b_S, S \in \mathcal{S}\})$ is called a *(generalized) potential representation* for p. Initially we take $b_S \equiv 1$, and the a's as described earlier.

If a graph is triangulated it possesses a junction tree, but this need not be unique. Any junction tree will serve our purposes. Figure 3.5 shows a junction tree formed from triangulating the moral CHILD graph in Figure 3.2. It has the property that if a node v is contained in any two cliques C_i and C_j, then it is contained in all the cliques in the unique path between C_i and C_j (see Section 4.3). For example, *n2: Disease?* is contained in $C_3, C_6, C_7, C_8, C_{12}$ and C_{13}, which form a connected sub-tree. Evidence on any node can then be passed to the rest of the network in a unique way.

3.3 The inference process

We now come to the point of all of this elaborate representation of the probability distribution of a probabilistic network — using it for inference.

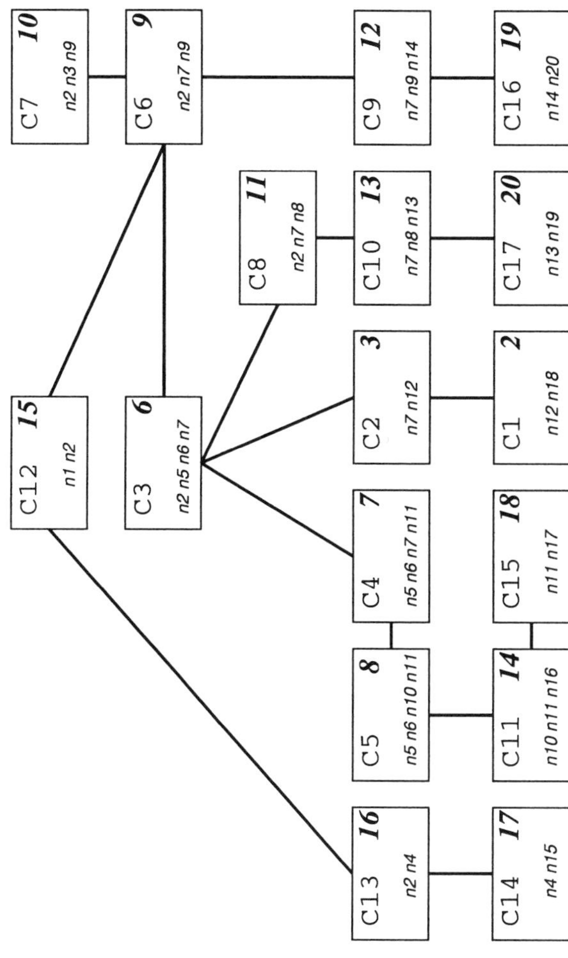

FIGURE 3.5. Junction tree of cliques derived from perfect ordering of the CHILD nodes. The members of each clique are shown; the highest node-label obtained in the ordering is shown in the top right-hand-corner, while the corresponding ordering of the cliques is shown in the top left-hand corner.

3.3.1 The clique-marginal representation

We work with a generalized potential representation on the junction tree, satisfying (3.5). The computational algorithms discussed in Chapter 6 below proceed by modifying the individual potential functions in a sequence of steps, which can be interpreted as local message-passing, but in such a way that (3.5) holds at all times. After all steps have been completed, the final potential functions will have a special form containing the desired information. Thus, the propagation algorithm finishes with every potential function being the marginal density for the relevant set of variables. Thus, (3.5) holds for this *marginal representation* of p:

$$p(x_V) = \frac{\prod_{C \in \mathcal{C}} p_C(x_C)}{\prod_{S \in \mathcal{S}} p_S(x_S)}. \tag{3.6}$$

3.3.2 Incorporation of evidence

Suppose that we observe 'evidence' $\mathcal{E} : X_A = x_A^*$. Define a new function p^* by

$$p^*(x) = \begin{cases} p(x), & \text{if } x_A = x_A^* \\ 0, & \text{otherwise.} \end{cases} \tag{3.7}$$

Then $p^*(x) = p(x, \mathcal{E}) = P(\mathcal{E})p(x \mid \mathcal{E})$, where $p(\cdot \mid \mathcal{E})$ is the density of the conditional distribution, given $\mathcal{E}$. We can rewrite (3.7) as

$$p^*(x) = p(x) \prod_{v \in A} l_v(x_v), \tag{3.8}$$

where $l_v(x_v)$ is 1 if $x_v = x_v^*$, and 0 otherwise. Thus, l_v is the *likelihood function* based on the partial evidence $X_v = x_v^*$.

If we have any representation for p satisfying (3.4) or (3.5) we immediately obtain such a representation for p^*, by associating each $v \in A$ with any one clique containing v and replacing each $a(C)$ by

$$a_C^*(x_C) = a_C(x_C) \prod \{l_v : v \text{ is assigned to } C\} \tag{3.9}$$

taking an empty product as unity.

The fact that p^* is proportional to, rather than equal to, a probability density function can be turned to advantage, as we shall see later. In particular, if we apply a routine for finding the marginal representation directly to p^*, it will give us

$$p^*(x_U) = P(\mathcal{E})p(x_U \mid \mathcal{E})$$

for any clique or separator U. Then the *normalizing constant* for any such U (i.e., the sum of all the values of $p^*(x_U)$) will be just $P(\mathcal{E})$. This provides a means of calculating the joint probability for specified states of any collection of variables, even when the corresponding set of nodes is not complete in $\mathcal{K}$. Finally, on performing the normalization, we obtain $p(x_U \mid \mathcal{E})$, and so shall have transmitted the effect of the evidence to every clique. In summary, we may obtain the marginal representation conditional on the evidence $\mathcal{E}$, for which

$$p(x_V \mid \mathcal{E}) = \frac{\prod_{C \in \mathcal{C}} p_C(x_C \mid \mathcal{E})}{\prod_{S \in \mathcal{S}} p_S(x_S \mid \mathcal{E})}. \tag{3.10}$$

By further marginalizing on the individual cliques, (3.10) can be used to calculate efficiently the marginal probabilities of individual variables given evidence $\mathcal{E}$. This is a common application and is illustrated in the next section.

3.4 Bayesian networks as expert systems

Figures 3.6 and 3.7 show a sequence of screen dumps taken from the HUGIN system (Andersen et al. 1989), illustrating the propagation algorithm in use on CHILD. Figure 3.6 shows the status of the network after observing the evidence '*LVH report?* = yes'; the marginal distribution on the observed node is now degenerate on the response 'yes', while the influence of that observation has been propagated through the underlying junction tree by a series of local message-passing operations. Finally the software displays the marginal posterior distributions on each of the nodes — we note that the fact that LVH is an accepted feature of PAIVS has raised this disease to be the most likely true diagnosis. The computations involved are described in more detail in Chapter 6.

After two more findings have been added, *X-ray report?* = oligaemic and *Lower body* $O_2 < 5$, the posterior probability that the disease is PAIVS is 0.76 (Figure 3.7). The status of the intermediate nodes gives some explanation of the findings: the X-ray shows oligaemia because of the low lung flow, although the lung vessels appear normal. There is likely to be complete cardiac mixing, with left-to-right (aorta to pulmonary artery) flow in the arterial duct providing the only source of blood to the lungs. Such depiction of the effects of changing evidence are particularly valuable for rapid sensitivity analysis, and the potential for an explanation facility is clear.

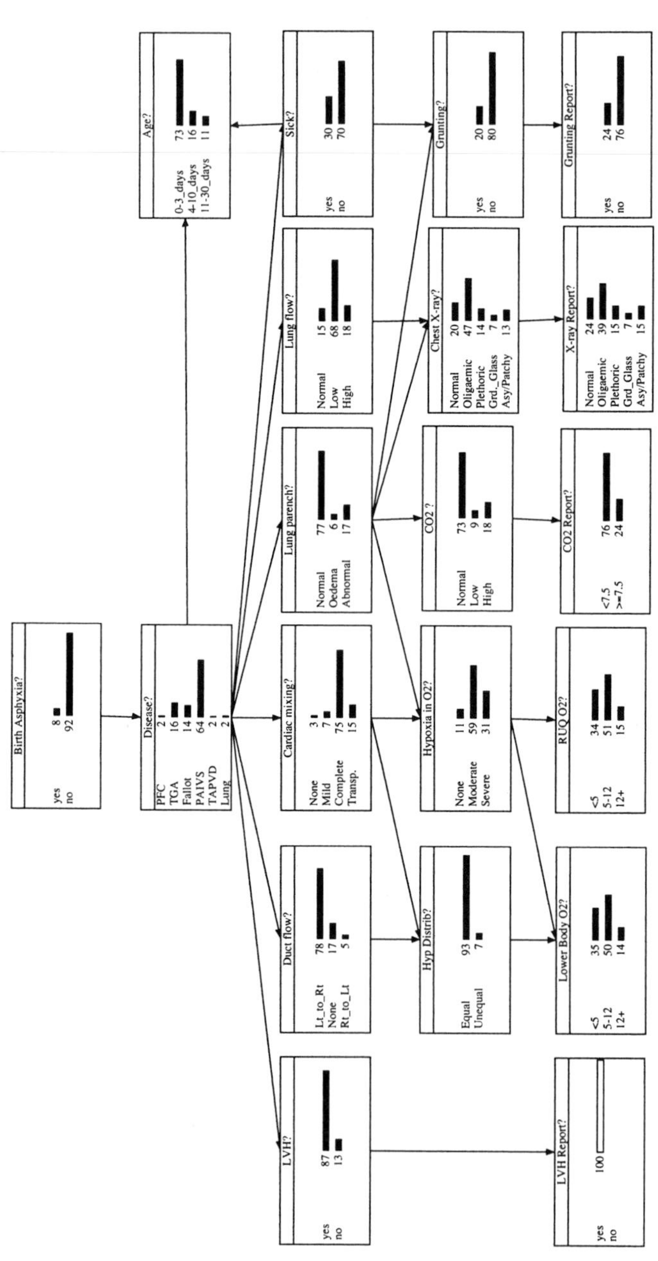

FIGURE 3.6. Conditional probability distributions on all nodes after propagation of evidence *LVH report?* = yes. The numbers and the lengths of the bars represent the current probability e.g., 64 percent belief that *PAIVS* is the true diagnosis. For observed evidence, e.g., *LVH report?* = yes, the bar is hollow.

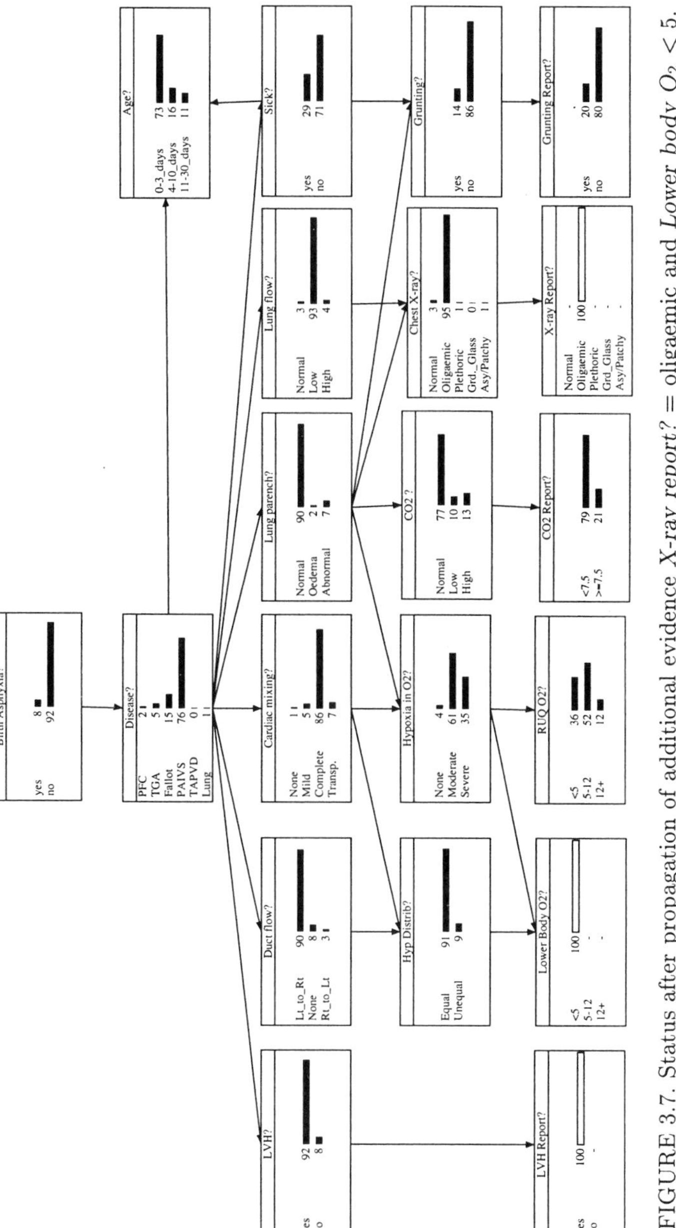

FIGURE 3.7. Status after propagation of additional evidence *X-ray report?* = oligaemic and *Lower body* $O_2 < 5$.

3.5 Background references and further reading

3.5.1 Structuring the graph

A number of early probabilistic expert systems involved constructing very large graphs (see, e.g., Henrion et al. (1991)). Examples include a probabilistic reconstruction (Shwe et al. 1991) of the QMR/INTERNIST system for general medical diagnosis (Miller et al. 1982). This system involved 4,500 vertices and over 40,000 edges. The resulting graph is very 'wide and shallow', in that symptoms are generally related directly to a large number of potential diagnoses. The MUNIN network (Andreassen et al. 1989), representing some of the muscles and nerves necessary for interpreting electromyographic data, had over 1,000 vertices each with up to 21 states, whose structuring was simplified by the repetition of similar subgraphs. This idea has been formalized in Koller and Pfeffer (1997). The PATHFINDER system for the diagnosis of lymph vertex pathology concerned over 60 diseases and required the specification of over 75,000 subjective probabilities (Heckerman et al. 1991). This has been converted to a commercial system, INTELLIPATH.

More recent applications have involved a range of structuring issues. For example, temporal networks and tracking problems require an evolving but repetitive structure (Kjærulff 1995; Nicholson and Brady 1994). Neural networks have been reconstructed as probabilistic networks (Neal 1996; Frey 1998), which again produce shallow but densely connected graphs.

A major application area has been troubleshooting and diagnostics, with the Decision Theory and Adaptive Systems Group at Microsoft Research taking a leading role. The background for this work is discussed in Heckerman et al. (1995a), and a number of applications have been implemented in Microsoft software. The March 1995 issue of *The Communications of the ACM* contains useful articles on structuring systems for software debugging (Burnell and Horvitz 1995) and information retrieval (Fung and Del Favero 1995); Microsoft Research's work in this area is briefly described on their Web page (see Appendix C).

An alternative to using expert knowledge to structure the graph is to use data-analytic techniques or other methods for structural learning; these are discussed in Chapter 11.

3.5.2 Specifying the probability distribution

In large systems there are so many numerical assessments required that it is unreasonable to expect each to be individually specified. A variety of simplifying models for the conditional probabilities has been exploited: these include assuming an underlying continuous mathematical model (MUNIN), two-stage assessments in which large sets of conditional probabilities are first assessed to be equal and then a single common value elicited (PATHFINDER), and 'noisy–or gates' (QMR) (Shwe et al. 1991). In the final example, up to 80 diseases may be parents of a symptom, but it is assumed that any single one of the diseases is sufficient to cause the symptom, and this causation occurs independently of other diseases present. The conditional distribution in this competing-risk model then only requires the specification of as many parameters as there are parents.

There has also been research into incomplete specification of the full joint distribution, using say upper and lower probabilities (Walley 1990), in which resulting intervals of probabilities may be computed using linear (van der Gaag 1991) or non-linear (Andersen and Hooker 1994) programming techniques. Alternatively, if the number of assessments made is insufficient to specify a joint distribution uniquely, it has been suggested that the distribution be completed by maximum entropy arguments (Nilsson 1986).

4
Graph Theory

The power and appeal of probabilistic networks stems from the *pictorial representation* they provide of the structural inter-relationships and dependencies between the variables of a problem, and the fact that these pictures have a formal definition as *graphs*. Many people find this form easy to understand and manipulate. But the graphs also serve as a precise and compact way of communicating these relations to a computer, paving the way for the use of efficient computational algorithms.

Problems that can be usefully tackled by probabilistic networks are those for which the graphs are relatively sparse. Such structures exhibit many independence relationships, thereby facilitating local inference, involving only a few variables at any one time. This chapter collects together a number of ideas and results from graph theory, and primarily contains definitions, notation, properties, and algorithms. Readers may prefer to skip this chapter on first reading and refer back to it only to clarify unfamiliar terms.

4.1 Basic concepts

Graph theory can be developed as a purely abstract mathematical subject. However, much of the immediate power of graph theory when applied to probabilistic expert systems lies in its ability to present a *visual* summary of expert knowledge or opinion about some subject. Accordingly, we shall develop graphical ideas and theorems making liberal use of pictorial representations. There are many variants of definitions and notations in use, hence the need here to state explicitly those that will be used throughout the book.

We define a *graph* $\mathcal{G}$ to be a pair $\mathcal{G} = (V, E)$, where V is a finite set of *vertices*, also called *nodes*, of $\mathcal{G}$, and E is a subset of the set $V \times V$ of ordered pairs of vertices, called the *edges* or *links* of $\mathcal{G}$. Thus, as E is a set, the graph $\mathcal{G}$ has no multiple edges. We further require that E consist of pairs of distinct vertices so that there are no loops.

If both ordered pairs (α, β) and (β, α) belong to E, we say that we have an *undirected* edge between α and β, and write $\alpha \sim \beta$ (or $\alpha \sim_{\mathcal{G}} \beta$ to indicate the relevant graph $\mathcal{G}$; similar elaborations may be made to the other notation introduced below). We also say that α and β are *neighbours*, α is a neighbour of β, or β is a neighbour of α. The set of neighbours of a vertex β is denoted by $\mathrm{ne}(\beta)$.

If $(\alpha, \beta) \in E$ but $(\beta, \alpha) \notin E$, we call the edge *directed*, and write $\alpha \to \beta$. We also say that α is a *parent* of β, and that β is a *child* of α. The set of parents of a vertex β is denoted by $\mathrm{pa}(\beta)$, and the set of children of a vertex α by $\mathrm{ch}(\alpha)$. The *family* of β, denoted $\mathrm{fa}(\beta)$, is $\mathrm{fa}(\beta) = \{\beta\} \cup \mathrm{pa}(\beta)$.

If $(\alpha, \beta) \in E$ or $(\beta, \alpha) \in E$ we say that α and β are *joined*. Then $\alpha \not\sim \beta$ indicates that α and β are not joined, i.e., both $(\alpha, \beta) \notin E$ and $(\beta, \alpha) \notin E$. We also write $\alpha \not\to \beta$ if $(\alpha, \beta) \notin E$.

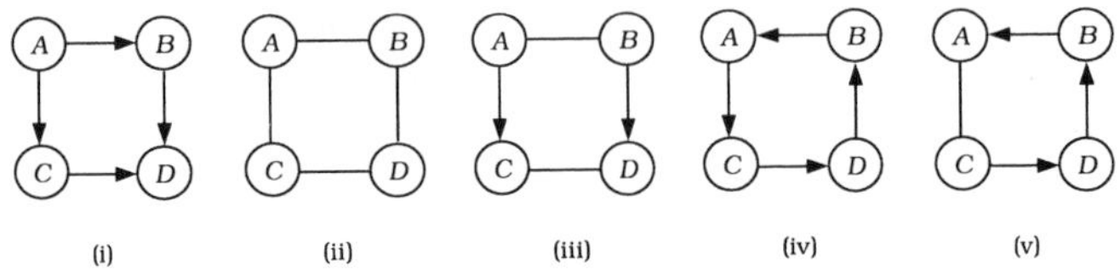

FIGURE 4.1. Examples of valid graphs on four vertices.

Figure 4.1 illustrates some graphs permitted by our definitions. Vertices are visually represented by (possibly) labelled circles, directed edges by arrows, and undirected edges by lines. In (iii) $A \sim B$ and $A \to C$, but $A \not\sim D$. In contrast, Figure 4.2 shows some 'graphs' which fail our definition, but which may pass definitions used by other authors for different purposes.

If $A \subset V$, the expressions $\mathrm{pa}(A)$, $\mathrm{ne}(A)$ and $\mathrm{ch}(A)$ will denote the collection of parents, children, and neighbours, respectively, of the elements of A, but exclude any element in A:

$$\begin{aligned} \mathrm{pa}(A) &= \textstyle\bigcup_{\alpha \in A} \mathrm{pa}(\alpha) \setminus A \\ \mathrm{ne}(A) &= \textstyle\bigcup_{\alpha \in A} \mathrm{ne}(\alpha) \setminus A \\ \mathrm{ch}(A) &= \textstyle\bigcup_{\alpha \in A} \mathrm{ch}(\alpha) \setminus A. \end{aligned}$$

Referring to Figure 4.1(iii), A is a parent of C, and thus C a child of A. Also, B is a parent of D, so that D is a child of B. In addition, C and

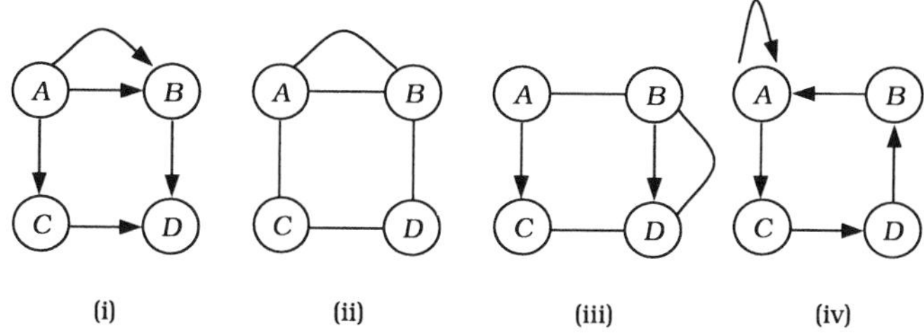

FIGURE 4.2. Examples of invalid graphs on four vertices. Examples (i), (ii) and (iii) exhibit illegal multiple edges of various types; example (iv) exhibits a loop as the node A is connected to itself.

D are neighbours, as are A and B. Finally, we have that $\text{pa}(\{A, B\}) = \emptyset$, while $\text{pa}(\{C, D\}) = \{A, B\}$.

If all the edges of a graph are directed, we say that it is a *directed graph.* Conversely, if all the edges of a graph are undirected, we say that it is an *undirected graph.* Referring again to Figure 4.1, graphs (i) and (iv) are directed graphs and (ii) is an undirected graph; neither of graphs (iii) nor (v) specialize to either of these two categories.

The *boundary* $\text{bd}(\alpha)$ of a vertex α is the set of parents and neighbours of α; the boundary $\text{bd}(A)$ of a subset $A \subset V$ is the set of vertices in $V \setminus A$ that are parents or neighbours to vertices in A, i.e., $\text{bd}(A) = \text{pa}(A) \cup \text{ne}(A)$. The *closure* of A is $\text{cl}(A) = A \cup \text{bd}(A)$. Hence, in Figure 4.1(i), $\text{bd}(A) = \emptyset$, while in (iii), $\text{bd}(C) = \{A, D\}$.

The *undirected version* $\mathcal{G}^{\sim}$ of a graph $\mathcal{G}$ is the undirected graph obtained by replacing the directed edges of $\mathcal{G}$ by undirected edges. For example, Figure 4.1(ii) is the undirected version of the other four graphs.

We call $\mathcal{G}_A = (A, E_A)$ a *subgraph* of $\mathcal{G} = (V, E)$ if $A \subseteq V$ and $E_A \subseteq E \cap (A \times A)$. Thus, it may contain the same vertex set but possibly fewer edges. If, in addition, $E_A = E \cap (A \times A)$, we say that $\mathcal{G}_A$ is the subgraph of $\mathcal{G}$ *induced* by the vertex set A. This is illustrated in Figure 4.3.

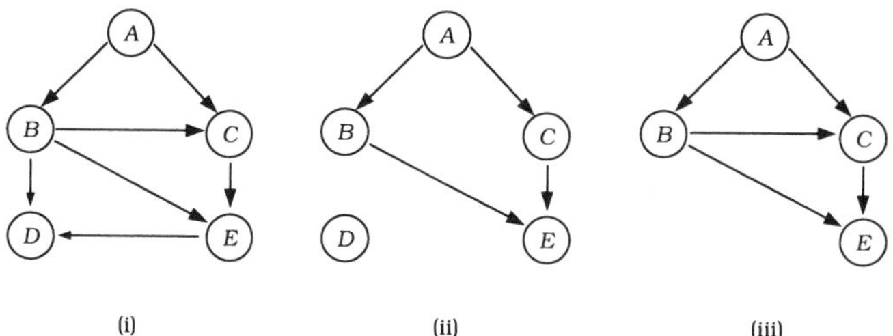

FIGURE 4.3. A graph (i), a subgraph (ii), and a vertex induced subgraph (iii).

A graph is called *complete* if every pair of vertices is joined. Figure 4.4 shows the complete undirected graph on five vertices. We say that a subset of vertices of $\mathcal{G}$ is *complete* if it induces a complete subgraph. A complete subgraph which is maximal (with respect to $\subseteq$) is called a *clique*. In all graphs of Figure 4.1 there are four cliques: $\{A, B\}$, $\{A, C\}$, $\{C, D\}$, and $\{B, D\}$. In Figure 4.3(i) there are three cliques: $\{A, B, C\}$, $\{B, C, E\}$, and $\{B, D, E\}$.

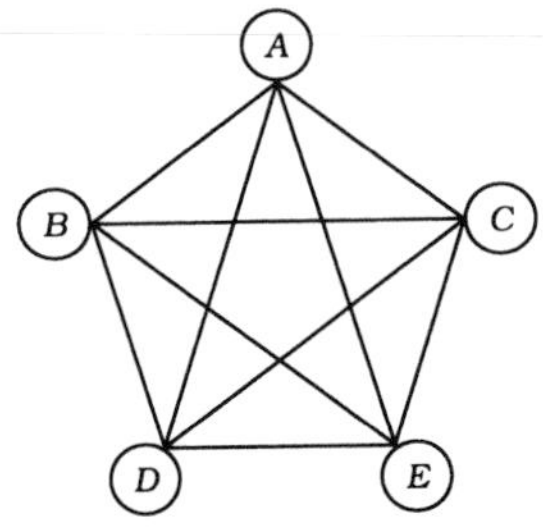

FIGURE 4.4. The complete undirected graph on five vertices.

A *path* of length n from α to β is a sequence $\alpha = \alpha_0, \ldots, \alpha_n = \beta$ of distinct vertices such that $(\alpha_{i-1}, \alpha_i) \in E$ for all $i = 1, \ldots, n$. Thus, a path can never cross itself and movement along a path never goes against the directions of arrows.

If the path of length n from α to β given by the sequence $\alpha = \alpha_0, \ldots, \alpha_n = \beta$ is such that for at least one $i \in \{1, \ldots, n\}$ there is a directed edge $\alpha_{i-1} \to \alpha_i$, we say that the path is *directed*.

We write $\alpha \mapsto \beta$ if there is a path from α to β, and say that α *leads to* β. If $\alpha \mapsto \beta$ and $\beta \mapsto \alpha$ we say that α and β *connect*, and write $\alpha \rightleftharpoons \beta$. This is clearly an equivalence relation which induces equivalence classes $[\alpha]$, where

$$\beta \in [\alpha] \Leftrightarrow \alpha \rightleftharpoons \beta.$$

We call the equivalence classes the *strong components* of $\mathcal{G}$. If $\alpha \in A \subseteq V$, the symbol $[\alpha]_A$ denotes the strong component of α in $\mathcal{G}_A$. In Figure 4.1(iii) the strong components are $\{A, B\}$ and $\{C, D\}$.

A graph $\mathcal{G}$ is said to be *connected* if there is a path between every pair of vertices in its undirected version $\mathcal{G}^{\sim}$. Any graph can be decomposed into a union of its *connected components*. The connected components are the strong components of $\mathcal{G}^{\sim}$.

An *n-cycle* is a path of length n with the modification that the end points are identical. Similarly a *directed n-cycle* is a directed path with the modification that the end points are identical. We say that a graph is *acyclic* if it does not possess any cycles.

A directed graph which is acyclic is called a *directed acyclic graph*, or DAG. A graph that has no directed cycles is called a *chain graph*. Thus, undirected graphs and directed acyclic graphs are both special cases of chain graphs.

For example, in Figure 4.1 graphs (i) and (iv) are directed graphs, but (iv) has a directed cycle and so is not a directed acyclic graph, whereas (i) is a DAG. Similarly, neither of the graphs (iv) and (v) are chain graphs, as they contain directed cycles, whereas graphs (i), (ii), and (iii) are chain graphs.

A *trail* of length n from α to β is a sequence $\alpha = \alpha_0, \ldots, \alpha_n = \beta$ of distinct vertices such that $\alpha_{i-1} \rightarrow \alpha_i$, or $\alpha_i \rightarrow \alpha_{i-1}$, or $\alpha_{i-1} \sim \alpha_i$ for all $i = 1, \ldots, n$. Thus, movement along a trail could go against the direction of the arrows, in contrast to the case of a path. In other words, a trail in $\mathcal{G}$ is a sequence of vertices that form a path in the undirected version $\mathcal{G}^{\sim}$ of $\mathcal{G}$.

If $\mathcal{K}$ is a chain graph, let $\mathcal{K}^{\not\to}$ denote the same graph but with the directed edges removed. Then each connected component of $\mathcal{K}^{\not\to}$ is called a *chain component* of $\mathcal{K}$. The strong components of a chain graph $\mathcal{K}$ are exactly its chain components. In fact, a graph is a chain graph if and only if its strong components induce undirected subgraphs.

As a special case, each node of a DAG $\mathcal{D}$ forms a chain component of $\mathcal{D}$. Figure 4.5 shows a six-vertex chain graph having five chain components.

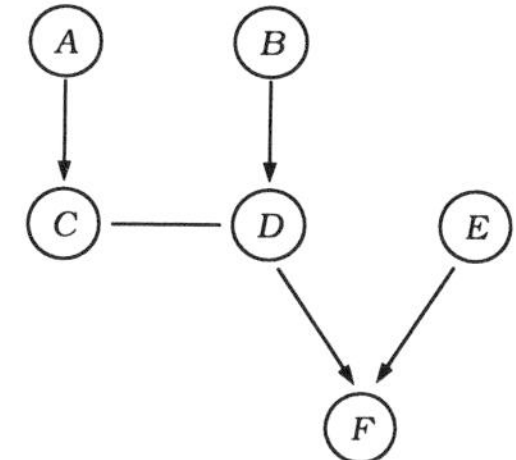

FIGURE 4.5. A six-vertex chain graph. The chain components are A, B, $\{C, D\}$, E, and F.

It is always possible to *well-order* the nodes of a DAG by a linear ordering or numbering such that, if two nodes are connected, the edge points from the lower to the higher of the two nodes with respect to the ordering. For example, the graph in Figure 4.3(i) has the unique well-ordering (A, B, C, E, D). Note that a DAG may not have a unique well-ordering. If a DAG is well-ordered, the *predecessors* of a node α, denoted by $\mathrm{pr}(\alpha)$, are those nodes that have a lower number than α.

A simple method to construct such a well-ordering is the following:

Algorithm 4.1 [Topological sort]

- Begin with all vertices unnumbered.
- Set counter $i := 1$.
- While any vertices remain:

- Select any vertex that has no parents;
- number the selected vertex as i;
- delete the numbered vertex and all its adjacent edges from the graph;
- increment i by 1. □

Alternatively, we can use the dual version of this algorithm, which recursively selects and deletes childless vertices, while numbering downward.

The above algorithm, or its dual, extends to well-ordering the chain components of a chain graph as follows. Let $\mathcal{K}$ be a chain graph having the set of chain components K. Then, instead of selecting a parentless node for deletion, one selects a parentless chain component, that is a chain component none of whose nodes have parents. The result is a well-ordering of the chain components. One possible well-ordering of the chain components of Figure 4.5 is $(A, B, \{C, D\}, E, F)$; yet another is $(E, B, A, \{C, D\}, F)$.

Given a chain graph, the set of vertices α such that $\alpha \mapsto \beta$ but not $\beta \mapsto \alpha$ is the set $\mathrm{an}(\beta)$ of the *ancestors* of β, and the *descendants* $\mathrm{de}(\alpha)$ of α are the vertices β such that $\alpha \mapsto \beta$ but not $\beta \mapsto \alpha$. The *nondescendants* $\mathrm{nd}(\alpha)$ of α is the set $V \setminus (\mathrm{de}(\alpha) \cup \alpha)$. If $\mathrm{bd}(\alpha) \subseteq A$ for all $\alpha \in A$, we say that A is an *ancestral* set. The symbol $\mathrm{An}(A)$ denotes the smallest ancestral set containing A. Note that in general $\mathrm{An}(A) \neq A \cup_{\alpha \in A} \mathrm{an}(\alpha)$. Thus, in Figure 4.5 the set of ancestors of F consists of all remaining nodes. The ancestors of C are $\{A, B\}$, F is the only descendant of C, $\{A, B, C, D\}$ is an ancestral set, and $\mathrm{An}(C) = \{A, B, C, D\}$. Node E has no ancestors, and $\{E\}$ is an ancestral set. The set of nondescendants of D is $\{A, B, C, E\}$.

A subset $C \subseteq V$ is said to be an (α, β)-*separator* if all trails from α to β intersect C. The subset C is said to *separate* A from B if it is an (α, β)-separator for every $\alpha \in A$ and $\beta \in B$. An (α, β)-separator C is said to be *minimal* if no proper subset of C is itself an (α, β)-separator. In Figure 4.5 the set $\{C, D\}$ is an (A, F)-separator; moreover, both C and D are each minimal (A, F)-separators. In Figure 4.3(i) both $\{B, C\}$ and $\{B, E\}$ are minimal (A, D)-separators.

An important class of graphs is that of the trees. We say that a graph $\mathcal{G}$ is a *tree* if it is connected and its undirected version $\mathcal{G}^{\sim}$ has no cycles; thus, there is a unique trail in a tree between any two vertices. We use the symbol $\mathcal{T}$ to denote a tree graph. A *rooted* tree is a tree with a designated vertex ρ called the *root*. A *leaf* of a tree is a node that is joined to at most one other node. A tree that has more than one node thus has at least two leaves. The *diameter* of a tree is the length of longest trail between two leaf nodes. A *forest* is a graph having no cycles, that is, its connected components are all trees. The graph of Figure 4.5 is a tree.

Given a chain graph $\mathcal{K}$, we define the *moral graph* of $\mathcal{K}$ to be the undirected graph $\mathcal{K}^m$ obtained from $\mathcal{K}$ by first adding undirected edges between all pairs of vertices that have children in a common chain component and that are not already joined, and then forming the undirected version of the resulting graph.

For the special case in which $\mathcal{K}$ is a DAG, this process of *moralization* involves adding undirected edges between all pairs of parents of each vertex which are not already joined, and then making all edges undirected. Figure 4.6 shows the moral graph of Figure 4.5, obtained by adding the two undirected edges $A \sim B$ (common parents of the chain component $\{C, D\}$) and $D \sim E$ (the parents of the chain component F), and then forming the undirected version. This moralization procedure is an important first step in constructing the inference engine for a probabilistic network specified by a chain graph.

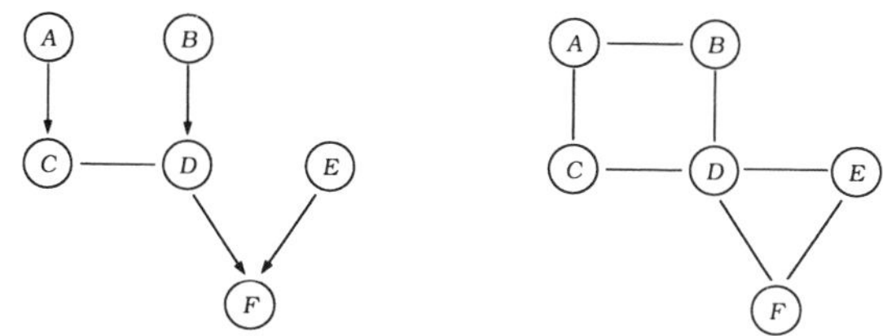

FIGURE 4.6. The graph of Figure 4.5 and its moral graph.

4.2 Chordal and decomposable graphs

An important type of graph is the *decomposable graph*, which we characterize below. This is basic to the analysis of probabilistic networks. The qualitative and quantitative expertise encoded within a probabilistic network can be transformed into a decomposable graph, using well-defined graphical algorithms and without loss of information, and then further into an associated *junction tree* which supports efficient computational algorithms. This section deals with the properties of decomposable graphs. Junction trees and their relation to decomposable graphs are discussed in Section 4.3 below, while Section 4.4 describes algorithms for constructing a junction tree.

Let $\mathcal{G}$ be an undirected graph with vertex set V. Recall that in this case an n-cycle in $\mathcal{G}$ is a sequence $(\alpha_0, \alpha_1, \dots, \alpha_n)$ of vertices in V, distinct except that $\alpha_0 = \alpha_n$, and such that $\alpha_i \sim \alpha_{i+1}$ for all i. Let σ be an n-cycle in $\mathcal{G}$. A *chord* of this cycle is a pair (α_i, α_j) of non-consecutive vertices in σ such that $\alpha_i \sim \alpha_j$ in $\mathcal{G}$. The undirected graph $\mathcal{G}$ is called *chordal* or *triangulated* if every one of its cycles of length ≥ 4 possesses a chord. A definition such as this is a so-called 'forbidden path' definition, which has several consequences. For example, the property is stable under taking vertex-induced subgraphs, i.e., if $\mathcal{G}$ is chordal and $A \subset V$, then $\mathcal{G}_A$ is also chordal. The moral graph of Figure 4.6 is clearly not chordal. Figure 4.7 shows two possible chordal graphs obtained from the moral graph in Figure 4.6 by adding one undirected edge.

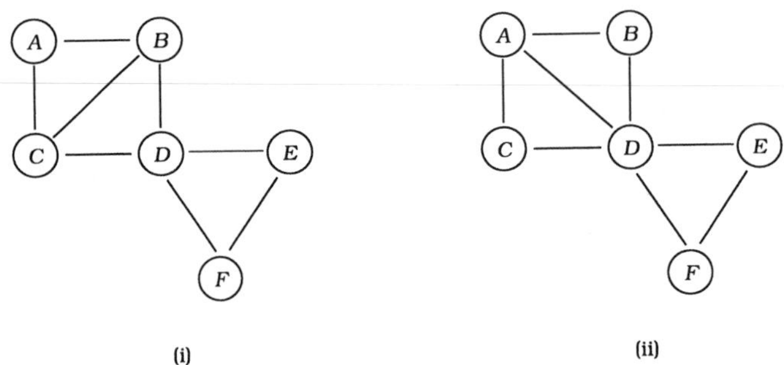

FIGURE 4.7. Two chordal graphs which can be derived from the moral graph in Figure 4.6, by either (i) adding the edge $B \sim C$, or (ii) adding the edge $A \sim D$.

An important concept that forms the basis of localizing computation in a probabilistic expert system is that of a decomposition of a graph, as defined below.

Definition 4.2 [DECOMPOSITION]
A triple (A, B, C) of disjoint subsets of the vertex set V of an undirected graph $\mathcal{G}$ is said to form a *decomposition* of $\mathcal{G}$, or to *decompose* $\mathcal{G}$, if $V = A \cup B \cup C$, and the following two conditions hold:

1. C separates A from B;

2. C is a complete subset of V. □

Note that we allow any of the sets A, B, and C to be empty. If both A and B are non-empty, we say that the decomposition is *proper*.

Definition 4.3 [DECOMPOSABLE GRAPH]
We say that an undirected graph $\mathcal{G}$ is *decomposable* if either: (i) it is complete, or (ii) it possesses a proper decomposition (A, B, C) such that both subgraphs $\mathcal{G}_{A\cup C}$ and $\mathcal{G}_{B\cup C}$ are decomposable. □

Note that this is a recursive definition, which is permissible because the decomposition (A, B, C) is required to be proper, so that each of $\mathcal{G}_{A\cup C}$ and $\mathcal{G}_{B\cup C}$ has fewer vertices than the original graph $\mathcal{G}$.

There is a strong connection between decomposability and chordality, as captured by the following theorem:

Theorem 4.4 *The following conditions are equivalent for an undirected graph $\mathcal{G}$:*

1. *$\mathcal{G}$ is decomposable;*

2. *$\mathcal{G}$ is chordal;*

3. *Every minimal (α, β)-separator is complete.*

Proof. The result is well-known (Berge 1973; Golumbic 1980). The present proof is taken from Lauritzen (1996).

We proceed by induction on the number of vertices $|V|$ of $\mathcal{G}$. The result is trivial for a graph with no more than three vertices since the three conditions are then all automatically fulfilled. So assume the result to hold for all graphs with $|V| \leq n$ and consider a graph $\mathcal{G}$ with $n+1$ vertices.

First we show $1 \Rightarrow 2$. Suppose that $\mathcal{G}$ is decomposable. If it is complete, it is obviously chordal. Otherwise it has a decomposition (A, B, C) into decomposable subgraphs $\mathcal{G}_{A \cup C}$ and $\mathcal{G}_{B \cup C}$, both with fewer vertices. By the inductive hypothesis these are chordal. Thus, the only possibility for a chordless cycle is one that intersects both A and B. But because C separates A from B, such a cycle must intersect C at least twice. But then it contains a chord because C is complete.

Now we show $2 \Rightarrow 3$. Assume that $\mathcal{G}$ is chordal and let C be a minimal (α, β)-separator. If C has only one vertex, it is complete. If not, it contains at least two vertices, γ_1 and γ_2 say. Since C is a minimal separator, there will be paths from α to β via γ_1 and back via γ_2. The sequence

$$(\alpha, \ldots, \gamma_1, \ldots, \beta, \ldots, \gamma_2, \ldots, \alpha)$$

forms a cycle, with the modification that it can have repeated points. These, and chords other than a link between γ_1 and γ_2, can be used to shorten the cycle, still leaving at least one vertex in the component $[\alpha]_{V \setminus C}$ and one in $[\beta]_{V \setminus C}$, where these symbols denote the connected components of the graph $\mathcal{G}_{V \setminus C}$ containing α and β respectively. This produces eventually a cycle of length at least 4, which must have a chord, whereby we get that $\gamma_1 \sim \gamma_2$. Repeating the argument for all pairs of vertices in C gives that C is complete.

Finally we show that $3 \Rightarrow 1$. Suppose that every minimal (α, β)-separator is complete. If $\mathcal{G}$ is complete there is nothing to show. Otherwise it has at least two non-adjacent vertices, α and β. Assume that the result has been established for every proper subgraph of $\mathcal{G}$. Let C be a minimal (α, β)-separator and partition the vertex set into $[\alpha]_{V \setminus C}$, $[\beta]_{V \setminus C}$, C and D (where D is the set of remaining vertices). Then, since C is complete, the triple (A, B, C), where $A = [\alpha]_{V \setminus C} \cup D$, and $B = [\beta]_{V \setminus C}$, forms a decomposition of $\mathcal{G}$. But each of the subgraphs $\mathcal{G}_{A \cup C}$ and $\mathcal{G}_{B \cup C}$ must be decomposable. For if C_1 is a minimal (α_1, β_1)-separator in $\mathcal{G}_{A \cup C}$, it is also a minimal separator in $\mathcal{G}$ and therefore complete by assumption. The inductive assumption implies that $\mathcal{G}_{A \cup C}$ is decomposable, and similarly with $\mathcal{G}_{B \cup C}$. Thus, we have decomposed $\mathcal{G}$ into decomposable subgraphs. □

The smallest graph that is not decomposable is a 4-cycle, as displayed in Figure 4.1(ii).

A directed acyclic graph for which the parents of every node form a complete set is called *perfect*. For example, Figure 4.3(i) is a perfect directed graph. If $\mathcal{G}$ is an undirected graph, then a numbering of its vertices, $(v_1, \dots, v_k)$ say, is called *perfect* if the neighbours of any node that have lower numbers, i.e., $\text{ne}(v_j) \cap \{v_1, \dots, v_{j-1}\}$, induce a complete subgraph.

For example, in the graph of Figure 4.7(i), $(A_1, B_2, C_3, D_4, E_5, F_6)$ is a perfect numbering, but this is not the case for $(A_1, B_2, C_3, E_4, F_5, D_6)$. For, in the latter numbering, the previously numbered neighbours of D, i.e., (B, C, E, F), do not induce a complete graph. Note that any vertex numbering of a complete undirected graph is perfect.

Given a well-ordered perfect directed graph $\mathcal{G}$, its undirected version $\mathcal{G}^{\sim}$ is a chordal graph for which the ordering $(v_1, v_2, \dots, v_k)$ constitutes a perfect numbering. This is easily seen by induction using the fact that for all j the triple (W_j, V_{j-1}, S_j) forms a decomposition of $\mathcal{G}^{\sim}_{V_j}$, where $V_j = \{v_1, \dots, v_j\}$, $W_j = \text{cl}^{\sim}(v_j) \cap V_j$, and $S_j = W_j \cap V_{j-1}$. Here $\text{cl}^{\sim}$ denotes closure relative to the undirected graph $\mathcal{G}^{\sim}$.

Conversely, given an undirected graph $\mathcal{G}$ and a perfect numbering of its vertices, $(v_1, \dots, v_k)$, one can construct a perfect directed graph simply by directing the edges from lower to higher numbered vertices. It follows that the graph $\mathcal{G}$ must be chordal for such a perfect numbering to exist. More precisely, the following result holds true:

Theorem 4.5 *An undirected graph is chordal if and only if it admits a perfect numbering.*

Proof. See Lauritzen (1996), Proposition 2.17. □

It is worth noting the slightly stronger result that if $\mathcal{G}$ is chordal and v is an arbitrary node of $\mathcal{G}$, then a perfect numbering of $\mathcal{G}$ exists with $v_1 = v$: see Algorithm 4.9 below.

4.3 Junction trees

In this section we summarize some of the important properties of junction trees and their relationship to decomposable graphs.

Let $\mathcal{C}$ be a collection of subsets of a finite set V and $\mathcal{T}$ a tree with $\mathcal{C}$ as its node set. Then $\mathcal{T}$ is said to be a *junction tree* if any intersection $C_1 \cap C_2$ of a pair C_1, C_2 of sets in $\mathcal{C}$ is contained in every node on the unique path in $\mathcal{T}$ between C_1 and C_2. Equivalently, for any vertex v in $\mathcal{G}$, the set of subsets in $\mathcal{C}$ containing v induces a connected subtree of $\mathcal{T}$. Junction trees also appear under other names in the literature, e.g., *join trees* in relational databases (see Section 4.5).

Now let $\mathcal{G}$ be an undirected graph, and $\mathcal{C}$ the family of its cliques. If $\mathcal{T}$ is a junction tree with $\mathcal{C}$ as its node set, we say that $\mathcal{T}$ is a junction tree (of cliques) for the graph $\mathcal{G}$. We have

Theorem 4.6 *There exists a junction tree $\mathcal{T}$ of cliques for the graph $\mathcal{G}$ if and only if $\mathcal{G}$ is decomposable.*

Proof. The theorem clearly holds if $\mathcal{G}$ contains at most two cliques. Suppose that the theorem holds for all graphs with at most k cliques and let $\mathcal{G}$ have $k+1$ cliques.

Assume $\mathcal{T}$ is a junction tree of cliques for $\mathcal{G}$. Take C_1 and C_2 adjacent in $\mathcal{T}$. On cutting the link $C_1 \sim C_2$, $\mathcal{T}$ separates into two subtrees, $\mathcal{T}_1$ and $\mathcal{T}_2$. Let V_i be the union of the nodes in $\mathcal{T}_i$ for $i = 1, 2$, and let $\mathcal{G}_i = \mathcal{G}_{V_i}$. The nodes in $\mathcal{T}_i$ are then the cliques of $\mathcal{G}_i$, and $\mathcal{T}_i$ is a junction tree for $\mathcal{G}_i$. By the inductive hypothesis, $\mathcal{G}_1$ and $\mathcal{G}_2$ are both decomposable. Thus, we are done if we can show that $S := V_1 \cap V_2$ is complete and separates V_1 from V_2. Suppose $v \in V_1 \cap V_2$. Then there exists a clique C_i' of $\mathcal{G}_i$ for each $i = 1, 2$, with $v \in C_i'$. Clearly the path in $\mathcal{T}$ joining C_1' and C_2' passes through both C_1 and C_2. Therefore, $v \in C_1 \cap C_2$ and so we must have $V_1 \cap V_2 \subseteq C_1 \cap C_2$. Since clearly $C_1 \cap C_2 \subseteq V_1 \cap V_2$, we must have that $S = C_1 \cap C_2$ and that S is complete.

Now take $u \in V_1 \setminus S$ and $v \in V_2 \setminus S$, and suppose there exists a path $u, w_1, w_2, \dots, w_k, v$ with each $w_i \notin S$. Then there exists a clique C containing the complete set $\{u, w_1\}$. Clearly $C \subseteq V_1$, so $w_1 \in V_1$, whence $w_1 \in V_1 \setminus S$. Repeat the argument to deduce $w_2 \in V_1 \setminus S, \dots, v \in V_1 \setminus S$. This is a contradiction, hence S separates V_1 from V_2, and (V_1, V_2, S) is a decomposition of $\mathcal{G}$. We have now decomposed $\mathcal{G}$ into subgraphs that possess junction trees and thus are decomposable by the inductive assumption.

Conversely, assume that $\mathcal{G}$ is decomposable, and let (W_1, W_2, S) be a decomposition of $\mathcal{G}$ into proper decomposable subgraphs $\mathcal{G}_{V_1}$, $\mathcal{G}_{V_2}$, where $V_i = W_i \cup S$. Then at least one of V_1 and V_2 — say V_1 — has the form $\bigcup_{C \in \mathcal{C}_1} C$, with $\mathcal{C}_1 \subset \mathcal{C}$; and then we can, if necessary, redefine $V_2 = \bigcup_{C \in \mathcal{C}_2} C$ (with $\mathcal{C}_2 = \mathcal{C} \backslash \mathcal{C}_1$) and still have a decomposition. Let $C_i \in \mathcal{C}_i$ satisfy $S \subseteq C_i$. By hypothesis, we have a junction tree $\mathcal{T}_i$ for $\mathcal{G}_i$ where, as before, $\mathcal{G}_i = \mathcal{G}_{V_i}$. Form $\mathcal{T}$ by linking C_1 in $\mathcal{T}_1$ to C_2 in $\mathcal{T}_2$.

Let $v \in V$. If $v \notin V_2$, then all cliques containing v are in $\mathcal{C}_1$, and so connected in $\mathcal{T}_1$, hence in $\mathcal{T}$. If $v \notin V_1$, then similarly for $\mathcal{T}_2$. Otherwise $v \in S$. The cliques in $\mathcal{C}_i$ containing v are connected in $\mathcal{T}_i$, and include C_i. Since C_1 and C_2 are connected in $\mathcal{T}$, the result follows. □

The above proof demonstrates that an intersection $S = C_1 \cap C_2$ between two neighbouring nodes in a junction tree of cliques separates the decomposable graph $\mathcal{G}$ (in fact, is a minimal separator). We therefore call S the *separator* associated with the edge between C_1 and C_2 of the junction tree; we use the term separator also in the case where the nodes of the junction tree are not all cliques. It is possible that distinct edges may have identical separators. The set of all separators, including any such repetitions, will be denoted by $\mathcal{S}$. When $\mathcal{G}$ admits more than one junction tree of cliques, it can be shown that $\mathcal{S}$ will be the same for all of them.

The separators are often displayed as labels on the edges of a junction tree. They play an important role in the propagation algorithms discussed in Chapters 6 to 8.

A clique $C^* \in \mathcal{C}$ is called *extremal* if, with $V_2 = \bigcup_{C \in \mathcal{C} \setminus \{C^*\}} C$, the triple $(C^* \setminus V_2, V_2 \setminus C^*, C^* \cap V_2)$ is a decomposition of $\mathcal{G}$. We have:

Corollary 4.7 *If a chordal graph $\mathcal{G}$ has at least two cliques, it has at least two extremal cliques.*

Proof. Any junction tree of $\mathcal{G}$ has at least two leaves. □

Figure 4.8 shows junction trees constructed from the chordal graphs of Figure 4.7, where the separators are displayed on the links as rectangular. There are two possible junction tree structures for Figure 4.7(ii), the difference not being in their cliques but in the way they are connected.

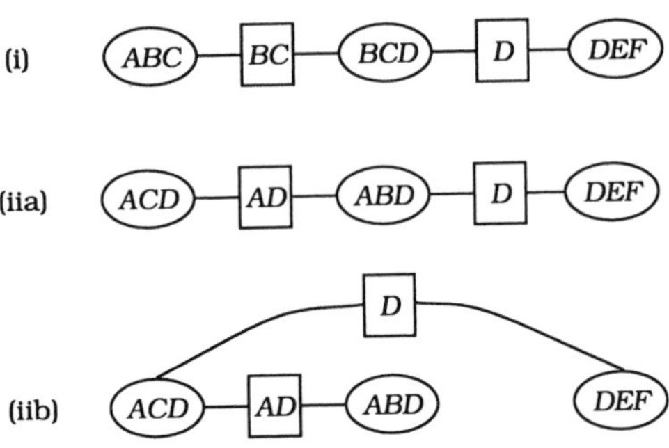

FIGURE 4.8. Junction trees of the chordal graphs of Figure 4.7.

A sequence $(C_1, C_2, \dots, C_k)$ of sets is said to have the *running intersection property* if, for all $1 < j \leq k$, there is an $i < j$ such that $C_j \cap (C_1 \cup \cdots \cup C_{j-1}) \subseteq C_i$.

The cliques of a decomposable graph can be ordered to satisfy this property simply by well-ordering the junction tree. Conversely, if the cliques have been ordered to satisfy the running intersection property, a junction tree (of cliques) can be built using the following algorithm.

Algorithm 4.8 [JUNCTION TREE CONSTRUCTION]
From the cliques $(C_1, \dots, C_p)$ of a chordal graph ordered to have running intersection property:

1. Associate a node of the tree with each clique C_i.

2. For $i = 2, \dots, p$, add an edge between C_i and C_j where j is any one value in $\{1, \dots, i-1\}$ such that

$$C_i \cap (C_1 \cup \cdots \cup C_{i-1}) \subseteq C_j. \qquad \square$$

4.4 From chain graph to junction tree

Advances in the computational analysis of probabilistic networks have come about through the realization that the joint distribution of a probabilistic network can be represented and manipulated efficiently using a junction tree derived from the original graph. This section collects together some of the algorithms for effecting this transformation.

Suppose that a probabilistic network has a chain graph structure $\mathcal{K}$ (we include as a possibility that $\mathcal{K}$ may be a directed or undirected graph). In Chapter 6 we shall see that the first stage in passing to the inference structure is to form the moral graph $\mathcal{K}^m$. The moral graph is undirected, but it may not be a chordal graph. Tarjan and Yannakakis (1984) gave the following efficient algorithm, and proved its correctness, for deciding whether a given undirected graph $\mathcal{G} = (V, E)$ is chordal or not; they also showed that it can be implemented to run in $O(n + e)$ time where $n = |V|$ is the number of nodes and $e = |E|$ the number of edges:

Algorithm 4.9 [MAXIMUM CARDINALITY SEARCH]

- Set **Output**:= '$\mathcal{G}$ is chordal'.
- Set counter $i := 1$.
- Set $L = \emptyset$.
- For all $v \in V$, set $c(v) := 0$.
- While $L \neq V$:
  - Set $U := V \setminus L$.
  - Select any vertex v maximizing $c(v)$ over $v \in U$ and label it i.
  - If $\Pi_{v_i} := \mathrm{ne}(v_i) \cap L$ is not complete in $\mathcal{G}$:
    Set **Output**:= '$\mathcal{G}$ is not chordal'.
  - Otherwise, set $c(w) = c(w) + 1$ for each vertex $w \in \mathrm{ne}(v_i) \cap U$.

– Set $L = L \cup \{v_i\}$.

– Increment i by 1.

• Report **Output**. □

At each stage, L consists of all previously labelled vertices. The algorithm recursively labels vertices in such a way as to maximize the cardinality of the set of previously labelled neighbours. If at any stage this set is not complete, $\mathcal{G}$ is not chordal. The process could be aborted at this stage.

If $\mathcal{G}$ passes the maximum cardinality search, the vertex numbering found will be perfect, as for any node v the set Π_v of its previously numbered neighbours will be complete, and thus $\mathcal{G}$ must be chordal. The converse is also true:

Theorem 4.10 *If $\mathcal{G}$ is chordal, then maximum cardinality search will provide a perfect numbering of $\mathcal{G}$.*

Proof. See Tarjan and Yannakakis (1984). □

If a graph is chordal and its vertices have been numbered by maximum cardinality search, its cliques can be identified in a simple fashion using the algorithm described below, which simultaneously provides an ordering of the cliques having the running intersection property.

Algorithm 4.11 [FINDING THE CLIQUES OF A CHORDAL GRAPH]
Starting from a numbering $(v_1, \dots, v_k)$ obtained by maximum cardinality search, we can find the cliques of a chordal graph as follows. Denote the cardinality of Π_{v_i} by π_i. Call node v_i a *ladder node* if $i = k$, or if $i < k$ and $\pi_{i+1} < 1 + \pi_i$. Let the jth ladder node, in ascending order, be λ_j, and define $C_j = \{\lambda_j\} \cup \Pi_{\lambda_j}$. □

Theorem 4.12 *There is a one-to-one correspondence between the ladder nodes and the cliques of $\mathcal{G}$, the clique associated with ladder node λ_j being C_j. The clique ordering $(C_1, C_2, \dots)$ will possess the running intersection property.*

Proof. Again we may suppose that $\mathcal{G}$ is connected. We argue by induction. If $|V| = 1$ there is nothing to prove. Suppose the algorithm works for $|V| \leq n$, and consider a case having $|V| = n + 1$. Let $v^* = v_{n+1}$, $\Pi^* = \Pi_{v_{n+1}} = \mathrm{bd}(v^*)$, $\pi^* = \pi_{n+1} = |\Pi^*|$. The first n nodes numbered induce a subgraph $\mathcal{G}'$ on $V' = V \setminus \{v^*\}$, which can itself be regarded as having been numbered by the same algorithm. Consequently, by the inductive hypothesis we can suppose that the algorithm has supplied a clique-ordering for $\mathcal{G}'$, $(C'_1, \dots, C'_p)$ say, with the running intersection property.

Since, by Theorem 4.10, Π^* is complete, there exists a clique C'_m of $\mathcal{G}'$ such that $\Pi^* \subseteq C'_m$. Let v_i be its corresponding ladder node. We distinguish two cases, according as whether (i) $\Pi^* \neq C'_m$, or (ii) $\Pi^* = C'_m$. Note that if $m < p$ we must be in case (i), since otherwise the maximum cardinality property would have selected v^* over v_{i+1}.

In case (i), $C_{p+1} := \{v^*\} \cup \Pi^*$ does not properly contain any clique of $\mathcal{G}'$. Taking $C_j = C'_j$ for $j = 1, \ldots, p$, it is easily seen that the algorithm behaves as asserted for the full graph $\mathcal{G}'$, delivering cliques $(C_j : j = 1, \ldots, p+1)$ having the running intersection property.

Otherwise, if we are in case (ii), we must have $p = m$. It readily follows that, in the final numbering of V, v_n is no longer a ladder node, while $v^* = v_{n+1}$ is; and that the algorithm applied to the full graph $\mathcal{G}'$ has again behaved as asserted, delivering cliques $C_j = C'_j$ for $j < p$, $C_p = C'_p \cup \{v^*\}$, having the running intersection property. □

Algorithm 4.11 is essentially the same as ordering the cliques by their largest numbered nodes in a maximum cardinality ordering, as described, for example, by Leimer (1989). However, Algorithm 4.11 can be carried out 'online' during the execution of Algorithm 4.9. Then, if $\mathcal{G}$ is chordal, one pass of this combined algorithm will not only verify the fact, but also identify its cliques, together with a running intersection ordering for them.

Note that Algorithm 4.11 need not work for an arbitrary perfect numbering if it is not generated by maximum cardinality search. A counterexample is given by the graph in Figure 4.9.

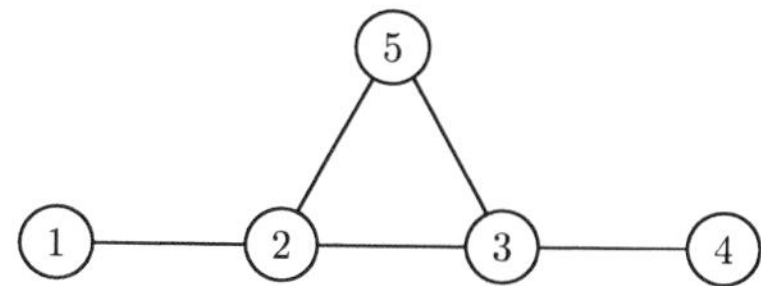

FIGURE 4.9. The numbering of the vertices is perfect, but the cliques, numbered as $(\{1,2\}, \{3,4\}, \{2,3,5\})$, do not have the running intersection property. This numbering could not have been generated by maximum cardinality search (which at least would reverse 4 and 5).

4.4.1 *Triangulation*

If the graph $\mathcal{G} = (V, E)$ is not chordal, it can always be made so by adding extra edges F in a suitable way to form $\mathcal{G}' = (V, E')$, where $E' = E \cup F$. The edges in F are referred to as *fill-in* edges. If $\mathcal{G}'$ is chordal, we refer to it as a *triangulation* of $\mathcal{G}$.

In general, given any ordering, say $(v_1, \ldots, v_k)$, of the nodes of an undirected graph $\mathcal{G}$, one can triangulate $\mathcal{G}$ by recursively examining each node v_j in turn in reverse order, beginning with v_k, and joining those pairs of neighbours that appear earlier in the ordering and are not already joined. The end result is a chordal graph $\mathcal{G}'$. Clearly the given ordering $(v_1, \ldots, v_k)$

is then a perfect numbering for the triangulation $\mathcal{G}'$ of $\mathcal{G}$. The problem of obtaining a good triangulation is thus one of finding a good ordering, but the general problem of finding optimal triangulations for undirected graphs is *NP-hard* (Yannakakis 1981), so heuristic algorithms must be developed.

Kjærulff (1992) examined various algorithms for triangulating a non-chordal graph. For problems in which large cliques are unavoidable the method of simulated annealing performs well. Although using simulated annealing to find a triangulation of a graph may be time-consuming, for any given probabilistic network it only needs to be performed once, so it can be a worthwhile investment of time for some problems. Kjærulff (1992) also looked at a number of heuristic algorithms that involve selecting the next node v on the basis of some *optimality criterion* $c(v)$, for example, either maximizing or minimizing some cost or utility function which depends upon the node being selected. The basic algorithm is described by Olmsted (1983) and Kong (1986) and runs as follows for an undirected graph $\mathcal{G}$ with k vertices.

Algorithm 4.13 [ONE-STEP LOOK AHEAD TRIANGULATION]

- Start with all vertices unnumbered, set counter $i := k$.
- While there are still some unnumbered vertices:
  - Select an unnumbered vertex v to optimize the criterion $c(v)$.
  - Label it with the number i, i.e., let $v_i := v$.
  - Form the set C_i consisting of v_i and its unnumbered neighbours.
  - Fill in edges where none exist between all pairs of vertices in C_i.
  - Eliminate v_i and decrement i by 1. □

Note that this algorithm operates with a numbering strategy opposite to that of maximum cardinality search. The quality of the triangulation, with regard to computational efficiency in applications to probabilistic networks, will depend upon the optimality criterion $c(v)$ used to select vertices.

For models with discrete random variables, selecting $c(v_j)$ to be the cardinality of the joint state space for the variables in the set C_j usually yields good results. Another possibility, which does not depend upon the interpretation of the variables, is to take $c(v)$ to be the number of fill-in edges required if v were to be selected for labelling. For other methods and comparisons between them, see Kjærulff (1992).

Although the maximum cardinality search algorithm is efficient for testing the chordality of a graph, as a numbering method for generating a chordal graph from a non-chordal graph it tends to introduce many more fill-in edges than are necessary, which in turn leads to larger than necessary cliques and reduces the efficiency of the algorithms for probabilistic computations.

4.4.2 Elimination tree

For a given numbering $(v_1, \dots, v_k)$ of the nodes of a graph $\mathcal{G}$ one can associate the sequence of sets $(C_1, \dots, C_k)$ defined by Algorithm 4.13. By construction, each set C_j has the following properties: it contains v_j; the indices of any remaining nodes in C_j are smaller than j; and v_j is not found in any earlier set, i.e., $v_j \notin C_l$ for all $l < j$. These sets are called the *elimination sets* induced by the numbering, and they may be used to form a tree structure called an *elimination tree* (Cowell 1994); this can be useful as an intermediate step to forming a junction tree of cliques and as the basis for the propagation algorithms (see Chapters 6 and 8).

Algorithm 4.14 [ELIMINATION TREE CONSTRUCTION]

1. Associate a node of the tree with each set C_i.

2. For $i = 1, \dots, k$, if C_i contains more than one vertex, add an edge between C_i and C_j where j is the largest index of a vertex in $C_i \setminus \{v_i\}$.

□

It is a simple matter to see that the sequence $(C_1, \dots, C_k)$ has the running intersection property and that the elimination tree is therefore a junction tree of sets. However, the elimination tree is generally not a junction tree of cliques of the triangulated graph $\mathcal{G}'$, because although the sequence $(C_1, \dots, C_k)$ will contain the cliques of $\mathcal{G}'$, it will also contain some subsets of the cliques. Figure 4.10 shows the elimination tree, using the elimination ordering obtained by ordering the nodes numerically, derived from the graph of Figure 4.9. The notation 5:23 reflects that when node 5 is eliminated its boundary, bd(5), is 23.

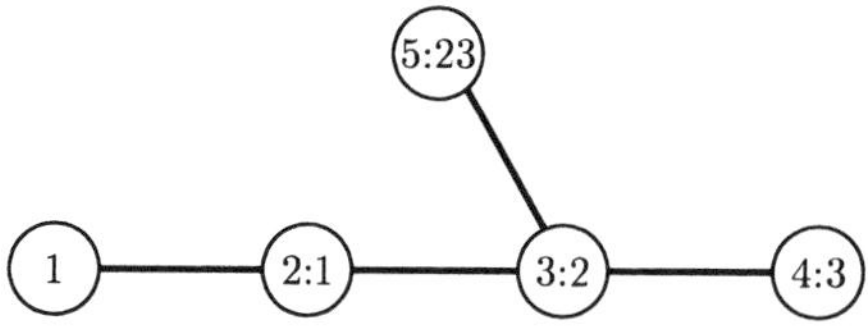

FIGURE 4.10. Elimination tree obtained from the chordal graph of Figure 4.9 using the given numbering.

Theorem 4.15 *The cliques of the triangulated graph $\mathcal{G}'$ are contained in the set of elimination sets $(C_1, \dots, C_k)$.*

Proof. Let C be any clique of the triangulated graph $\mathcal{G}'$ formed by the triangulation algorithm. Then it possesses a greatest-numbered vertex v. At the stage when v was eliminated, it must have been a neighbour of all the other vertices in the clique; for if there is a vertex w for which this is not true, at no stage subsequent to eliminating v could the edge $v \sim w$ be added because the algorithm only adds edges between unnumbered vertices, and at all later stages v remains numbered. Hence, we may identify C with the elimination set formed by eliminating v because the algorithm adds sufficient edges to ensure all pairs of unnumbered neighbours of v are joined. □

One can thus find the cliques of $\mathcal{G}'$ simply by deleting redundant elimination sets, i.e., sets that are contained in other sets. However, this will in general destroy the running intersection property of the sequence. For example, deleting the elimination set $\{2, 3\}$ of the graph in Figure 4.9 (with elimination tree displayed in Figure 4.10) will destroy the running intersection property of the sequence $(\{1\}, \{1, 2\}, \{2, 3\}, \{3, 4\}, \{2, 3, 5\})$.

However, Lemma 4.16 below, due to Leimer (1989) (see also Lemma 2.13 of Lauritzen (1996)) shows how to modify the ordering when a redundant subset is to be deleted. We omit the proof.

Lemma 4.16 *Let $C_1, \ldots, C_k$ be a sequence of sets having the running intersection property. Assume that $C_t \subseteq C_p$ for some $t \neq p$ and that p is minimal with this property for fixed t. Then:*

(i) *If $t > p$, then $C_1, \ldots, C_{t-1}, C_{t+1}, \ldots, C_k$ has the running intersection property.*

(ii) *If $t < p$, then $C_1, \ldots, C_{t-1}, C_p, C_{t+1}, \ldots, C_{p-1}, C_{p+1}, \ldots, C_k$ has the running intersection property.*

Note that the condition $t < p$ is always true for the sequence of elimination sets generated by a vertex numbering because C_p contains v_p by construction, and no lower numbered elimination set can contain v_p because v_p remains numbered when these sets are formed.

We can remove those sets in $(C_1, \ldots, C_k)$ that are proper subsets of others until there is no redundancy, by repeatedly applying Lemma 4.16. The result is an ordering of the cliques of $\mathcal{G}'$ having the running intersection property. These can now be joined up to form a junction tree using Algorithm 4.8. Alternatively, knowing that $\mathcal{G}'$ is now chordal, maximum cardinality search combined with Algorithm 4.11 can be used to find the cliques and order them with the running intersection property, enabling the junction tree of cliques to be built.

4.5 Background references and further reading

There are many general textbooks on graph theory: Harary (1972) and Berge (1973) are standard references.

Chordal graphs are well-studied objects which appear under a variety of names, including triangulated and decomposable graphs, and also *rigid circuit graphs* (Dirac 1961). They are extensively dealt with in Golumbic (1980). Chain graphs were introduced by Lauritzen and Wermuth (1984); see also Lauritzen and Wermuth (1989).

The notion of a graph decomposition has deep connections to many areas of mathematics (Lauritzen et al. 1984; Diestel 1987, 1990), including the four-colour problem (Wagner 1937), measure theory (Kellerer 1964a, 1964b; Vorob'ev 1962, 1963), the solution of systems of linear equations (Parter 1961; Rose 1970, 1973), game theory (Vorob'ev 1967), and relational databases (Beeri et al. 1981, 1983).

The notion of a junction tree has appeared under an abundance of names. The first explicit identification seems to be in relational databases, where it has been known as a *join tree* (Maier 1983); the terms *k-tree* (Arnborg et al. 1987), *Markov tree* and *hypertree* (Shenoy and Shafer 1990), or simply *clique tree* have also been used.

There is an extensive literature concerned with algorithms for manipulating decomposable graphs in an efficient way. It includes other algorithms for checking decomposability of a graph and finding their cliques (Rose et al. 1976; Gavril 1972), for constructing optimal junction trees for given decomposable graphs (Jensen and Jensen 1994), and for constructing optimal decompositions of a non-chordal graph into its indecomposable components (Tarjan 1985; Leimer 1993). There is also some efficiency to be gained by constructing junction trees of special types (Almond 1995; Shenoy 1997). For recent results on triangulation algorithms, see Becker and Geiger (1996), Larrañaga et al. (1997), and Meilă and Jordan (1997).

5

Markov Properties on Graphs

The Markov properties of graphs provide a theoretical foundation of localized computation for inference in probabilistic networks. General chain graphs and their specializations — directed and undirected graphs — each have different types of Markov properties. A common theoretical tool to understanding these properties is the notion of conditional independence.

5.1 Conditional independence

In Chapter 2 we saw simple examples of probabilistic networks, in which the factorization of joint distributions is expressed by directed graphs, and inference consists of reversing arrows. Our aim is to develop tools for manipulating the probability distributions of variables in models that factorize over graphical structures having more complicated topologies, and to enable efficient inference to be performed for such models.

As a precursor to this we need to introduce the notion of *conditional independence*, which will allow us to justify the local computations developed for our inference process later in the book. For simplicity, we mainly confine ourselves to distributions of discrete variables, each having a finite number of states. We will let V denote the index set of a collection of variables $(X_v), v \in V$ taking values in probability spaces $\mathcal{X}_v, v \in V$. The probability spaces can be quite general, just sufficiently well-behaved to ensure the existence of regular conditional probabilities. For A being a typical subset of

V, we let $\mathcal{X}_A = \times_{v \in A} \mathcal{X}_v$ and further $\mathcal{X} = \mathcal{X}_V$. Typical elements of $\mathcal{X}_A$ are denoted by $x_A = (x_v)_{v \in A}$ and so on.

Thus, let $X, Y, Z, \ldots$ denote random variables with a joint distribution P, having density p with respect to a product measure. Note that each variable may itself be a random vector. We can consider, for any possible value y of Y, the conditional distribution of X given $Y = y$, denoted by $D(X \mid Y = y)$. This is defined if $p(y) > 0$, in which case we may call y a *possible value* of Y. Thus, if A denotes a set of possible values for X, then $D(X \mid Y = y)$ attaches to A the conditional probability value $P(X \in A \mid Y = y)$.

Definition 5.1 [CONDITIONAL INDEPENDENCE]

We say X is *conditionally independent of Y given Z*, and write $X \perp\!\!\!\perp Y \mid Z$, if, for any possible pair of values (y, z) for (Y, Z), we have $D(X \mid Y = y, Z = z) = D(X \mid Z = z)$, i.e., for any A, $P(X \in A \mid Y, Z) = P(X \mid Z)$. □

As a special case we note that the expression $X \perp\!\!\!\perp Y$ means that we have $D(X \mid Y = y) = D(X)$ (the marginal distribution of X), for any possible value y of Y; we say that X and Y are (marginally) *independent.*

Suppose for simplicity that all variables are discrete. Similar properties hold for continuous quantities. Let $p(x, y \mid z)$ denote $P(X = x, Y = y \mid Z = z)$, and let $a(x, z)$, for example, denote unspecified functions of x, z, etc. Then $X \perp\!\!\!\perp Y \mid Z$ if and only if any of the following equivalent conditions holds:

$$
\begin{array}{lll}
\text{C1a}: & p(x \mid y, z) \equiv p(x \mid z) & \text{if } p(y, z) > 0 \\
\text{C1b}: & p(x \mid y, z) \text{ has the form } a(x, z) & \text{if } p(y, z) > 0 \\
\text{C2a}: & p(x, y \mid z) \equiv p(x \mid z) p(y \mid z) & \text{if } p(z) > 0 \\
\text{C2b}: & p(x, y \mid z) \text{ has the form } a(x, z) b(y, z) & \text{if } p(z) > 0 \\
\text{C3a}: & p(x, y, z) \equiv p(x \mid z) p(y \mid z) p(z) & \\
\text{C3b}: & p(x, y, z) \equiv p(x, z) p(y, z) / p(z) & \text{if } p(z) > 0 \\
\text{C3c}: & p(x, y, z) \text{ has the form } a(x, z) b(y, z) &
\end{array}
$$

The ternary relation $X \perp\!\!\!\perp Y \mid Z$ has the following properties:

$$
\begin{array}{llll}
\text{P1}: \text{If } X \perp\!\!\!\perp Y \mid Z & & \text{then} & Y \perp\!\!\!\perp X \mid Z \\
\text{P2}: \text{If } X \perp\!\!\!\perp Y \mid Z & \text{and} \quad U \text{ is a function of } X & \text{then} & U \perp\!\!\!\perp Y \mid Z \\
\text{P3}: \text{If } X \perp\!\!\!\perp Y \mid Z & \text{and} \quad U \text{ is a function of } X & \text{then} & X \perp\!\!\!\perp Y \mid (Z, U) \\
\text{P4}: \text{If } X \perp\!\!\!\perp Y \mid Z & \text{and} \quad X \perp\!\!\!\perp W \mid (Y, Z) & \text{then} & X \perp\!\!\!\perp (W, Y) \mid Z
\end{array}
$$

Another property sometimes holds, viz.:

$$
\text{P5}: \text{If } X \perp\!\!\!\perp Y \mid (Z, W) \quad \text{and} \quad X \perp\!\!\!\perp Z \mid (Y, W) \quad \text{then} \quad X \perp\!\!\!\perp (Y, Z) \mid W,
$$

but only under additional assumptions, essentially that there be no non-trivial logical relationships between Y and Z. This is true when the density

p is strictly positive. For if $p(x, y, z, w) > 0$ and both $X \perp\!\!\!\perp Y \mid (Z, W)$ and $X \perp\!\!\!\perp Z \mid (Y, W)$ hold, then by C3c

$$p(x, y, z, w) = g(x, y, w)\, h(y, z, w) = k(x, z, w)\, l(y, z, w)$$

for suitable strictly positive functions g, h, k, l. Thus, for all z we must have

$$g(x, y, w) = \frac{k(x, z, w)\, l(y, z, w)}{h(y, z, w)}.$$

Hence, choosing a fixed $z = z_0$ we have $g(x, y, w) = \pi(x, w)\rho(y, w)$, where $\pi(x, w) = k(x, z_0, w)$ and $\rho(y, w) = l(y, z_0, w)/h(y, z_0, w)$. Thus, $p(x, y, z, w) = \pi(x, w)\rho(y, w)h(y, z, w)$, and hence $X \perp\!\!\!\perp (Y, Z) \mid W$.

If properties (P1) to (P4) are regarded as axioms with "is a function of" replaced by a suitable partial order, then it is possible to develop an *abstract calculus of conditional independence* which applies to other mathematical systems than probabilistic conditional independence. Any such model of these abstract axioms has been termed a *semi-graphoid* by Pearl (1988) or, if (P5) is also satisfied, a *graphoid*. A range of examples is described by Dawid (1998). Important statistical applications include *meta* conditional independence, which generalizes the concept of a *cut* in a parametric statistical family (Barndorff-Nielsen 1978); and *hyper* conditional independence, which imposes, in addition, corresponding independence properties on a prior distribution over the parameters. A detailed description and discussion may be found in Dawid and Lauritzen (1993). Further application areas of interest in artificial intelligence include concepts of conditional independence for belief functions (Shafer 1976) and various purely logical structures such as, e.g., embedded multi-valued dependencies (Sagiv and Walecka 1982) and natural conditional functions (Spohn 1988; Studený 1995). Purely mathematical examples include orthogonality of linear spaces and the various separation properties in graphs that form the basis of this chapter. The last properties form the reason for Pearl's nomenclature. Virtually all the general results on conditional independence which can be shown to hold for probability distributions can be reinterpreted in these alternative models, and remain valid.

Now let $(X_v), v \in V$ be a collection of random variables, and let $\mathcal{B}$ be a collection of subsets of V. For $B \in \mathcal{B}$, let $a_B(x)$ denote a non-negative function of x depending only on $x_B = (x_v)_{v \in B}$.

Definition 5.2 [HIERARCHICAL DISTRIBUTION]
A joint distribution P for X is $\mathcal{B}$-hierarchical if its probability density p factorizes as

$$p(x) = \prod_{B \in \mathcal{B}} a_B(x).$$

□

Note that if all of the functions a are strictly positive, then P is $\mathcal{B}$-hierarchical if and only if it satisfies the restrictions of a hierarchical log–linear model with generating class $\mathcal{B}^*$ (Christensen 1990), where $\mathcal{B}^*$ is obtained from $\mathcal{B}$ by removing sets that are subsets of other sets in $\mathcal{B}$.

Example 5.3 Let $V = \{A, B, C\}$ and $\mathcal{B} = \{\{A, B\}, \{B, C\}\}$; then the density factorizes as $p(x_A, x_B, x_C) = a(x_A, x_B)b(x_B, x_C)$. □

Example 5.4 Now let $V = \{A, B, C\}$ and $\mathcal{B} = \{\{A, B\}, \{B, C\}, \{A, C\}\}$; then the density factorizes as $p(x_A, x_B, x_C) = a(x_A, x_B)b(x_B, x_C)c(x_A, x_C)$. □

In Example 5.3 the factorization is equivalent to having $X_A \perp\!\!\!\perp X_C \mid X_B$, but Example 5.4 shows that not all factorizations have representations in terms of conditional independence. We can ask what conditional independence properties are implicit in a hierarchical distribution? To answer this we form an undirected graph $\mathcal{G}$ with node set V, in which we join nodes u and v if and only if they occur together within any subset in $\mathcal{B}$. Figure 5.1 shows the graphs obtained in this way for the Examples 5.3 and 5.4.

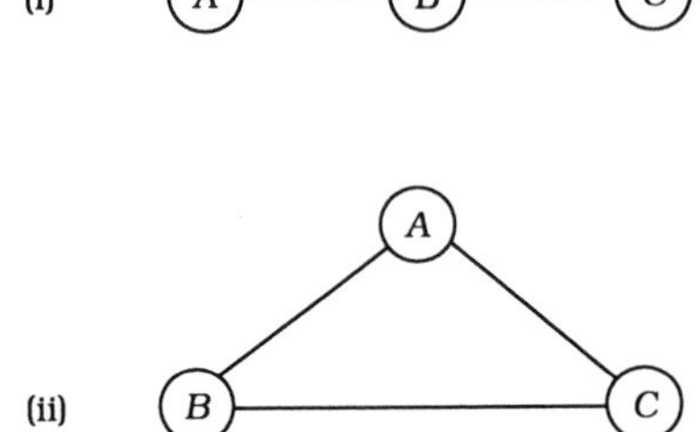

FIGURE 5.1. Undirected graphs formed from the hierarchical distributions of (i) Example 5.3 and (ii) Example 5.4.

Clearly any subset in $\mathcal{B}$ is a complete subset of $\mathcal{G}$. However, in the process of forming $\mathcal{G}$, other complete sets not belonging to $\mathcal{B}$ may be introduced, for example, $\{A, B, C\}$ in Example 5.4. Now let $\mathcal{C}$ denote the collection of cliques of $\mathcal{G}$. Since any subset in $\mathcal{B}$ is contained within some clique in $\mathcal{C}$, it follows that every $\mathcal{B}$-hierarchical distribution is also $\mathcal{C}$-hierarchical. Thus, in discussing the conditional independence properties of hierarchical distributions, we are led to consider the cliques of an underlying graph and, as we shall see, separation properties in such graphs.

5.2 Markov fields over undirected graphs

Let $\mathcal{G} = (V, E)$ be an undirected graph, and consider a collection of random variables $(X_v), v \in V$. Let P be probability measure on $\mathcal{X}_v$ which *factorizes* according to $\mathcal{G}$, i.e., there exist non-negative functions ϕ_A defined on $\mathcal{X}_A$ for only complete subsets A, and a product measure $\mu = \otimes_{v \in V} \mu_v$ on $\mathcal{X}$, such that the density p of P with respect to μ factorizes in the form

$$p(x) = \prod_A \phi_A(x_A).$$

In other words, P factorizes if and only if P is $\mathcal{A}$-hierarchical, for $\mathcal{A}$ the class of complete subsets of $\mathcal{G}$. The functions ϕ_A are referred to as *factor potentials* of P. They are not uniquely determined because there is arbitrariness in the choice of the measure μ, and also groups of potentials can be multiplied together or split up in different ways. One can without loss of generality assume that only the cliques of $\mathcal{G}$ appear in the sets $\mathcal{A}$, i.e., that

$$p(x) = \prod_{C \in \mathcal{C}} \psi_C(x_C), \tag{5.1}$$

where $\mathcal{C}$ is the set of cliques of $\mathcal{G}$, or, in other words, that P is $\mathcal{C}$-hierarchical. If P factorizes as above, we also say that P has property (F).

Associated with the graph $\mathcal{G}$ is a range of Markov properties, different in general. Write $A \perp\!\!\!\perp B \mid C$ if $X_A \perp\!\!\!\perp X_B \mid X_C$ under P. A probability measure P on $\mathcal{X}$ is said to obey:

(P) *the pairwise Markov property*, relative to $\mathcal{G}$, if for any pair (α, β) of non-adjacent vertices,

$$\alpha \perp\!\!\!\perp \beta \mid V \setminus \{\alpha, \beta\};$$

(L) *the local Markov property*, relative to $\mathcal{G}$, if for any vertex $\alpha \in V$,

$$\alpha \perp\!\!\!\perp V \setminus \mathrm{cl}(\alpha) \mid \mathrm{bd}(\alpha);$$

(G) *the global Markov property*, relative to $\mathcal{G}$, if for any triple (A, B, S) of disjoint subsets of V such that S separates A from B in $\mathcal{G}$,

$$A \perp\!\!\!\perp B \mid S.$$

Note that, if we write $A \perp\!\!\!\perp_{\mathcal{G}} B \mid S$ to denote that S separates A from B in $\mathcal{G}$, replace "function" by "subset" in (P2) and (P3), and similarly (Z, U) by $Z \cup U$, etc., then the subsets of V constitute a graphoid under $\perp\!\!\!\perp_{\mathcal{G}}$, and the various Markov definitions relate properties of probabilistic conditional independence $\perp\!\!\!\perp$ to corresponding obvious properties of graph separation $\perp\!\!\!\perp_{\mathcal{G}}$.

In the terminology defined above we have that

$$(\mathrm{F}) \Rightarrow (\mathrm{G}) \Rightarrow (\mathrm{L}) \Rightarrow (\mathrm{P}), \tag{5.2}$$

but in general the properties are different. Note that (5.2) only depends on the properties (P1) to (P4) of conditional independence. If P admits a strictly positive density p with respect to μ, (P5) can also be used and then all the properties are equivalent. This is a consequence of the theorem below, due to Pearl and Paz (1987) (see also Pearl (1988)).

Theorem 5.5 (Pearl and Paz) *If a probability distribution on $\mathcal{X}$ is such that* (P5) *holds for all pairwise disjoint subsets, then*

$$(\mathrm{G}) \iff (\mathrm{L}) \iff (\mathrm{P}).$$

Proof. We need to show that (P) implies (G), so assume that S separates A from B in $\mathcal{G}$ and that (P) as well as (P5) hold. The proof is then reverse induction on the number of vertices $n = |S|$ in S. If $n = |V| - 2$, then both A and B consist of one vertex, and the required conditional independence follows from (P).

So assume $|S| = n < |V| - 2$, and that the independence $A \perp\!\!\!\perp B \mid S$ holds for all S with more than n elements. We first assume that $A \cup B \cup S = V$, implying that at least one of A and B has more than one element, A, say. If $\alpha \in A$ then $S \cup \{\alpha\}$ separates B from $A \setminus \{\alpha\}$, and also $S \cup A \setminus \{\alpha\}$ separates B from α. Thus, by the inductive hypothesis

$$B \perp\!\!\!\perp A \setminus \{\alpha\} \mid S \cup \{\alpha\} \text{ and } B \perp\!\!\!\perp \alpha \mid S \cup A \setminus \{\alpha\}.$$

Now (P5) gives $A \perp\!\!\!\perp B \mid S$.

If $A \cup B \cup S \subset V$ we choose $\alpha \in V \setminus (A \cup B \cup S)$. Then $S \cup \{\alpha\}$ separates A and B, implying $A \perp\!\!\!\perp B \mid S \cup \{\alpha\}$. Further, either $A \cup S$ separates B from $\{\alpha\}$ or $B \cup S$ separates A from $\{\alpha\}$. Assuming the former gives $B \perp\!\!\!\perp \{\alpha\} \mid A \cup S$. Using (P5) we derive $A \perp\!\!\!\perp B \mid S$. The latter case is similar. □

The global Markov property (G) is important because it gives a general criterion for deciding when two groups of variables A and B are conditionally independent given a third group of variables S.

In the case where all state spaces are discrete and P has a positive density, we can show that (P) implies (F), and thus that all Markov properties are equivalent. More precisely, we have the classical result:

Theorem 5.6 *A probability distribution P on a discrete sample space with strictly positive density satisfies the pairwise Markov property if and only if it factorizes.*

Proof. See Lauritzen (1996). □

In general, without positivity assumptions on the density, the global Markov property (G) may not imply the factorization property (F). An example was given by Moussouris (1974) for the graph being a four-cycle (see also Lauritzen (1996)).

When we use the term *Markov probability distribution* on an undirected graph $\mathcal{G}$ without further qualification, we shall always mean one that factorizes, hence satisfies all of the properties. The set of such probability distributions is denoted by $M(\mathcal{G})$. When (A, B, S) forms a decomposition of $\mathcal{G}$ the Markov property is decomposed accordingly:

Proposition 5.7 *Assume that (A, B, S) decomposes $\mathcal{G} = (V, E)$. Then P factorizes with respect to $\mathcal{G}$ if and only if both $P_{A\cup S}$ and $P_{B\cup S}$ factorize with respect to $\mathcal{G}_{A\cup S}$ and $\mathcal{G}_{B\cup S}$ respectively and the density p satisfies*

$$p(x) = \frac{p_{A\cup S}(x_{A\cup S})p_{B\cup S}(x_{B\cup S})}{p_S(x_S)}. \tag{5.3}$$

Proof. Suppose that P factorizes with respect to $\mathcal{G}$ such that

$$p(x) = \prod_{C\in\mathcal{C}} \psi_C(x).$$

Since (A, B, S) decomposes $\mathcal{G}$, all cliques are subsets of either $A \cup S$ or of $B \cup S$, so that

$$p(x) = \prod_{C\in\mathcal{A}} \psi_C(x) \prod_{C\in\mathcal{B}} \psi_C(x) = h(x_{A\cup S})k(x_{B\cup S}).$$

By direct integration we find

$$p(x_{A\cup S}) = h(x_{A\cup S})\bar{k}(x_S)$$

where

$$\bar{k}(x_S) = \int k(x_{B\cup S})\mu_B(dx_B),$$

and similarly with the other marginals. This gives (5.3) as well as the factorizations of both marginal densities. The converse is trivial. □

In the case of discrete sample spaces we further have, if we take $0/0 = 0$:

Proposition 5.8 *Assume that (A, B, S) decomposes $\mathcal{G} = (V, E)$ and the sample space is discrete. Then P is globally Markov with respect to $\mathcal{G}$ if and only if both $P_{A\cup S}$ and $P_{B\cup S}$ are globally Markov with respect to $\mathcal{G}_{A\cup S}$ and $\mathcal{G}_{B\cup S}$ respectively, and*

$$p(x) = \frac{p(x_{A\cup S})p(x_{B\cup S})}{p(x_S)}. \tag{5.4}$$

Proof. See Lauritzen (1996). □

When $\mathcal{G}$ is decomposable, recursive application of (5.3) shows that a distribution P is Markov with respect to $\mathcal{G}$ if and only if it factorizes as

$$p(x) = \frac{\prod_{C\in\mathcal{C}} p(x_C)}{\prod_{S\in\mathcal{S}} p(x_S)},$$

where $\mathcal{C}, \mathcal{S}$ are, respectively, the sets of cliques and separators of $\mathcal{G}$. The clique-marginals $\{P_C\}$ can be assigned arbitrarily, subject only to implying identical marginals over any common separators. Markov properties of decomposable graphs are studied by Dawid and Lauritzen (1993).

5.3 Markov properties on directed acyclic graphs

Before we proceed to the case of a general chain graph we consider the same set-up as in the previous section, except that now the graph $\mathcal{D}$ is assumed to be directed and acyclic.

We say that a probability distribution P admits a *recursive factorization* according to $\mathcal{D}$ if there exist (σ-finite) measures μ_v over $\mathcal{X}$ and non-negative functions $k^v(\cdot,\cdot), v \in V$, henceforth referred to as *kernels*, defined on $\mathcal{X}_v \times \mathcal{X}_{\mathrm{pa}(v)}$ such that

$$\int k^v(y_v, x_{\mathrm{pa}(v)})\mu_v(dy_v) = 1,$$

and P has density p with respect to the product measure $\mu = \otimes_{v \in V}\mu_v$ given by

$$p(x) = \prod_{v \in V} k^v(x_v, x_{\mathrm{pa}(v)}).$$

We then also say that P *has property* (DF). It is easy to show that if P admits a recursive factorization as above, then the kernels $k^v(\cdot, x_{\mathrm{pa}(v)})$ are in fact densities for the conditional distribution of X_v, given $X_{\mathrm{pa}(v)} = x_{\mathrm{pa}(v)}$, and thus

$$p(x) = \prod_{v \in V} p(x_v \mid x_{\mathrm{pa}(v)}). \tag{5.5}$$

Also it is immediate that if we form the (undirected) moral graph $\mathcal{D}^m$ (see Section 4.1) we have the following:

Lemma 5.9 *If P admits a recursive factorization according to the directed acyclic graph $\mathcal{D}$, it factorizes according to the moral graph $\mathcal{D}^m$ and therefore obeys the global Markov property relative to $\mathcal{D}^m$.*

Proof. The factorization follows from the fact that, by construction, the sets $\{v\} \cup \mathrm{pa}(v)$ are complete in $\mathcal{D}^m$ and we can therefore let $\psi_{\{v\}\cup\mathrm{pa}(v)} = k^v$. The remaining part of the statement follows from (5.2). □

This simple lemma has very useful consequences when constructing the inference engine in a probabilistic expert system (see Section 3.2.1 for an example of this). Also, using the local Markov property on the moral graph $\mathcal{D}^m$, we find that

$$v \perp\!\!\!\perp V \setminus v \mid \mathrm{bl}(v),$$

where $\mathrm{bl}(v)$ is the so-called *Markov blanket* of v. The Markov blanket is the set of neighbours of v in the moral graph $\mathcal{D}^m$. It can be found directly from the original DAG $\mathcal{D}$ as the set of v's parents, children, and children's parents:

$$\mathrm{bl}(v) = \mathrm{pa}(v) \cup \mathrm{ch}(v) \cup \{w : \mathrm{ch}(w) \cap \mathrm{ch}(v) \neq \emptyset\}. \tag{5.6}$$

The following result is easily shown:

Proposition 5.10 *If P admits a recursive factorization according to the directed acyclic graph $\mathcal{D}$ and A is an ancestral set, then the marginal distribution P_A admits a recursive factorization according to $\mathcal{D}_A$.*

Corollary 5.11 *Let P factorize recursively according to $\mathcal{D}$. Then*

$$A \perp\!\!\!\perp B \mid S$$

whenever A and B are separated by S in $(\mathcal{D}_{\mathrm{An}(A \cup B \cup S)})^m$, the moral graph of the smallest ancestral set containing $A \cup B \cup S$.

The property in Corollary 5.11 will be referred to as the *directed global Markov property* (DG), and a distribution satisfying it is a *directed Markov field* over $\mathcal{D}$. If we now reinterpret $\perp\!\!\!\perp_{\mathcal{D}}$ to denote the relation between subsets described in Corollary 5.11, the subsets of V again form a graphoid under $\perp\!\!\!\perp_{\mathcal{D}}$, and the global directed Markov property again relates probabilistic conditional independence $\perp\!\!\!\perp$ with graph separation $\perp\!\!\!\perp_{\mathcal{D}}$.

One can show that the global directed Markov property has the same role as the global Markov property does in the case of an undirected graph, in the sense that it gives the sharpest possible rule for reading conditional independence relations off the directed graph. The procedure is illustrated in the following example:

Example 5.12 Consider a directed Markov field on the first graph in Figure 5.2 and the problem of deciding whether $a \perp\!\!\!\perp b \mid S$. The moral graph of the smallest ancestral set containing all the variables involved is shown in the second graph of Figure 5.2. It is immediate that S separates a from b in this moral graph, implying $a \perp\!\!\!\perp b \mid S$. □

An alternative formulation of the global directed Markov property was given by Pearl (1986a) with a formal treatment in Verma and Pearl (1990). Recall that a trail in $\mathcal{D}$ is a sequence of vertices that forms a path in the undirected version $\mathcal{D}^\sim$ of $\mathcal{D}$, i.e., when the directions of arrows are ignored. A trail π from a to b in a directed acyclic graph $\mathcal{D}$ is said to be *blocked* by S if it contains a vertex $\gamma \in \pi$ such that either

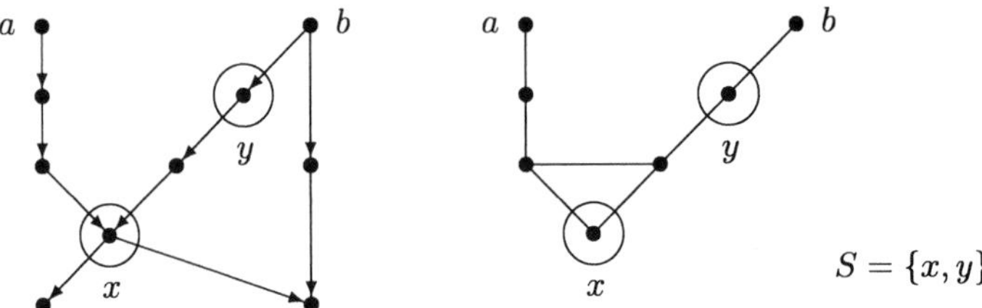

FIGURE 5.2. The directed global Markov property. Is $a \perp\!\!\!\perp b \mid S$? In the moral graph of the smallest ancestral set in the graph containing $\{a\} \cup \{b\} \cup S$, clearly S separates a from b, implying $a \perp\!\!\!\perp b \mid S$.

- $\gamma \in S$ and arrows of π do not meet head-to-head at γ, or
- γ and all its descendants are not in S, and arrows of π meet head-to-head at γ.

A trail that is not blocked by S is said to be *active*. Two subsets A and B are said to be *d-separated* by S if all trails from A to B are blocked by S. We then have the following result:

Proposition 5.13 *Let A, B, and S be disjoint subsets of a directed acyclic graph $\mathcal{D}$. Then S d-separates A from B if and only if S separates A from B in $(\mathcal{D}_{\mathrm{An}(A \cup B \cup S)})^m$.*

Proof. Suppose S does not *d*-separate A from B. Then there is an active trail from A to B such as, for example, the one indicated in Figure 5.3.

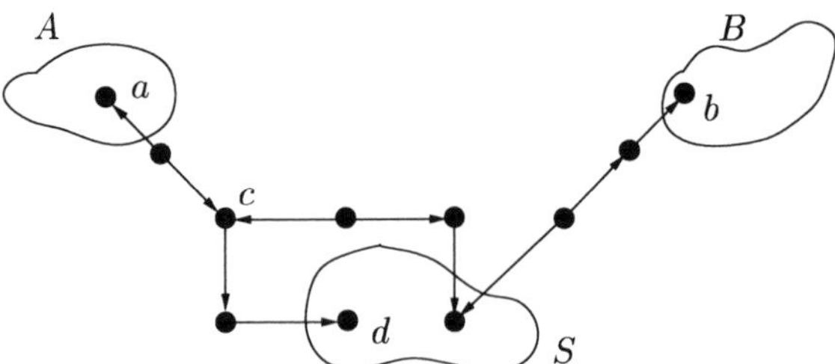

FIGURE 5.3. Example of an active trail from A to B. The path from c to d is not part of the trail, but indicates that c must have descendants in S.

All vertices in this trail must lie within $\mathrm{An}(A \cup B \cup S)$. Because if the arrows meet head-to-head at some vertex γ, either $\gamma \in S$ or γ has descendants in S. And if not, either of the subpaths away from γ either meets another arrow, in which case γ has descendants in S, or leads all the way to A or B. Each of these head-to-head meetings will give rise to a marriage in the moral graph, such as illustrated in Figure 5.4, thereby creating a trail from A to B in $(\mathcal{D}_{\mathrm{An}(A \cup B \cup S)})^m$, circumventing S.

Suppose conversely that A is not separated from B in $(\mathcal{D}_{\mathrm{An}(A\cup B\cup S)})^m$. Then there is a trail in this graph that circumvents S. The trail has pieces that correspond to edges in the original graph and pieces that correspond to marriages. Each marriage is a consequence of a meeting of arrows head-to-head at some vertex γ. If γ is in S or it has descendants in S, the meeting does not block the trail. If not, γ must have descendants in A or B since the ancestral set was smallest. In the latter case, a new trail can be created with one less head-to-head meeting, using the line of descent, such as illustrated in Figure 5.5.

Continuing this substitution process eventually leads to an active trail from A to B, and the proof is complete. □

We illustrate the concept of d-separation by applying it to the query of Example 5.12. As Figure 5.6 indicates, all trails between a and b are blocked by S, whereby the global Markov property gives that $a \perp\!\!\!\perp b \mid S$.

Geiger and Pearl (1990) show that the criterion of d-separation cannot be improved, in the sense that, for any given directed acyclic graph $\mathcal{D}$, one can find state spaces $\mathcal{X}_\alpha, \alpha \in V$ and a probability P such that

$$A \perp\!\!\!\perp B \mid S \iff S \text{ } d\text{-separates } A \text{ from } B. \tag{5.7}$$

Indeed, we can take each state space to be the real plane, with the overall distribution Gaussian. Meek (1995) proved a similar result for the case where the state spaces are all binary.

A variant on the d-separation criterion, well-suited to computation of separating sets, is the "Bayes-ball" algorithm of Shachter (1998).

To complete this section we say that P obeys the *local directed Markov property* (DL) if any variable is conditionally independent of its non-descendants, given its parents

$$v \perp\!\!\!\perp \mathrm{nd}(v) \mid \mathrm{pa}(v).$$

A seemingly weaker requirement, the *ordered directed Markov property* (DO), replaces all non-descendants of v in the above condition by the predecessors $\mathrm{pr}(v)$ of v in some given well-ordering of the nodes:

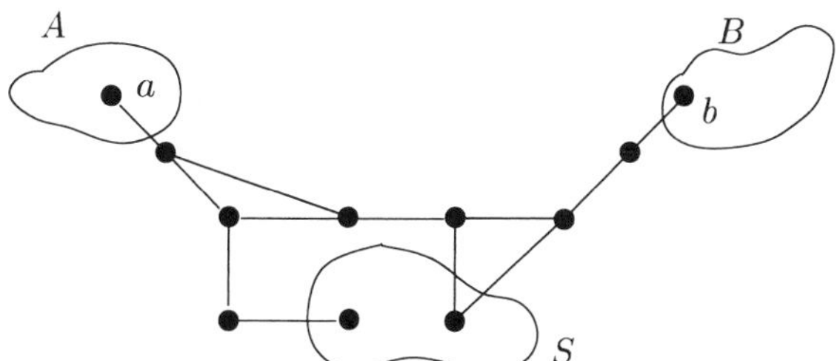

FIGURE 5.4. The moral graph corresponding to the active trail in $\mathcal{D}$.

$$v \perp\!\!\!\perp \mathrm{pr}(v) \mid \mathrm{pa}(v).$$

In contrast with the undirected case we have that all the four properties (DF), (DL), (DG), and (DO) are equivalent just assuming existence of the density p. This is stated formally as:

Theorem 5.14 *Let $\mathcal{D}$ be a directed acyclic graph. For a probability distribution P on $\mathcal{X}$ which has density with respect to a product measure μ, the following conditions are equivalent:*

(DF) *P admits a recursive factorization according to $\mathcal{D}$;*

(DG) *P obeys the global directed Markov property, relative to $\mathcal{D}$;*

(DL) *P obeys the local directed Markov property, relative to $\mathcal{D}$;*

(DO) *P obeys the ordered directed Markov property, relative to $\mathcal{D}$.*

Proof. That (DF) implies (DG) is Corollary 5.11. That (DG) implies (DL) follows by observing that $\{v\} \cup \mathrm{nd}(v)$ is an ancestral set and that $\mathrm{pa}(v)$ obviously separates $\{v\}$ from $\mathrm{nd}(v) \setminus \mathrm{pa}(v)$ in $(\mathcal{D}_{\{v\}\cup \mathrm{nd}(v)})^m$. It is trivial that (DL) implies (DO), since $\mathrm{pr}(v) \subseteq \mathrm{nd}(v)$. The final implication is shown by induction on the number of vertices $|V|$ of $\mathcal{D}$. Let v_0 be the last vertex of $\mathcal{D}$. Then we can let k^{v_0} be the conditional density of X_{v_0}, given $X_{V\setminus\{v_0\}}$, which by (DO) can be chosen to depend on $x_{\mathrm{pa}(v_0)}$ only. The marginal distribution of $X_{V\setminus\{v_0\}}$ trivially obeys the ordered directed Markov property and admits a factorization by the inductive assumption. Combining this factorization with k^{v_0} yields the factorization for P. This completes the proof. □

Since the four conditions in Theorem 5.14 are equivalent, it makes sense to speak of a *directed Markov field* as one where any of the conditions is satisfied. The set of such distributions for a directed acyclic graph $\mathcal{D}$ is denoted by $M(\mathcal{D})$.

In the particular case when the directed acyclic graph $\mathcal{D}$ is perfect (see Section 4.2) the directed Markov property on $\mathcal{D}$ and the factorization prop-

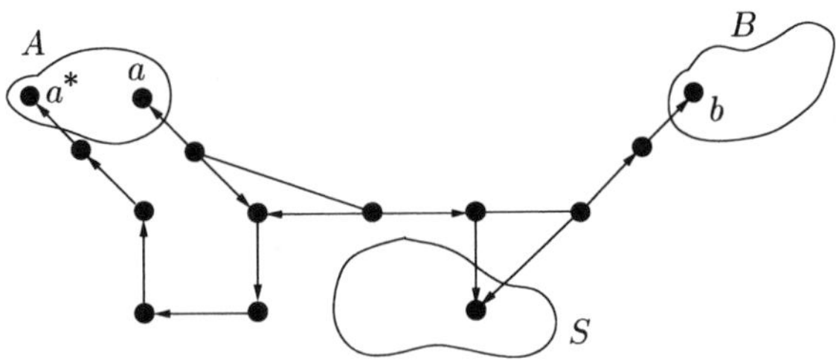

FIGURE 5.5. The trail in the graph $(\mathcal{D}_{\mathrm{An}(A\cup B\cup S)})^m$ makes it possible to construct an active trail in $\mathcal{D}$ from A to B.

erty on its undirected version $\mathcal{D}^{\sim}$ coincide (note that in this case $\mathcal{D}^{\sim}$ is decomposable):

Proposition 5.15 *Let $\mathcal{D}$ be a perfect directed acyclic graph and $\mathcal{D}^{\sim}$ its undirected version. Then a distribution P is directed Markov with respect to $\mathcal{D}$ if and only if it factorizes according to $\mathcal{D}^{\sim}$.*

Proof. That the graph is perfect means that $\mathrm{pa}(\alpha)$ is complete for all $\alpha \in V$. Hence, $\mathcal{D}^m = \mathcal{D}^{\sim}$. From Lemma 5.9 it then follows that any $P \in M(\mathcal{D})$ also factorizes with respect to $\mathcal{D}^{\sim}$.

The reverse inclusion is established by induction on the number of vertices $|V|$ of $\mathcal{D}$. For $|V| = 1$ there is nothing to show. For $|V| = n+1$ let $P \in M(\mathcal{D}^{\sim})$ and find a terminal vertex $\alpha \in V$. This vertex has $\mathrm{pa}_{\mathcal{D}}(\alpha) = \mathrm{bd}_{\mathcal{D}^{\sim}}(\alpha)$ and, since $\mathcal{D}$ is perfect, this set is complete in both graphs as well. Hence, $(V \setminus \{\alpha\}, \{\alpha\}, \mathrm{bd}(\alpha))$ is a decomposition of $\mathcal{D}^{\sim}$ and Proposition 5.7 gives the factorization

$$p(x) = p(x_{V\setminus\{\alpha\}})p(x_{\mathrm{cl}(\alpha)})/p(x_{\mathrm{bd}(\alpha)}) = p(x_{V\setminus\{\alpha\}})k^{\alpha}(x_{\alpha}, x_{\mathrm{pa}(\alpha)}),$$

say, where $\int k^{\alpha}(y_{\alpha}, x_{\mathrm{pa}(\alpha)})\mu_{\alpha}(dy_{\alpha}) = 1$, and the first factor factorizes according to $\mathcal{D}^{\sim}_{V\setminus\{\alpha\}}$. Using the inductive assumption on this factor gives the full recursive factorization of P. □

5.4 Markov properties on chain graphs

In this section we deal with general chain graphs $\mathcal{K} = (V, E)$. We further assume that all probability measures have positive densities, implying that all five of the basic properties of conditional independence (P1) to (P5) hold. Again there is a pairwise, a local, and a global Markov property. More precisely we say that a probability P satisfies:

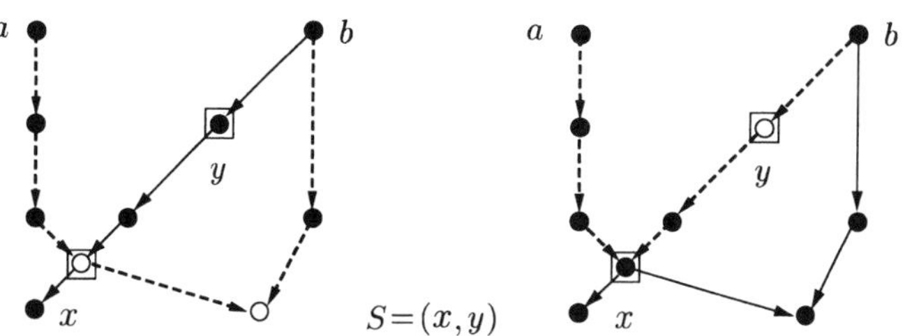

FIGURE 5.6. Illustration of Pearl's d-separation criterion. There are two trails from a to b, drawn with broken lines. Both are blocked, but different vertices γ, indicated with open circles, play the role of blocking vertices.

(CP) the *pairwise chain Markov property*, relative to $\mathcal{K}$, if for any pair (α, β) of non-adjacent vertices with $\beta \in \mathrm{nd}(\alpha)$,

$$\alpha \perp\!\!\!\perp \beta \mid \mathrm{nd}(\alpha) \setminus \{\alpha, \beta\};$$

(CL) the *local chain Markov property*, relative to $\mathcal{K}$, if for any vertex $\alpha \in V$,

$$\alpha \perp\!\!\!\perp \mathrm{nd}(\alpha) \setminus \mathrm{bd}(\alpha) \mid \mathrm{bd}(\alpha);$$

(CG) the *global chain Markov property*, relative to $\mathcal{K}$, if for any triple (A, B, S) of disjoint subsets of V such that S separates A from B in $(\mathcal{K}_{\mathrm{An}(A \cup B \cup S)})^m$, the moral graph of the smallest ancestral set containing $A \cup B \cup S$, we have

$$A \perp\!\!\!\perp B \mid S.$$

Studený and Bouckaert (1998) have introduced a definition of *c-separation* of A and B by S in a chain graph, which extends d-separation for directed acyclic graphs and is equivalent to the separation property used in the global chain Markov property above.

Once again, the graph separation property $\perp\!\!\!\perp_{\mathcal{K}}$ described in (CG) engenders a graphoid structure on subsets of V, and the Markov properties relate $\perp\!\!\!\perp$ and $\perp\!\!\!\perp_{\mathcal{K}}$. We note that these Markov properties unify the properties for the directed and undirected cases. For in the undirected case $\mathrm{nd}(\alpha) = V$ and $\mathcal{K} = (\mathcal{K}_{\mathrm{An}(A \cup B \cup S)})^m$, and in the directed case $\mathrm{bd}(\alpha) = \mathrm{pa}(\alpha)$.

When interpreting the conditional independence relationships in a chain graph, it is occasionally more straightforward to use the following approach, an extension of the ordered directed Markov property for directed acyclic graphs: since the graph is a chain graph, the vertex set can be partitioned as $V = V(1) \cup \cdots \cup V(T)$ such that each of the sets $V(t)$ only has undirected edges between its vertices, and any directed edges point from vertices in sets with lower number to those with higher number. Such a partition is called a *dependence chain*. The set of *concurrent* variables of $V(t)$ is defined to be the set $C(t) = V(1) \cup \cdots \cup V(t)$. Then P satisfies the *block-recursive Markov property* (CB) if for any pair (α, β) of non-adjacent vertices we have

$$\alpha \perp\!\!\!\perp \beta \mid C(t^*) \setminus \{\alpha, \beta\},$$

where t^* is the smallest t that has $\{\alpha, \beta\} \subseteq C(t)$. It appears that this property depends on the particular partitioning, but it can be shown (Frydenberg 1990) that — if P satisfies (P5) — it is equivalent to any of the above.

Theorem 5.16 *Assume that P is such that* (P5) *holds for subsets of V. Then*

$$\text{(CG)} \iff \text{(CL)} \iff \text{(CP)} \iff \text{(CB)}.$$

Proof. See Frydenberg (1990). □

Example 5.17 As an illustration of this, consider the graph in Figure 5.7 and the question of deciding whether $3 \perp\!\!\!\perp 8 \mid \{2,5\}$. The smallest ancestral set containing these variables is the set {1,2,3,4,5,6,7,8}. The moral graph of this adds an edge between 3 and 4, because these both have children in the chain component {5,6}. Thus, the graph in Figure 5.8 appears.

Since there is a path between 3 and 8 circumventing 2 and 5 in this graph, we cannot conclude that $3 \perp\!\!\!\perp 8 \mid \{2,5\}$.

If we instead consider the question whether $3 \perp\!\!\!\perp 8 \mid 2$, the smallest ancestral set becomes {1,2,3,4,7,8}, no edge has to be added between 3 and 4, and Figure 5.9 reveals that $3 \perp\!\!\!\perp 8 \mid 2$. □

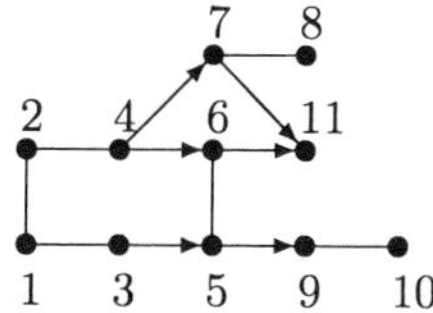

FIGURE 5.7. A chain graph with chain components {1,2,3,4}, {5,6}, {7,8}, {9,10}, {11}. Is $3 \perp\!\!\!\perp 8 \mid \{2,5\}$? Is $3 \perp\!\!\!\perp 8 \mid 2$?

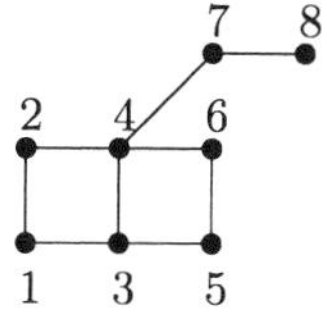

FIGURE 5.8. Moral graph of smallest ancestral set in the graph of Figure 5.7 containing {2,3,5,8}. A connection between 3 and 4 has been introduced since these both have children in the same chain component $\{5,6\}$. We cannot conclude $3 \perp\!\!\!\perp 8 \mid \{2,5\}$.

One way of constructing a distribution that satisfies the chain graph Markov property is through factorization. For example, if $V(1), \ldots, V(T)$ is a dependence chain of $\mathcal{K}$ or the chain components of $\mathcal{K}$, then any distribution P with density p with respect to a product measure μ will factorize as

$$p(x) = \prod_{t=1}^{T} p\left(x_{V(t)} \mid x_{C(t-1)}\right)$$

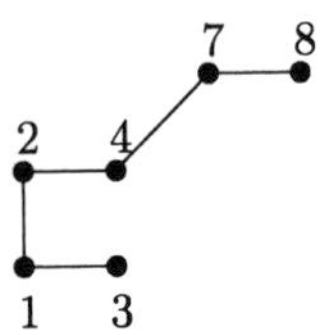

FIGURE 5.9. Moral graph of smallest ancestral set in the graph of Figure 5.7 containing {2,3,8}. We conclude that $3 \perp\!\!\!\perp 2 \mid 8$.

where $C(t) = V(1) \cup \cdots \cup V(t)$ are the concurrent variables, as usual. If $B(t) = \mathrm{pa}(V(t)) = \mathrm{bd}(V(t))$ and p is Markov with respect to $\mathcal{K}$, the factorization reduces to

$$p(x) = \prod_{t=1}^{T} p\left(x_{V(t)} \mid x_{B(t)}\right). \tag{5.8}$$

This factorization is essentially identical to the factorization for directed Markov densities due to the chain graph forming a directed acyclic graph of its chain components. But the factorization does not reveal all conditional independence relationships. To describe the remainder, let $\mathcal{K}^*(t)$ be the undirected graph with vertex set $V(t) \cup B(t)$ and α adjacent to β in $\mathcal{K}^*(t)$ if either $(\alpha, \beta) \in E$ or $(\beta, \alpha) \in E$ or if $\{\alpha, \beta\} \subseteq B(t)$, i.e., $B(t)$ is made complete in $\mathcal{K}^*(t)$ by adding all missing edges between these, and directions on existing edges are ignored. We cannot expect factorization results to be more general for chain graphs than for undirected graphs, since the chain graphs contain these as special cases. But if all variables are discrete, there is a result analogous to Theorem 5.6.

Theorem 5.18 *A probability distribution on a discrete sample space with strictly positive density p satisfies the pairwise chain graph Markov property with respect to $\mathcal{K}$ if and only if it factorizes as*

$$p(x) = \prod_{t=1}^{T} \frac{p\left(x_{V(t) \cup B(t)}\right)}{p\left(x_{B(t)}\right)}, \tag{5.9}$$

and each of the numerators factorizes on the graph $\mathcal{K}^*(t)$.

Proof. See Lauritzen (1996). □

Corollary 5.19 *If the density p of a probability distribution factorizes as in (5.9), it also factorizes according to the moral graph $\mathcal{K}^m$ and therefore obeys the undirected global Markov property relative to $\mathcal{K}^m$.*

Proof. By construction, sets that are complete in $\mathcal{K}^*(t)$ are also complete in $\mathcal{K}^m$. □

An equivalent formulation of the factorization (5.9) is

$$p(x) = \prod_{t=1}^{T} p\left(x_{V(t)} \mid x_{B(t)}\right), \tag{5.10}$$

where each factor further factorizes according to $\mathcal{K}^*(t)$. This is true because $B(t)$ is complete in $\mathcal{K}^*(t)$.

Alternatively, each of the factors in (5.10) must further factorize as

$$p\left(x_{V(t)} \mid x_{B(t)}\right) = Z^{-1}\left(x_{B(t)}\right) \prod_{A \in \mathcal{A}(t)} \phi_A(x_A), \tag{5.11}$$

where $\mathcal{A}(t)$ are the complete subsets of $\mathcal{K}^*_{V(t)\cup B(t)}$ and

$$Z\left(x_{B(t)}\right) = \sum_{x_{V(t)}} \prod_{A \in \mathcal{A}(t)} \phi_A(x_A).$$

5.5 Current research directions

There has been much recent research activity developing and extending the links between graphical structures and conditional independence properties which have been the subject of this chapter. Some of this concentrates on logical issues, such as *strong completeness*, i.e., whether a probability distribution (perhaps of a special form, e.g., Gaussian or having a positive density) exists displaying all and only the conditional properties displayed by a given graphical representation (Geiger and Pearl 1990, 1993; Studený and Bouckaert 1998). Studený (1997) and Andersson et al. (1998) give good overviews of recent issues and advances in graphical descriptions of conditional independence. We briefly describe some of the major current research themes below.

5.5.1 Markov equivalence

There may be more than one graph representing the same conditional independence relations. A focus of recent attention has been how to characterize such *Markov equivalence* of graphs and, if possible, nominate a natural representative of any equivalence class. Issues of equivalence of seemingly distinct representations need careful attention when attempting statistical model selection on the basis of data (Heckerman et al. 1995b).

Extending a result of Verma and Pearl (1991) for directed acyclic graphs, Frydenberg (1990) showed that two chain graphs are Markov equivalent if and only if they have the same skeleton (i.e., undirected version), and the same complexes, where a *complex* is a subgraph, induced by a set of nodes $\{v_1, v_2, \ldots, v_k\}$ with $k \geq 3$, whose edge set consists of $v_1 \to v_2$, $v_{k-1} \leftarrow v_k$, and $v_i \sim v_{i+1}$ for $2 \leq i \leq k-2$. In any class of Markov equivalent chain graphs there is a unique 'largest' one, having the maximum number of undirected edges; its arrows are present in every other member of the class. If we restrict attention to directed acyclic graphs, there is no natural representative of an equivalence class within the class, but it can be characterized by its *essential graph* (Andersson et al. 1996b), the chain graph (with the same skeleton) in which an edge has an arrow if and only if at least one member of the equivalence class has that arrow, and none has the reverse arrow (see Section 11.5).

5.5.2 *Other graphical representations*

Alternative graphical representations of conditional independence have been considered. Cox and Wermuth (1996) allow their graph edges to be dashed as well as solid lines and arrows, introducing an appropriately modified semantics to relate the graph structure to marginal and conditional independence properties of Gaussian distributions. This approach is related to that of Andersson et al. (1996a), who use chain graphs, but with a new 'AMP' semantics, based on a graph separation property different from that considered here, which they term 'LWF.' The AMP semantics is related to the interpretation of structural equation models (Bollen 1988). It gives rise to many questions similar to those for the LWF approach: equivalence of different descriptions of graphical separation, Markov equivalence of distinct graphs, etc. However, it does not correspond in a simple fashion to a factorization property of the joint density, an aspect that is crucial for computational efficiency.

Further graphical representations include possibly cyclic directed graphs using essentially the same moralization semantics as in the acyclic case; Markov equivalence and related issues have been addressed by Richardson (1996). An extension to 'reciprocal graphs,' a generalization of chain graphs, has been studied by Koster (1996).

Starting from any graphical Markov criterion, we can also consider the effects of collapsing out over unobservable variables, or conditioning on 'selection variables,' thus broadening still further the range of conditional independence structures that may be represented (Cox and Wermuth 1996). Again, issues of equivalence, etc., need addressing (Spirtes and Richardson 1997).

5.6 Background references and further reading

Dawid (1979, 1980b) proposed the axioms of conditional independence, without any graphical connections, and showed how they could be developed as a unifying concept within probability and statistics. Applications within theoretical statistics include: sufficiency and ancillarity; nuisance parameters (Dawid 1980a); Simpson's paradox; optional stopping, selection and missing data effects (Dawid 1976; Dawid and Dickey 1977); invariance (Dawid 1985); and model-building (Dawid 1982). An overview is given by Dawid (1998).

Graphical representations of probability models have a long history. Directed models can be traced back to the path analysis of Wright (1921, 1923, 1934), and undirected models to the work of Bartlett (1935) on interactions in contingency tables. The latter was taken up by Darroch et al. (1980) and has led to intensive investigation of graphical models in statistics, well summarized by Whittaker (1990) and Lauritzen (1996). The connections between undirected graphs and conditional independence were first made in the unpublished work of Hammersley and Clifford (1971). Statistical use of directed graphs came into its own with the introduction of influence diagrams (Howard and Matheson 1984), but it was the application by Pearl (1986b) (see also Pearl (1988)) to probability calculations in graphical networks which initiated the recent explosion of interest in directed graphical representations. Their Markov properties were explored by Pearl (1986a) and Verma and Pearl (1990) using d-separation, while Lauritzen et al. (1990) introduced the moralization criterion. A detailed study of chain graph representations can be found in Frydenberg (1990).

6
Discrete Networks

Probabilistic networks with only discrete random variables are the simplest probabilistic networks admitting an exact analysis. Our starting point is a probabilistic network model, from which we have constructed a representation of the joint distribution factorizing over a junction tree of cliques or sets. Since the cliques in the tree will constitute the largest collections of variables that need to be handled simultaneously, they need to be manageable, otherwise we run into problems with computational limitations. Thus, the premise underlying the construction of a probabilistic network is that, while the overall problem is too large to allow the application of naïve techniques for calculating and updating probabilities (cf. Chapter 2), the individual cliques in the triangulated moral graph are of manageable size. In particular, we assume that we can perform (e.g., by brute force) any desired operations, such as marginalization or conditioning, within any clique, although not necessarily directly for the full network. Our aim is to use the network structure to extend such calculations to the complete set of variables.

We start by giving a simple illustration of such localized calculation. Then, after defining the fundamental operations, we describe in detail the central algorithm for propagating information through a junction tree so as to calculate marginal and conditional distributions. Several variations on this algorithm are described, including localized calculation of the most probable overall configuration or configurations, simulation with and without replacement, and extraction of moments of suitable functions. We illustrate in detail the various stages of the procedure using a chain graph variation of the ASIA example described in Section 2.9.

6.1 An illustration of local computation

We first illustrate the general principles of local computation using a simple example. Suppose that a strictly positive joint density for three discrete variables X, Y, Z factorizes as

$$p(x, y, z) = f(x, y)\frac{1}{h(y)}g(y, z). \tag{6.1}$$

This may be represented on a junction tree with two cliques XY and YZ, and separator Y. Note that a factorization of the form (6.1) is possible if and only if $X \perp\!\!\!\perp Z \mid Y$ (see Section 5.1).

The marginal density $p(x, y)$ is obtained by summing over z:

$$\begin{aligned} p(x, y) &= \sum_z p(x, y, z) \\ &= \sum_z f(x, y)\frac{1}{h(y)}g(y, z) = f(x, y)\frac{1}{h(y)}\sum_z g(y, z). \end{aligned} \tag{6.2}$$

Thus, if we define

$$h^*(y) = \sum_z g(y, z) \tag{6.3}$$

and

$$f^*(x, y) = f(x, y)\frac{h^*(y)}{h(y)}, \tag{6.4}$$

we have

$$f^*(x, y) = p(x, y). \tag{6.5}$$

The calculation of $p(x, y)$ by means of (6.3) and (6.4) may be described as effected by passing a local *message* or *flow* from YZ to XY through their separator Y. The factor $h^*(y)/h(y)$ is termed the *update ratio.*

We now have:

$$\begin{aligned} p(x, y, z) &= f(x, y)\frac{1}{h(y)}g(y, z) \\ &= f(x, y)\frac{h^*(y)}{h(y)}\frac{1}{h^*(y)}g(y, z) \\ &= f^*(x, y)\frac{1}{h^*(y)}g(y, z) \qquad \text{by (6.4).} \end{aligned} \tag{6.6}$$

Passage of the flow has thus resulted in a new representation for $p(x, y, z)$, similar in the form to (6.1), in which one of the factors, $f^*(x, y)$, is now the marginal density $p(x, y)$.

Next we can pass a message in the other direction. We first calculate, parallel to (6.3),

$$\begin{aligned} h^{\dagger}(y) &= \sum_{x} f^{*}(x, y) \\ &= p(y) \qquad \text{by (6.5).} \end{aligned}$$

Parallel to (6.4), we calculate

$$g^{\dagger}(y, z) = g(y, z) \frac{h^{\dagger}(y)}{h^{*}(y)},$$

and then we shall have $g^{\dagger}(y, z) = p(y, z)$. Finally, parallel to (6.6), we have the overall representation

$$\begin{aligned} p(x, y, z) &= f^{*}(x, y) \frac{1}{h^{\dagger}(y)} g^{\dagger}(y, z) \\ &= p(x, y) \frac{1}{p(y)} p(y, z). \end{aligned}$$

By the passage of these two flows, we have thus transformed the original representation into a new representation involving only marginal densities.

The above calculation has been illustrated for a junction tree of only two cliques, but, as we shall show, the basic idea extends to arbitrary junction trees or elimination trees.

6.2 Definitions

We consider a probabilistic network constructed over a set of nodes, or vertices, V, the variable X_v associated with node v having a finite set $\mathcal{X}_v$ of possible states. The nodes are organized into a chain graph $\mathcal{K}$ — this includes the important special cases of undirected and directed acyclic graphs. To each chain-component k of $\mathcal{K}$ is attached a table specifying the conditional probabilities $p(x_k \mid x_{\mathrm{pa}(k)})$. As described in Sections 3.1.2 and 5.4, the structure of the chain graph may lead to further factorization of the conditional probabilities $p(x_k \mid x_{\mathrm{pa}(k)})$ in terms of potentials (see (6.8) below). If $\mathrm{pa}(k) = \emptyset$, the table consists of unconditional probabilities.

Let $U \subseteq V$. The *space* of U is the Cartesian product of the state sets of the nodes of U, and will be denoted by $\mathcal{U}$ or $\mathcal{X}_U$. The space of V may be denoted by $\mathcal{V}$ or $\mathcal{X}$. We shall also, with some misuse of notation, denote the space of $U \cup W$ by $\mathcal{U} \cup \mathcal{W}$, and that of $W \setminus U$ by $\mathcal{W} \setminus \mathcal{U}$.

A *discrete potential*, or more briefly, a *potential*, on U is a mapping from $\mathcal{U}$ to the non-negative real numbers $\mathbb{R}_0$. In particular, the table of conditional probabilities $p(x_k \mid x_{\mathrm{pa}(k)})$ forms a potential on $k \cup \mathrm{pa}(k)$, with the additional property of being normalized to sum to unity, for each fixed parent configuration, when summed over the states in k. The product of all such terms yields the joint probability distribution over $\mathcal{X}$.

6.2.1 Basic operations

Definition 6.1 [Extension]
Let $U \subseteq W \subseteq V$ and let ϕ be a potential on U. The potential ϕ is extended to W in the following way. Let $x \in \mathcal{W}$. Then we define $\phi(x) = \phi(y)$, where y is the projection of x onto $\mathcal{U}$. □

When it is clear from the context, we shall not distinguish between a function and its various extensions. Alternatively, all potentials can be considered as defined on the entire space V by first extending them to this space. A potential on U or extended from U may be denoted by ϕ_U. Such a potential depends on x through x_U only.

Definition 6.2 [Multiplication]
Multiplication is defined in the obvious way. For potentials ϕ and ψ on U and W respectively, we define their *product* $\phi\psi$ on $U \cup W$ by

$$(\phi\psi)(x) = \phi(x)\psi(x),$$

where ϕ and ψ on the right-hand side have first been extended to $U \cup W$. □

Definition 6.3 [Addition]
Addition is defined analogously. For potentials ϕ and ψ on U and W respectively, we define their *sum* $\phi + \psi$ on $U \cup W$ by

$$(\phi + \psi)(x) = \phi(x) + \psi(x),$$

where ϕ and ψ on the right hand side have first been extended to $U \cup W$. □

Definition 6.4 [Division]
Division is likewise defined in the obvious way, except that special care has to be taken when dividing by zero:

$$(\phi/\psi)(x) = \begin{cases} \phi(x)/\psi(x) & \text{if } \psi(x) \neq 0, \\ 0 & \text{otherwise.} \end{cases} \tag{6.7}$$

□

The reason for this definition is that intermediate expressions may sometimes arise during local computation that involve a division by 0, even when the overall joint probability is well-defined.

Definition 6.5 [Marginalization]
Let $W \subseteq U \subseteq V$, and let ϕ be a potential on U. The expression $\sum_{U \setminus W} \phi$ denotes the *margin* (or, more fully, the *sum-margin*) of ϕ on W, and is

defined by, for $x \in \mathcal{W}$,

$$\left(\sum_{U \setminus W} \phi\right)(x) = \sum_{z \in \mathcal{U} \setminus \mathcal{W}} \phi(z.x),$$

where $z.x$ is the element in $\mathcal{W}$ with projections x to $\mathcal{U}$ and z to $\mathcal{W} \setminus \mathcal{U}$. □

6.3 Local computation on the junction tree

Before a junction tree can be used, it must first be initialized to provide a local representation of the overall distribution. Then evidence, if any, is entered into the junction tree as described below in Section 6.3.9. After this is done, local computation can be performed to yield marginal or conditional distributions of interest.

6.3.1 Graphical specification

The initial graphical representation of the problem will involve a translation of qualitative judgments of causal relationships, associations, and conditional independence relationships relevant to the specific applied context, into Markov properties of a suitable undirected, directed acyclic, or general chain graph $\mathcal{K}$. This vital first step is as much art as science, and we shall have little to add here.

Let $\mathcal{T}$ denote a junction tree of sets or cliques for the triangulated moralized graph $\mathcal{K}^{mt}$ of $\mathcal{K}$. For definiteness we shall talk principally about a tree of cliques, although other junction trees, for example, an elimination tree, can be used in an essentially identical way. We may suppose that $\mathcal{T}$ is connected, since the disconnected case can easily be handled by considering each component separately. Denote the cliques of $\mathcal{T}$ by $\mathcal{C}$; we shall also refer to these cliques as *belief universes* or simply *universes* — they constitute the units of modularity in our scheme of local computation.

6.3.2 Numerical specification and initialization

As described toward the end of Section 5.4, the density $p(\cdot)$ of the probability distribution P in the model with chain graph $\mathcal{K}$ and set of chain components K factorizes first as

$$p(x) = \prod_{k \in K} p_k(x_k \mid x_{\mathrm{pa}(k)}), \tag{6.8}$$

and then still further in the form

$$p(x) \propto \prod_{k \in K} Z^{-1}\left(x_{\mathrm{pa}(k)}\right) \prod_{A \in \mathcal{A}_k} \psi_A(x_A), \tag{6.9}$$

where in (6.9) $\mathcal{A}_k$ denotes the set of maximal subsets of $k \cup \mathrm{pa}(k)$ that are complete in the moralized version $\mathcal{K}^m$ of $\mathcal{K}$ and contain at least one child variable in k. The factors $\{p_k(x_k \mid x_{\mathrm{pa}(k)})\}$ and $\{\psi_A(x_A)\}$ are part of the initial external numerical specification of the problem.

A factorization over $\mathcal{T}$ of the density of the joint distribution P on $\mathcal{T}$ can be constructed as follows. To each clique $C \in \mathcal{C}$ we associate a potential ϕ_C, and to each separator $S \in \mathcal{S}$ on the edge connecting two cliques in $\mathcal{T}$ we associate a potential ϕ_S. We initialize all the potentials on the tree, $\{\phi_C, C \in \mathcal{C}\}$ and $\{\phi_S, S \in \mathcal{S}\}$, to have value unity. We then take each factor in (6.9) and multiply it into the potential on any one clique of $\mathcal{T}$ which contains all the vertices labelling its arguments. The moralization (Section 4.1) ensures that there is always one such clique; if there are several possibilities it does not matter which is chosen. After each factor has been so multiplied we shall have

$$p(x) = \frac{\prod_{C \in \mathcal{C}} \phi_C(x_C)}{\prod_{S \in \mathcal{S}} \phi_S(x_S)}, \tag{6.10}$$

in which $\phi_S \equiv 1$.

6.3.3 Charges

A collection of non-negative potential functions $\Phi = \{\phi_A, A \in \mathcal{C} \cup \mathcal{S}\}$ will be termed a *charge* on $\mathcal{T}$.

For any such charge, the expression on the right-hand side of (6.10), calculated using Definitions 6.2 and 6.4, defines its *contraction*. Whenever (6.10) holds, we may call Φ a *(generalized potential) representation* of P. The representation constructed in Section 6.3.2 is termed the *initial representation* and its potentials the *initial potentials*.

The computational algorithm to be introduced operates by transforming one representation for $p^{\mathcal{E}}(\cdot) := p(\cdot \,\&\, \mathcal{E})$ to another (where $\mathcal{E}$ is the available evidence) in a series of simple steps, starting from the initial (or any other convenient) representation of P, modified by the incorporation of $\mathcal{E}$ as described in Section 6.3.9 below, and finishing with the *marginal representation* of $p^{\mathcal{E}}$ in which, for each $A \in \mathcal{C} \cup \mathcal{S}$, ϕ_A is equal to $\sum_{V \setminus A} p^{\mathcal{E}}$, the joint probability function for the variables in A and the evidence $\mathcal{E}$.

6.3.4 Flow of information between adjacent cliques

The full algorithm comprises a sequence of messages, or flows, passing along the edges of $\mathcal{T}$, each such flow affecting the potentials on exactly one clique and one separator.

Let C_1 and C_2 be adjacent cliques in $\mathcal{T}$, and S_0 the separator on the edge joining them. We describe the structure of a flow (more fully, a *sum-flow*) from C_1 (the *source*) to C_2 (the *sink*). Suppose that, prior to this flow, we have a charge $\Phi = (\{\phi_C, C \in \mathcal{C}\}, \{\phi_S, S \in \mathcal{S}\})$. Then the effect of the flow is specified to be the replacement of this charge by a new charge $\Phi^* = (\{\phi^*_C, C \in \mathcal{C}\}, \{\phi^*_S, S \in \mathcal{S}\})$, where the new potentials on S_0 and C_2 are given (using Definitions 6.2, 6.4, and 6.5) by

$$\phi^*_{S_0} = \sum_{C_1 \setminus S_0} \phi_{C_1}, \tag{6.11}$$

and

$$\phi^*_{C_2} = \phi_{C_2}\, \lambda_{S_0}, \tag{6.12}$$

where

$$\lambda_{S_0} = \phi^*_{S_0} / \phi_{S_0}. \tag{6.13}$$

All other potentials are unaltered. We term λ_{S_0} the *update ratio* carried by the flow along S_0 into C_2. A charge Φ is unaffected by the passage of any sum-flow if and only if it is *sum-consistent*, i.e., $\sum_{C \setminus S} \phi_C = \phi_S$ for any $C \in \mathcal{C}$ and neighbouring $S \in \mathcal{S}$.

Lemma 6.6 *Passage of a flow does not affect the contraction of a charge.*

Proof. Let f and f^* be the respective contractions of Φ and Φ^*, the charges before and after passage of the flow. We have to show $f^*(x) = f(x)$ for all $x \in \mathcal{X}$. We consider three different cases:

1. $\lambda_{S_0}(x_{S_0}) > 0$. Clearly $f^*(x) = f(x)$ in this case.

2. $\phi_{S_0}(x_{S_0}) = 0$. Then $f(x) = 0$ by definition, while $f^*(x) = 0$ because $\phi^*_{C_2}(x_{C_2}) = 0$.

3. $\phi_{S_0}(x_{S_0}) > 0$ but $\phi^*_{S_0}(x_{S_0}) = 0$. In this case we have $f^*(x) = 0$ by definition; while $f(x) = 0$ also because, ϕ_{C_1} being non-negative, $\sum\{\phi_{C_1}(y_{C_1}) : y_{S_0} = x_{S_0}\} = 0 \Rightarrow \phi_{C_1}(x_{C_1}) = 0$.

Thus, in all three cases $f^* = f$, as was to be proved. □

We now consider the effect of passing flows in a particular sequence along the edges of $\mathcal{T}$.

6.3.5 *Active flows*

A *subtree* $\mathcal{T}'$ of the junction tree $\mathcal{T}$ is a connected set of vertices (cliques) of $\mathcal{T}$, together with the edges in $\mathcal{T}$ between them. A clique C is a *neighbour* of a subtree $\mathcal{T}'$ if the corresponding vertex of $\mathcal{T}$ is not a vertex of $\mathcal{T}'$, but is connected to a vertex of $\mathcal{T}'$ by an edge of $\mathcal{T}$.

A *schedule* is an ordered list of directed edges of $\mathcal{T}$, specifying which flows are to be passed and in what order. Relative to such a schedule, a flow is called *active* if, before it is sent, the source has itself already received active flows from all its neighbours in $\mathcal{T}$, with the possible exception of the sink; and it is the first flow in the list with this property. This is a recursive but effective definition. Note that the first active flow must originate in a leaf of $\mathcal{T}$, i.e., a clique in the tree with only one neighbour.

A schedule is *full* if it contains an active flow in each direction along every edge of $\mathcal{T}$. It is *active* if it contains only active flows, and *fully active* if it is both full and active. If from any full schedule we omit the inactive flows, we obtain a fully active schedule.

Proposition 6.7 *For any tree $\mathcal{T}$, there exists a fully active schedule.*

Proof. The result is vacuously true if $\mathcal{T}$ consists of only one clique. Otherwise, let C_0 be a leaf of the tree $\mathcal{T}$, and $\mathcal{T}_0$ the subtree of $\mathcal{T}$ obtained by removing C_0 and the associated edge S_0. By induction we may suppose a fully active schedule to exist for $\mathcal{T}_0$. Then adjoining a flow along S_0 out of C_0 at the beginning, and a flow along S_0 into C_0 at the end, we obtain such a schedule for $\mathcal{T}$. □

6.3.6 Reaching equilibrium

Let $\mathcal{T}'$ be a subtree of $\mathcal{T}$, with vertices $\mathcal{C}' \subseteq \mathcal{C}$ and edges $\mathcal{S}' \subseteq \mathcal{S}$. The *base* of $\mathcal{T}'$ is $U' := \bigcup\{C : C \in \mathcal{C}'\}$, the collection of variables associated with $\mathcal{T}'$. If $\Phi = (\{\phi_C, C \in \mathcal{C}\}, \{\phi_S, S \in \mathcal{S}\})$ is a charge for $\mathcal{T}$, its *restriction* to $\mathcal{T}'$ is $\Phi_{\mathcal{T}'} := (\{\phi_C, C \in \mathcal{C}'\}, \{\phi_S, S \in \mathcal{S}'\})$, and its *potential on* $\mathcal{T}'$ is the contraction of $\Phi_{\mathcal{T}'}$, which is thus a function on $\mathcal{U}'$.

With respect to a given schedule of flows, a subtree $\mathcal{T}'$ is *live* at a certain stage of the schedule if it has already received active flows from all its neighbours. This property is then retained throughout the remainder of the schedule.

Theorem 6.8 *Suppose we start with a representation $\Phi^0 = (\{\phi^0_C, C \in \mathcal{C}\}, \{\phi^0_S, S \in \mathcal{S}\})$ for a function f that factorizes on $\mathcal{T}$ and progressively modify it by passing a sequence of flows according to some schedule. Then whenever $\mathcal{T}'$ is live, the potential on $\mathcal{T}'$ is the sum-margin $f_{U'}$ of f on U'.*

Proof. By Lemma 6.6, we shall at all stages have a representation for f on $\mathcal{T}$.

The result holds trivially if $\mathcal{T}' = \mathcal{T}$. Otherwise, assuming $\mathcal{T}'$ to be live, let C^* be the last neighbour to have passed a flow (active or not) into $\mathcal{T}'$, and let $\mathcal{T}^*$ be the subtree obtained by adding C^* and the associated edge S^* to $\mathcal{T}'$. Let $\mathcal{C}^*$, $\mathcal{S}^*$, and U^* be the cliques, separators, and base of $\mathcal{T}^*$ respectively. By the junction tree property of $\mathcal{T}$, the separator associated with the edge S^* joining C^* to $\mathcal{T}'$ is $S^* = C^* \cap U'$. Then $\mathcal{C}^* = \mathcal{C}' \cup \{C^*\}$, $\mathcal{S}^* = \mathcal{S}' \cup \{S^*\}$, and $U^* = U' \cup C^*$.

By induction, we may suppose the result to hold for $\mathcal{T}^*$, so that, if $\Phi = (\{\phi_C, C \in \mathcal{C}\}, \{\phi_S, S \in \mathcal{S}\})$ is the overall representation for f just before the last flow from C^* into $\mathcal{T}'$,

$$f_{U^*} = \frac{\phi_{C^*}\,\alpha_{U'}}{\phi_{S^*}}, \tag{6.14}$$

where $\alpha_{U'} = \prod_{C \in \mathcal{C}'} \phi_C / \prod_{S \in \mathcal{S}'} \phi_S$ is the potential on $\mathcal{T}'$. After the flow, $\alpha_{U'}$ is replaced by

$$\alpha^*_{U'} = \lambda_{S^*}\,\alpha_{U'}, \tag{6.15}$$

with λ_{S^*} defined by (6.13).

But if we marginalize (6.14) over $U^* \setminus U' = C^* \setminus S^*$, noting that, by Definition 6.4, $\phi_{S^*}(x_{S^*}) = 0 \Rightarrow \phi_{U^*}(x_{U^*}) = 0$, we obtain

$$\phi_{U'}(x_{U'}) = \lambda_{S^*}\,\alpha_{U'}, \tag{6.16}$$

and the result follows. □

Corollary 6.9 *Whenever a clique C is live, its potential is f_C.*

Corollary 6.10 *Any time after active flows have passed in both directions across an edge in $\mathcal{T}$, the potential for the associated separator S is f_S.*

Corollary 6.11 *Any time after active flows have passed in both directions across an edge in $\mathcal{T}$ between cliques C and D with associated separator S, the tree is sum-consistent along S,* i.e.,

$$\sum_{D \setminus S} \phi_D = \phi_S = \sum_{C \setminus S} \phi_C.$$

We can now deduce our principal result, showing the effectiveness of the flow propagation routine for calculating margins on all cliques and separators.

Corollary 6.12 *After passage of a full schedule of flows, the resulting charge is the marginal charge Φ_f of f.*

After such a full schedule of flows has passed the tree will be sum-consistent. We may also say that the system has reached equilibrium.

By Lemma 6.6, the charge at equilibrium must be a representation for f. Consequently, we have the following corollary.

Corollary 6.13 *Suppose that f factorizes on $\mathcal{T}$. Then Φ_f is a representation for f, and thus*

$$f = \frac{\prod_{C \in \mathcal{C}} f_C}{\prod_{S \in \mathcal{S}} f_S}. \tag{6.17}$$

6.3.7 *Scheduling of flows*

Generally there is no point in passing inactive flows: while not affecting the correctness of the algorithm, they introduce unnecessary computation.

There are various ways of constructing a fully active schedule of flows, the differences between them lying not so much in the results they deliver as in the control mechanisms they employ. The inductive proof of Proposition 6.7 provides one scheduling routine: this yields a schedule, which we may term *palindromic*, in which an initial sequence of directed edges is followed by its reversal in both order and direction. Jensen et al. (1990a) propose a two-phase schedule. This requires the selection of an arbitrary clique $C_0 \in \mathcal{C}$ to be the *root-clique*. An initial *collection* phase involves passing active flows only along edges directed toward C_0; this is then followed by a *distribution* phase, in which, starting from C_0, active flows are passed back toward the periphery. After collection, the potential for C_0 will be f_{C_0}, and if this is all that is desired the distribution phase may be omitted. Any fully active schedule implicitly defines a two-phase routine, with root the first clique to become live, and may thus be divided into its collection and distribution phases.

As a variant of this and other scheduling routines, some flows can be passed in parallel, and a receiving node can process several incoming flows simultaneously by multiplying all their update ratios into its potential — which it can delay until required to calculate an active output flow. Yet another routine, similar to the one described by Shafer and Shenoy (1990) treats each clique as an autonomous processor, which at time t examines each of the edges incident on it to see if a new flow has been passed along it, and if so, multiplies the associated update ratio into its potential. At time $t + 1$ it passes a new flow along any edge if it has already received flows along all its other edges, at least one of these being new at time t. Update ratios remain attached to their directed edges until overwritten. This method is well-suited to implementation by parallel processing, object-oriented methods, or production systems. When executed in parallel, the number of cycles needed for all flows to pass is equal to the diameter of the junction tree.

6.3.8 *Two-phase propagation*

For certain generalizations of the algorithm, it is important to distinguish the two phases of collection to, and distribution from, the root-clique, since 'flow' may be defined differently in the two phases. Moreover, there may need to be restrictions on the choice of root-clique. Because of the importance of such problems, we now describe in more detail this two-phase control structure.

Collection phase

Let R be the selected root-clique. The collection phase is executed by R's calling COLLECT EVIDENCE in each of its neighbouring cliques, where the effect of this instruction is described by the following recursive construction.

Definition 6.14 [COLLECT EVIDENCE]
When COLLECT EVIDENCE is called in a clique U from a neighbour W, U calls COLLECT EVIDENCE in all its other neighbours, if any; and, after they have all responded by passing flows into U, U responds to the call by passing a flow into W. □

Note that evidence collection by distinct neighbours of W may be performed in parallel, since the portions of the junction tree affected are disjoint. The final flows back to W from those neighbours in which it has called COLLECT EVIDENCE may be effected as soon as each such neighbour is ready, or be delayed until the last is ready, all flows then passing into W simultaneously, a process termed ***absorption*** by Jensen et al. (1990a).

Distribution phase

After completion of the collection phase, the root-clique R will have absorbed all information available, and its potential will have become its sum-margin f_R. Now it must pass on its information to all remaining cliques by calling DISTRIBUTE EVIDENCE in each of its neighbours according to the following recursive procedure.

Definition 6.15 [DISTRIBUTE EVIDENCE]
When DISTRIBUTE EVIDENCE is called in a clique U from a neighbour W, a flow is passed from W to U, and U then calls DISTRIBUTE EVIDENCE in all its other neighbours, if any. □

After both phases have terminated, active flows will have passed in both directions between every pair of cliques, and the final charge will be the marginal representation of f.

6.3.9 *Entering and propagating evidence*

A probabilistic network typically is used dynamically. Initially the junction tree holds a representation of the overall (prior) distribution of all the variables. If the propagation algorithm is executed at this point, it will deliver the prior probability distribution in each clique and separator. When information is to be incorporated, it is first applied to modify the potentials, and these modifications are then propagated through the tree to yield the relevant new (posterior) probabilities.

By *evidence* we formally understand a function $\mathcal{E} : \mathcal{X} \rightarrow \{0, 1\}$. This evidence represents a statement that some elements of $\mathcal{X}$ (those assigned value 0) are *impossible.* We shall also call such $\mathcal{E}$ a *finding*. We restrict attention to cases in which the evidence function can written in a factored form as

$$\mathcal{E}(x) \equiv \prod_{v \in U} l_v(x_v), \tag{6.18}$$

for some set U of nodes. In particular, and most usefully, when the evidence consists of finding that, for each node $v \in U$, X_v is in a definite state, (6.18) will hold with

$$l_v(x_v) = \begin{cases} 1, & \text{if } x_v \text{ is the observed state of node } v \\ 0, & \text{otherwise.} \end{cases} \tag{6.19}$$

We shall not deal here with more complicated scenarios involving findings on a set of variables not contained in a clique, which cannot be factorized into the product of such findings.

If the prior joint probability function for the probabilistic network is p then the posterior joint probability function is

$$p(x \mid \mathcal{E}) \equiv \kappa\, p(x)\, \mathcal{E}(x),$$

where κ is a normalizing constant (the reciprocal of the prior probability of $\mathcal{E}$).

When evidence arrives in a probabilistic network, it is first entered into the junction tree. With $\mathcal{E}$ factoring as in (6.18), this is accomplished by, for each $v \in U$, multiplying the potential ϕ_C by l_v for some arbitrary clique C containing v. In effect, when we have observed the state of v, so that (6.19) holds, we simply set $\phi_C(x_C)$ to 0 for any configuration x_C involving a different state at v. The modified potentials now constitute a representation of $p^{\mathcal{E}}(x) \equiv p(x \,\&\, \mathcal{E}) \equiv p(x)\mathcal{E}(x) \propto p(x \mid \mathcal{E})$, i.e., the contraction of the resulting charge will now be equal to the joint probability of x and the evidence. The junction tree is now brought to equilibrium by passing a full schedule of flows, for example, using the operations of Collect Evidence and Distribute Evidence. At this point the resulting charge will be $(\{p^{\mathcal{E}}_C, C \in \mathcal{C}\}, \{p^{\mathcal{E}}_S, S \in \mathcal{S}\})$. Posterior probabilities can then be found by normalizing the resulting potentials to sum to 1. If the original tree was (connected and) normalized, the prior probability of the evidence can be obtained by selecting any clique or separator in it and taking the sum of the values of its potential after propagation.

We obtain the following formula for the joint posterior probability:

$$p(x \,\&\, \mathcal{E}) = \frac{\prod_{C \in \mathcal{C}} p(x_C \,\&\, \mathcal{E})}{\prod_{S \in \mathcal{S}} p(x_S \,\&\, \mathcal{E})}, \tag{6.20}$$

and on normalizing overall, or within each clique and separator, we obtain

$$p(x \mid \mathcal{E}) = \frac{\prod_{C \in \mathcal{C}} p(x_C \mid \mathcal{E})}{\prod_{S \in \mathcal{S}} p(x_S \mid \mathcal{E})}. \tag{6.21}$$

6.3.10 A propagation example

We illustrate the message-passing algorithm for calculating the marginal representation with the CHILD example. Consider two adjacent cliques from Figure 3.5: $C_{13} = \{n2, n4\}$ and $C_{14} = \{n4, n15\}$ and their separator $S = \{n4\}$. Initially, the potential tables for these are as in Table 6.1(a); these are given by $p(n15 \mid n4)$ for C_{14}, and $p(n4 \mid n2)$ for C_{13}. If we incorporate the evidence $\mathcal{E}$: "*LVH report?* = yes," then the potential for C_{14} changes to that shown in Table 6.1(b). Starting from these modified potentials, we now describe the passage of a flow from C_{14}, the source, to C_{13}, the sink, across S. This has two phases.

1. Within C_{14}, we sum out over variables not in S (*n15: LVH report?*) to obtain a new potential function over S. This is shown in Table 6.1(b).

2. The potential over C_{13} is now modified by multiplying each term by the associated update ratio, which is the ratio of the relevant value of the new potential over S to that of the old one. In this case, the update ratios are just the new potential values since the previous potentials were unity, and we obtain the potentials on C_{13} shown in Table 6.1(b).

The above routine applies when any flow is passed between adjacent cliques. The potentials on the separator and on the receiving clique are modified.

We need to schedule the flows in the full junction tree so that there will be exactly one flow in each direction along each edge. Assuming a suitable scheduling, Table 6.1(c) shows the potentials on C_{13}, C_{14}, and S, having incorporated the evidence "*LVH report?* = yes" just prior to the final flow, from C_{13} back to C_{14}. Passage of the flow now produces the new potential function on S shown in Table 6.1(d), and corresponding update ratios $(0.248/0.900 = 0.276, 0.036/0.050 = 0.720)$. These factors are then incorporated, in this case trivially, into the potential for C_{14}, yielding Table 6.1(d) as the final tables after propagation. At this point, for example, the potential ϕ_{13} for C_{13} satisfies $\phi_{13}(d, l) = P(\mathit{Disease?} = d, \mathit{LVH?} = l, \mathcal{E})$ and by normalization of this or any other potential table we find $p(\mathcal{E}) = 0.284$. Dividing each term by this marginal probability of the evidence observed gives us the final representation shown in Table 6.1(e). We can now see, for example, that $P(\mathit{LVH?} = \text{yes} \mid \mathcal{E}) = 0.872$.

6.4 Generalized marginalization operations

The previous sections have shown that the marginal distribution of a variable, conditional on some evidence (if any), can be found by local computation on the junction tree. The basic reason this can be done is because of

C_{13}:

Disease?	(a) *LVH?* Yes	(a) *LVH?* No	(b) *LVH?* Yes	(b) *LVH?* No	(c) *LVH?* Yes	(c) *LVH?* No	(d) *LVH?* Yes	(d) *LVH?* No	(e) *LVH?* Yes	(e) *LVH?* No
PFC	0.100	0.900	0.090	0.045	0.004	0.002	0.004	0.002	0.015	0.007
TGA	0.100	0.900	0.090	0.045	0.031	0.016	0.031	0.016	0.109	0.055
Fallot	0.100	0.900	0.090	0.045	0.026	0.013	0.026	0.013	0.091	0.045
PAIVS	0.900	0.100	0.810	0.005	0.180	0.001	0.180	0.001	0.634	0.004
TAPVD	0.050	0.950	0.045	0.048	0.002	0.002	0.002	0.002	0.008	0.008
Lung	0.100	0.900	0.090	0.045	0.004	0.002	0.004	0.002	0.016	0.008

S:

	(a) *LVH?* Yes	(a) *LVH?* No	(b) *LVH?* Yes	(b) *LVH?* No	(c) *LVH?* Yes	(c) *LVH?* No	(d) *LVH?* Yes	(d) *LVH?* No	(e) *LVH?* Yes	(e) *LVH?* No
	1.000	1.000	0.900	0.050	0.900	0.050	0.248	0.036	0.872	0.128

C_{14}:

LVH report?	(a) *LVH?* Yes	(a) *LVH?* No	(b) *LVH?* Yes	(b) *LVH?* No	(c) *LVH?* Yes	(c) *LVH?* No	(d) *LVH?* Yes	(d) *LVH?* No	(e) *LVH?* Yes	(e) *LVH?* No
Yes	0.900	0.050	0.900	0.050	0.900	0.050	0.248	0.036	0.872	0.128
No	0.100	0.950	0	0	0	0	0	0	0	0

TABLE 6.1. Propagation of evidence through cliques C_{13} and C_{14} with separator S of junction tree: (a) initial potentials, (b) after incorporation of evidence "*LVH report?* = yes," (c) after propagation through the rest of the network and back to C_{13}, (d) final potentials, (e) marginal tables after normalization.

the commutation of the summation operations with the factorization over the tree structure, which allows terms not involved in the summation of leaf variables to be taken through the product. For example if we have a real-valued function f which factorizes as

$$f(a,b,c,d,e) = \psi(ab)\psi(bcd)\psi(bde),$$

then we can write

$$\sum_e f(a,b,c,d,e) = \sum_e \psi(ab)\psi(bcd)\psi(bde) = \psi(ab)\psi(bcd)\left(\sum_e \psi(bde)\right).$$

However, operations other than summation also have this commutation property. This was exploited by Shenoy and Shafer (1990), who gave a set of abstract axioms under which a form of local computation is possible. Axioms validating the specific propagation scheme described in this chapter were given by Lauritzen and Jensen (1997). We describe below some useful applications of this more general approach to propagation in expert systems.

6.4.1 Maximization

Suppose that a probability function p, factorizing on $\mathcal{T}$, is represented by a charge $(\{\phi_C, C \in \mathcal{C}\}, \{\phi_S, S \in \mathcal{S}\})$, and we wish to calculate the most probable configuration of all the variables. This may be effected by a simple modification of the main propagation routine; we merely substitute for the marginalization definition of Definition 6.5 the following definition:

Definition 6.16 [Max-Marginalization]
Let $W \subseteq U \subseteq V$ and ϕ be a potential on U. The expression $\mathbf{M}_{U\setminus W}\,\phi$ denotes the *max-margin* of ϕ on W, defined by

$$\left(\mathbf{M}_{U\setminus W}\,\phi\right)(x) = \max_{z\in\mathcal{U}\setminus\mathcal{W}} \phi(z.x).$$

□

We shall call a flow, defined exactly as in Section 6.3.4 but with $\mathbf{M}$ replacing $\sum$ in (6.11), a *max-flow*. A fully active schedule of such flows constitutes a *max-propagation*. It may then be seen that all the considerations and results of the previous sections continue to apply, so long as we everywhere replace sum-marginalization by max-marginalization. After propagating, with this new definition for marginalization, the tree will reach equilibrium, and the resulting potential for each clique C (respectively separator S), denoted by $p_C^{\max}$ (respectively $p_S^{\max}$), will now have the interpretation that $p_C^{\max}(x_C)$ (respectively $p_S^{\max}(x_S)$) is the probability of the joint configuration of the collection of all variables X which is most probable under the restriction that $X_C = x_C$ (respectively $X_S = x_S$). For this *max-marginal* representation, we thus have

$$p(x) = \frac{\prod_{C \in \mathcal{C}} p_C^{\max}(x_C)}{\prod_{S \in \mathcal{S}} p_S^{\max}(x_S)}. \tag{6.22}$$

In particular, since the charge $(\{p_C^{\max}, C \in \mathcal{C}\}, \{p_S^{\max}, S \in \mathcal{S}\})$ is a valid potential representation of p, even after max-propagation a further sum-propagation will reconstruct the correct sum-margins.

More generally, we may also include evidence $\mathcal{E}$, and so obtain

$$p(x \,\&\, \mathcal{E}) = \frac{\prod_{C \in \mathcal{C}} p_C^{\max}(x_C \,\&\, \mathcal{E})}{\prod_{S \in \mathcal{S}} p_S^{\max}(x_S \,\&\, \mathcal{E})}. \tag{6.23}$$

In (6.22) or (6.23), the largest value in every clique and separator will be the same, and equal to the probability of the most probable configuration of the variables (consistent with the evidence). The states of the variables in such a configuration will typically differ from the collection of most probable states for each variable considered separately. We can find a most probable configuration by a single further distribute operation, as follows. We start from a representation in which the potential for some clique R is $p_R^{\max}$; for example, using max-marginalization, we may have performed Collect Evidence to R, or a full evidence propagation (consisting of Collect Evidence to R followed by Distribute Evidence from R). We then perform the following variation of Distribute Evidence, using clique R as root (for $U = R$ the references to W are ignored).

Definition 6.17 [Find Max-Configuration]

Each $U \in \mathcal{C}$ is given the action Find Max-Configuration. When Find Max-Configuration is called in U from a neighbour W, then a max-flow is passed from W to U, and U then finds a configuration of its variables to maximize its potential, breaking ties arbitrarily. This configuration, say $x_U^{\max}$, is then entered as additional evidence by setting all other potential table entries for U to zero, and finally U issues Find Max-Configuration in all its other neighbours. □

The junction tree property ensures that when the state of a variable is fixed in one clique, that variable assumes the same state in all other cliques. When Find Max-Configuration has terminated, it has found a most probable configuration, consistent with the evidence. This can be interpreted as the 'best explanation' of that evidence. For further details and proofs see Dawid (1992a).

6.4.2 *Degeneracy of the most probable configuration*

It is possible to find the degeneracy of the most probable configuration (i.e., the total number of distinct configurations that have the same maximum probability $p^{\max}(x \mid \mathcal{E}) = p^*$) by a simple device. (For many realistic applications there is unlikely to be any degeneracy, but it sometimes arises, for example, in genetic pedigree problems.) First one performs a max-propagation to obtain the max-marginal representation. Then one sets each value in each clique to either 1 or 0, depending on whether or not it had attained the maximum probability value. Thus,

$$I_C(x_C \mid \mathcal{E}) = \begin{cases} 1, & \text{if } p_C^{\max}(x_C \mid \mathcal{E}) = p^* \\ 0, & \text{otherwise,} \end{cases}$$

and sets $I_S(x_S \mid \mathcal{E}) \equiv 1$ for every separator S. Then

$$I(x \mid \mathcal{E}) = \frac{\prod_C I_C(x_C \mid \mathcal{E})}{\prod_S I_S(x_S \mid \mathcal{E})}$$

is a generalized potential representation of the indicator function of most probable configurations; sum-propagation on this will yield the degeneracy as the normalization.

6.4.3 *Simulation*

For some purposes we may wish to generate a random sample of configurations from the probability distribution of a probabilistic network model. Henrion (1988) proposed an algorithm called *probabilistic logic sampling* for DAGs to generate such samples. Let $(v_1, v_2, \ldots, v_n)$ be a well-ordering (see Section 4.1) of the nodes of the DAG $\mathcal{G}$, so that for every node v all parents $\mathrm{pa}(v)$ of v occur before v in the ordering. First suppose that there is no evidence. To sample a complete configuration one starts with node v_1. This has no parents and so one can sample X_1 directly from its specified marginal distribution $\pounds(X_1)$ to yield a state, say x_1^*. Next one samples a state for node v_2. If v_2 has no parents one samples from $\pounds(X_2)$, which is known from the specification of the model. Otherwise, if v_2 has a parent it can only be v_1, whose state has been fixed at x_1^*; hence, one samples from $\pounds(X_2 \mid X_1 = x_1^*)$. In either case we obtain a sampled state, x_2^* say. We then move onto v_3, etc. In general, at the jth stage we can sample directly from $\pounds(X_j \mid X_{\mathrm{pa}(j)})$ because the states of all parents have been fixed by previous stages and the required probabilities are known from the specification of the probabilistic network. One completes this process until all nodes have been sampled — this defines one complete configuration. The procedure is then repeated, until one has a sample of complete configurations of sufficient size. This algorithm is simple and efficient. But now suppose that there is evidence $\mathcal{E}$ on one or more nodes. One can still use the algorithm, but now a rejection step has to be inserted. Thus, suppose one is at stage j and that one has evidence that X_j is in state $x_j^\dagger$ say. One cannot just set $X_j^\dagger = x_j^\dagger$ because this would lead to a sample drawn from the wrong distribution. Instead, one samples directly from $\mathcal{L}(X_j \mid X_{\mathrm{pa}(j)})$ to obtain x_j^* say. Then, if this is equal to the actual evidence $x_j^\dagger$, one proceeds to the next step, otherwise one *rejects* the results of all computations on that sample and starts back at X_1 again. This rejection ensures that one is sampling from the correct distribution, but it can be very inefficient, with most of the time spent generating partial configurations which get rejected. The rejection step can be avoided by generating weighted samples from the DAG, but then one does not sample directly from the posterior distribution (see Shachter and Peot (1990) for a comparison of a number of algorithms of this type).

However, it is possible to generate samples directly and efficiently from the joint distribution on the junction tree by propagation, even with evidence $\mathcal{E}$ (Dawid 1992a). The first stage, after incorporating the evidence, is to perform a regular (sum-propagation) COLLECT EVIDENCE to an arbitrary root clique R. Then on the root the potential is proportional to the correct distribution for its variables X_R given the evidence, and one samples a configuration x_R^* directly from it. Next, one fixes the states of those variables, entering them as additional evidence $X_R = x_R^*$ and propagating, and then generates sample configurations for the neighbouring cliques of R. When each such clique has been sampled, one again enters and propagates the sampled values as additional evidence and samples from their neighbouring cliques further away from the root; and so on until one has a complete configuration.

We can describe this variation of DISTRIBUTE EVIDENCE as follows (again ignoring references to W when $U = R$).

Definition 6.18 [SAMPLE RANDOM-CONFIGURATION]

When SAMPLE RANDOM-CONFIGURATION is called in U from a neighbour W, U absorbs a flow from W and then picks at random a configuration of its variables from the distribution proportional to the potential on U. This configuration is then entered as additional evidence by setting all other table entries of U to zero, after which U issues SAMPLE RANDOM-CONFIGURATION in all its other neighbours. □

Again the junction tree property is vital here. When a configuration is selected in a clique, it defines a configuration in every one of its neighbouring separators, so that when it passes flows to its neighbouring cliques further away they too are sampled from the correct distribution, and no variable has its state fixed in two different cliques. When there is no evidence, the probabilistic logic sampling algorithm will generally be more efficient; but when there is evidence, sampling in the junction tree, as described in Definition 6.18, will soon become more efficient because one avoids rejection. Indeed, it is possible to implement sampling in the junction tree so that it becomes more efficient the more findings there are, in contrast to probabilistic logic sampling in which rejection rates increase sharply with increase in evidence.

Note the similarity of this to the method of finding the most probable configuration after a max-propagation.

6.4.4 Finding the M most probable configurations

An extension to the problem of finding the most probable configuration is to find the M most probable configurations for some given integer M. For tree-like Bayesian networks, Sy (1993) developed a computation scheme based upon an algorithm by Pearl (1988) for finding the most probable configuration in such networks. For more general networks, Seroussi and Golmard (1992) proposed a search method based on a bottom-up search of the junction tree, storing at each stage the M most probable configurations of the subtree examined. However the complexity of the algorithm is high. Nilsson (1994) proposed an iterative divide-and-conquer algorithm for finding the top M most probable configurations in a probabilistic network, which exploits the use of multiple max-propagations, the fact that one can find the top three most probable configurations directly by inspection of clique potentials after a max-propagation, and the fact that one can also propagate likelihood vectors. Later Nilsson (1998) presented a more efficient version of the algorithm which avoids multiple max-propagations. We describe the last method here.

Suppose we are given a junction tree, which for simplicity we assume to be connected, with a potential representation of the probability function p. We can order the cliques to have the running intersection property as $(C_0, C_1, \dots, C_k)$, with separators $(S_1, \dots, S_k)$ such that the separator S_j joins the clique C_j to the adjacent clique which is closer to the root C_0. We denote the corresponding ordered *residual sets* by $(R_0, R_1, \dots, R_k)$, where $R_j := C_j \setminus S_j$ for every $j > 0$, and we define $R_0 := C_0$. Let $\mathcal{X}$ denote the complete set of configurations, and for $L = 1, 2, \dots$, etc., let x^L denote the Lth most probable configuration. After having incorporated any initial evidence $\mathcal{E}$, we perform a max-propagation to put the potentials in the junction tree into their max-marginal form, and then extract the most probable configuration x^1 by a call to FIND MAX-CONFIGURATION. Denote the max-marginal representation by $(\{p_C^{\max}, C \in \mathcal{C}\}, \{p_S^{\max}, S \in \mathcal{S}\})$ and express the most probable configuration x^1 as $(r_0^1, \dots, r_k^1)$, where the r's lie in the configuration spaces of the residual sets.

We now partition the search space to exclude x^1. The second most probable configuration x^2 must differ from x^1 in the configuration of variables in at least one residual set. Hence, one considers $k+1$ distinct sets of 'virtual evidence,' of the form

$$\begin{aligned}
\mathcal{E}_0^1 &:= \mathcal{E} \ \& \ R_0 \neq r_0^1 \\
\mathcal{E}_1^1 &:= \mathcal{E} \ \& \ R_0 = r_0^1 \ \& \ R_1 \neq r_1^1 \\
\mathcal{E}_2^1 &:= \mathcal{E} \ \& \ R_0 = r_0^1 \ \& \ R_1 = r_1^1 \ \ \& \ R_2 \neq r_2^1 \\
&\ \vdots \\
\mathcal{E}_k^1 &:= \mathcal{E} \ \& \ R_0 = r_0^1 \ \& \ \dots \ \& \ R_{k-1} = r_{k-1}^1 \ \ \& \ R_k \neq r_k^1.
\end{aligned} \tag{6.24}$$

These constraints partition the search space $\mathcal{X} \setminus \{x^1\}$ in such a way that x^1 is incompatible with all of these possible evidences, but the second most probable configuration x^2 must be compatible with at least one (and, if x^2 is unique, only one) of the evidences in the set $\{\mathcal{E}_j^1, j = 0, \dots, k\}$; which one it is can be found by comparing their 'max-normalization' constants and choosing the largest. These normalization constants *could* be found using $k + 1$ junction tree initializations and max-collect operations. However, as Nilsson (1998) shows, it is not necessary to perform full initializations and max-collect operations for these evidences. This is because of the way the residual sets have been ordered. Thus, for example, for evidence $\mathcal{E}_j^1$ no new evidence over and above the initial evidence $\mathcal{E}$ has been entered in any clique having a label larger than j. Thus, all of these cliques and separators are initially in max-equilibrium, so if one were to perform a max-collect their contributions would cancel out in calculating the max-normalization. Furthermore, for the cliques and separators having labels i smaller than j, the evidence is such that for each clique C_i and separator S_i, only one configuration (which we denote by $x^*_{C_i}$ and $x^*_{S_i}$) is allowed. Thus, the max-normalization constant for evidence $\mathcal{E}_j^1$ has the simple form (Theorem 3 of Nilsson (1998)):

$$\frac{\prod_{i=0}^{j-1} p_{C_i}^{\max}(x^*_{C_i})}{\prod_{i=0}^{j-1} p_{S_i}^{\max}(x^*_{S_i})} \max p_{C_j}^{\max}(x_{C_j} \ \& \ \mathcal{E}_j^1). \tag{6.25}$$

To evaluate (6.25) requires a simple search in the jth clique for the last term, and that one value be picked out for each preceding clique and separator. Having found which evidence is consistent with x^2, one uses it to find the actual configuration by a call to FIND MAX-CONFIGURATION; this can be achieved without altering the clique and separator potentials.

Thus, it is simple to find x^2; suppose it came from the jth evidence, $\mathcal{E}_j^1$. Then either the third most probable configuration x^3 is consistent with one of the other evidences of (6.24), excluding $\mathcal{E}_j^1$, or it disagrees with both the x^1 and x^2 configurations in at least one place for some residual set having index $\geq j$ while agreeing for all residual sets having indices less than j. To find out which, we perform a further partition of the search space to obtain an extra set of $k - j + 1$ disjoint evidences of the form:

$$\begin{aligned}
\mathcal{E}_j^2 &:= \mathcal{E}_j^1 \;\&\; R_j \neq r_j^2 \\
\mathcal{E}_{j+1}^2 &:= \mathcal{E}_j^1 \;\&\; R_j = r_j^2 \;\&\; R_{j+1} \neq r_{j+1}^2 \\
&\;\vdots \\
\mathcal{E}_k^2 &:= \mathcal{E}_j^1 \;\&\; R_j = r_j^2 \;\&\; \ldots \;\&\; R_{k-1} = r_{k-1}^2 \;\&\; R_k \neq r_k^2.
\end{aligned} \tag{6.26}$$

Using arguments similar to those above the max-normalization constant for $\mathcal{E}_l^2$ reduces to:

$$\frac{\prod_{i=0}^{l-1} p_{C_i}^{\max}(x_{C_i}^*)}{\prod_{i=0}^{l-1} p_{S_i}^{\max}(x_{S_i}^*)} \max p_{C_l}^{\max}(x_{C_l} \;\&\; \mathcal{E}_l^2). \tag{6.27}$$

That evidence which has highest max-normalization among these or those of (6.24) (excluding $\mathcal{E}_j^1$ from the latter) will yield a set of evidence consistent with x^3; this evidence may then be used with a call to FIND MAX-CONFIGURATION to find x^3. Further partitions are required to find x^4, x^5, etc. Full proofs and details can be found in Nilsson (1998).

Regarding implementation: at each stage one keeps a list of all configurations found and also a list of candidate evidence partitions with their max-normalization weights. The method is implemented in XBAIES (see Appendix C) working on an elimination tree, so that each residual set consists of a single variable; this simplifies computation. For the CHILD network, without prior evidence, finding the top 500 configurations, which accounts for just over 18 percent of the total probability mass, requires approximately 2,600 possible evidences to be considered in total (actual numbers will depend upon the junction tree topology, choice of root, etc., which are not unique).

6.4.5 Sampling without replacement

We saw in Section 6.4.3 that sampling a random configuration (SAMPLE RANDOM-CONFIGURATION) is similar in many respects to finding the most probable configuration (FIND MAX-CONFIGURATION). One could use SAMPLE RANDOM-CONFIGURATION to generate many independent samples. Cowell (1997) showed that one can employ Nilsson's partitioning scheme, as described in Section 6.4.4, to draw *samples without replacement* correctly from the posterior distribution. The differences are: (1) sum-propagation rather than max-propagation is used to initialize the tree, and (2) sampling rather than maximization is performed in the candidate partition search stages. Apart from these differences, the essential steps are as for finding the top M configurations.

Specifically, suppose that we have put the junction tree into sum-marginal representation by sum-propagation of evidence. Then one samples one configuration by the method described in Section 6.4.3. Denote this first sample by x^1, writing it in terms of the residual sets as $x^1 = (r_0^1, \ldots, r_k^1)$. Then to sample a second configuration, in analogy to finding the second most probable configuration, one generates $k+1$ candidates using evidences given by (6.24), with the candidate defined by $\mathcal{E}_j^1$ being assigned weight (cf. (6.25)):

$$\frac{\prod_{i=0}^{j-1} p_{C_i}(x_{C_i}^*)}{\prod_{i=0}^{j} p_{S_i}(x_{S_i}^*)} \sum p_{C_j}(x_{C_j} \;\&\; \mathcal{E}_j^1). \tag{6.28}$$

This weight could alternatively be found using a sum-collect operation with evidence $\mathcal{E}_j^1$, but again, in direct analogy to the max-collect operation, by sum-consistency the cliques and separators having indices greater than j have their contributions cancel. Thus, to evaluate (6.28) requires a simple summation of terms in the jth clique consistent with $\mathcal{E}_j^1$ for the last term, and picking out one value for each preceding clique and separator. Next, from this set of weighted candidates one samples a candidate according to its weight and then completes the configuration by a sample-distribute operation, which ensures that the final configuration obtained is drawn from the distribution correctly.

Thus, assume that in this process the jth term was sampled. Then in analogy to drawing a single sample, the configuration is completed by SAMPLE RANDOM-CONFIGURATION using $\mathcal{E}_j^1$ (which can begin from the jth clique, as all previous cliques have completely specified residuals). Its probability can then be found easily. After completion, if a third configuration is required then this evidence $\mathcal{E}_j^1$ must also be partitioned to form new extra candidates. These candidates will have respective evidences as in (6.26), with weights given in analogy to (6.28).

6.4.6 Fast retraction

Suppose that we have evidence $\mathcal{E} = \{\mathcal{E}_v, v \in U^*\}$, with each item $\mathcal{E}_v$ of evidence represented by a function l_v that depends on a single node v; in what follows, $\mathcal{E}$ will be regarded as fixed. For $A \subseteq V$, let $\mathcal{E}_A = \{\mathcal{E}_v, v \in U^* \cap A\}$ and correspondingly $l_A(x_A) = \prod_{v \in U^* \cap A} l_v(x_v)$. We may as well assume that $U^* = V$ by defining $l_v(x_v) \equiv 1$ for $v \notin U^*$.

For purposes of monitoring the performance of a probabilistic network (see Chapter 10), it can be informative to compare each item of evidence $\mathcal{E}_v$ with the probabilistic prediction given by the system for X_v on the basis of the remaining evidence $\mathcal{E}_{V \setminus \{v\}}$, as expressed by the conditional distribution of X_v given $\mathcal{E}_{V \setminus \{v\}}$. Such comparisons can form the basis of an investigation of model adequacy, using general methods of assessing the success of probability forecasts, such as scoring rules or calibration plots (Dawid 1986).

By further within-clique conditioning the required distribution of X_v given $\mathcal{E}_{V\setminus\{v\}}$ can be calculated from that of X_C given $\mathcal{E}_{V\setminus C}$, C being any clique containing v. To find this, we need to *retract* the evidence $\mathcal{E}_C$. If we have evidence in k cliques, these conditional densities could be found by using k initializations and propagations, each propagating the evidence in the remaining $k-1$ cliques. However, under suitable conditions the same results can be found in just one propagation, as below. Further details and proofs, together with conditions, may be found in Cowell and Dawid (1992).

We describe how the basic propagation algorithm can be modified so as to calculate, simultaneously for every clique, the clique-density engendered by retraction of the evidence entered into that clique, under the assumption that the overall probability density p is strictly positive. Recall that in the standard set-up one multiplies the clique potentials by the evidence potentials on the variables in that clique, and then marginalizes the result to initiate the flow. The idea here is to pass the flow incorporating the evidence, but to do so without multiplying the clique potential by the evidence. This leads to another type of marginalization operation now defined.

Definition 6.19 [OUT-MARGINALIZATION]

Let $W \subseteq U \subseteq V$ and let ϕ be a potential on U. The expression $\sum_{V\setminus W}^{\mathcal{E}} \phi$, for evidence $\mathcal{E} = \{\mathcal{E}_v : v \in V\}$, is defined by:

$$\left(\sum_{U\setminus W}^{\mathcal{E}} \phi\right)(x) = \sum_{z \in \mathcal{U}\setminus\mathcal{W}} \phi(z.x) l_{U\setminus W}(z).$$

□

We term $\sum_{V\setminus W}^{\mathcal{E}} \phi$ the *out-margin* of ϕ on W, since it incorporates the evidence functions related to nodes outside W, but leaves out those associated with W. We shall also write $\phi_W^{*\mathcal{E}}$ for this out-margin. Note that, if $Y \subseteq W$ then $\sum_{W\setminus Y}^{\mathcal{E}} \sum_{V\setminus W}^{\mathcal{E}} \phi = \sum_{V\setminus Y}^{\mathcal{E}} \phi$. We have

$$p_A^{*\mathcal{E}}(x_A) = P(X_a = x_A \;\&\; \mathcal{E}_{V\setminus A}).$$

With this definition of marginalization, and for positive densities p, a full evidence propagation (a call of COLLECT EVIDENCE followed by DISTRIBUTE EVIDENCE) leads to the *out-marginal representation* $(\{p_C^{*\mathcal{E}}, C \in \mathcal{C}\}, \{p_S^{*\mathcal{E}}, S \in \mathcal{S}\})$, so that we have:

$$p(x) = \frac{\prod_{C\in\mathcal{C}} p_C^{*\mathcal{E}}(x_C)}{\prod_{S\in\mathcal{S}} p_S^{*\mathcal{E}}(x_S)} \tag{6.29}$$

(Cowell and Dawid 1992). This routine works because (1) we start with a generalized potential representation $(\{\phi_C, C \in \mathcal{C}\}, \{\phi_S, S \in \mathcal{S}\})$ of the joint density p without the evidence $\mathcal{E}$ incorporated, and (2) because of the positivity assumption on p, all flows in the separators are also strictly positive; hence, it follows that (3) the flows preserve the contraction p of the representation, while still transmitting the evidence around the tree. One could develop hybrid schemes in which evidence about some variables is incorporated into clique potentials, while evidence about other variables is treated by fast retraction.

The out-marginal representation has one other computationally useful feature. After propagating the evidence, it is still a representation of the prior p, not the posterior, which means that if we wish to examine different evidence we can propagate it without the need to reinitialize the junction tree potentials with the original conditional probabilities.

One could more properly describe Definition 6.19 as *out-sum-marginalization*: the manner in which fast retraction handles evidence is unrelated to the type of marginalization being performed. Thus, for example, one could define an *out-max-marginalization* operation as follows:

Definition 6.20 [OUT-MAX-MARGINALIZATION]

Let $W \subseteq U \subseteq V$ and let ϕ be a potential on U. The expression $\mathbf{M}^{\mathcal{E}}_{U \setminus W} \phi$, the *out-max-margin* of ϕ on W for the evidence $\mathcal{E}$, is defined by:

$$\left(\overset{\mathcal{E}}{\underset{U \setminus W}{\mathbf{M}}} \phi \right)(x) = \max_{z \in \mathcal{U} \setminus \mathcal{W}} \phi(z.x) l_{U \setminus W}(z).$$

□

Propagating using this type of marginalization operation allows one to calculate for every clique the maximum probabilities of a full configuration of those variables consistent with the evidence for variables not in that clique.

The assumption that p is strictly positive can be relaxed under some conditions, depending on the junction tree under consideration. The problem is that without this assumption the entries of some configurations in one or more separator potentials can become zero, leading to incorrect flows being sent in the distribute stage because of illegal divisions by zero. The approach of Shenoy and Shafer (1990) is not affected by such problems because their propagation scheme avoids division operations. Thus, in their scheme fast retraction can be applied without restriction.

6.4.7 *Moments of functions*

Suppose that we have a probability distribution P for a set of discrete variables $X = X_V$ whose density is expressed by (6.10), in terms of a generalized potential representation on a junction tree $\mathcal{T}$. Suppose also that we have a real-valued function $g : \mathcal{X} \to \mathbb{R}$. We may wish to find the expectation of g under P, perhaps given evidence $\mathcal{E}$: $\mathrm{E}_P(g \mid \mathcal{E}) = \sum_x g(x)p(x \mid \mathcal{E})$. For arbitrary functions g this may be difficult because we will not be able to exploit the decomposition of p on $\mathcal{T}$. However, if g decomposes *additively* on $\mathcal{T}$, that is, if we can express g in the form

$$g(x) = \sum_{C \in \mathcal{C}} g_C(x_C) - \sum_{S \in \mathcal{S}} g_S(x_S) \tag{6.30}$$

for functions g_C and g_S defined on the cliques and separators of $\mathcal{T}$, then it is straightforward to find the expectation $\mathrm{E}_P(g \mid \mathcal{E})$: one propagates the evidence on the junction tree to obtain the marginal representation $(\{p_C, C \in \mathcal{C}\}, \{p_S, S \in \mathcal{S}\})$, from which one can readily evaluate

$$\begin{aligned} \mathrm{E}_P(g \mid \mathcal{E}) &= \sum_x g(x) p(x \mid \mathcal{E}) \\ &= \sum_{C \in \mathcal{C}} \sum_{x_C} g_C(x_C) p_C(x_C \mid \mathcal{E}) - \sum_{S \in \mathcal{S}} \sum_{x_S} g_S(x_S) p_S(x_S \mid \mathcal{E}). \end{aligned}$$

Cowell (1992) showed that, if g decomposes as in (6.30), it is also possible to calculate the higher moments $\mathrm{E}_P(g^n \mid \mathcal{E})$ by local computation. There are two key steps. The first is to consider the *moment generating function* of g, the second to define a new type of message-passing in which polynomials, with potentials as coefficients, are transmitted. We summarize the method here, assuming for simplicity that there is no evidence.

Thus, suppose we wish to find the first n moments of g for some fixed integer n. Introducing a dummy parameter, or indeterminate, ϵ, the moment-generating function G is given by

$$G(\epsilon) := \sum_x p(x) e^{\epsilon g(x)}. \tag{6.31}$$

Then $G(\epsilon) = \sum_{i=0}^{\infty} (\mu_i / i!) \epsilon^i$, where $\mu_i = \mathrm{E}_P(g^n)$.

Now using the decomposition according to $\mathcal{T}$ of both the density p (6.10) and function g (6.30), we obtain

$$G(\epsilon) = \sum_x G_\epsilon(x), \tag{6.32}$$

where the function $G_\epsilon : \mathcal{X} \rightarrow \mathbb{R}$ is defined by

$$G_\epsilon(x) = \frac{\prod_{C \in \mathcal{C}} \phi_C(x_C) e^{\epsilon g_C(x_c)}}{\prod_{S \in \mathcal{S}} \phi_S(x_S) e^{\epsilon g_S(x_S)}}. \tag{6.33}$$

The moments of g up to order n are obtained by local propagations of polynomials of degree n, having potential coefficients, which leave G_ϵ invariant. Specifically:

1. Fix the highest degree n required for the moments and set $\epsilon^k = 0$ for all $k > n$ in all calculations.

2. Replace the exponential in each clique by its truncated power series

$$\phi_C^n(x_C) := \phi_C(x_C)e^{\epsilon g_C(x_C)} = \sum_{j=0}^{n} \phi_{C:j}(x_C)\epsilon^j, \tag{6.34}$$

where $\phi_{C:j}(x_C) := \phi_C(x_C)g_C^j(x_C)/j!$, and do similar expansions for the separators. Then to degree n we have

$$G_\epsilon(x) = \frac{\prod_{C\in\mathcal{C}} \sum_{j=0}^{n} \phi_{C:j}(x_C)\epsilon^j}{\prod_{S\in\mathcal{S}} \sum_{j=0}^{n} \phi_{S:j}(x_S)\epsilon^j}.$$

3. Addition and subtraction of the polynomial potentials ϕ_U^n, $U \subseteq V$, are defined termwise in the obvious manner. Multiplication of two or more such polynomial potentials is also defined in the obvious manner, taking care to delete any terms of degree ϵ^{n+1} and higher which arise.

4. A message sent from a clique C_1 to a neighbouring clique C_2 via the separator S is defined by the following operations:

   - the potential $\phi_S^n = \sum_{j=0}^{n} \phi_{S:j}\epsilon^j$ is replaced by the potential $\phi_S^{*n} := \sum_{j=0}^{n} \phi_{S:j}^*\epsilon^j$ where $\phi_{S:j}^*(x_S) = \sum_{C_1\setminus S} \phi_{C_1:j}(x_{C_1})$; and
   - the potential $\phi_{C_2}^n$ on C_2 is modified by the update ratio ϕ_S^{*n}/ϕ_S^n to yield a new potential $\phi_{C_2}^{*n} = \sum_{j=0}^{n} \phi_{C_2:j}^*\epsilon^j$, where the $\phi_{C_2:j}^*$ are chosen so that $\phi_{C_2}^{*n}\phi_S^n \equiv \phi_{C_2}^n\phi_S^{*n}$, with terms of degree ϵ^{n+1} and higher discarded.

     This defines the division of one polynomial potential by another. In practice, one would first find the term in ϵ^0:

     $$\phi_{C_2:0}^* = \phi_{C_2:0}\phi_{S:0}^*/\phi_{S:0}$$

     then solve for the term in ϵ^1:

     $$\phi_{C_2:1}^* = (\phi_{C_2:0}\phi_{S:1}^* + \phi_{C_2:1}\phi_{S:0}^* - \phi_{C_2:0}^*\phi_{S:1})/\phi_{S:0}$$

     etc., up to the term of degree ϵ^n.

   This message leaves the polynomial potential $G_\epsilon(x)$ unchanged.

5. Collect to any clique by passing these messages, and then further marginalize on this clique. The result is a polynomial in ϵ of degree n whose coefficient of ϵ^i is $\mu^i/i!$.

For the proof that this algorithm works, see Cowell (1992). Evidence, if any, can be incorporated into the clique potentials $\phi_C(x_C)$ at Stage 2, before the collection operation of Stage 4. Instead of treating the potential entries as polynomials of degree n, one could treat them as $(n+1)$-vectors. Vector potentials will be used to solve decision problems in Chapter 8.

Note that, for p of the form (6.10), $\log p \equiv \sum_{C\in\mathcal{C}} \log \phi_C - \sum_{S\in\mathcal{S}} \log \phi_S$ which is of the form (6.30), hence by this method one can calculate the moments of $\log p(X)$, in particular the *entropy* $\mathrm{E}_P\{-\log p(X)\}$. This can be applied to evaluating global monitors for complete data (see Section 10.4).

6.5 Example: CH-ASIA

We use a chain graph variation of the fictitious ASIA given on page 20 to illustrate the above concepts. We refer to this example as CH-ASIA.

6.5.1 *Description*

> *Cough?* (C) can be caused by *Tuberculosis?* (T), *Lung cancer?* (L), *Bronchitis?* (B), or a combination of them, or none of them. *Dyspnoea?* (D) is associated with cough and can be caused by bronchitis. An *X-ray?* investigation (X) does not distinguish between tuberculosis or lung cancer, and these two diseases also have similar risks for developing cough. A recent *Visit to Asia?* (A) could be a risk factor for tuberculosis, and *Smoking?* (S) increases the risk for both lung cancer and bronchitis.

6.5.2 *Graphical specification*

The model is a chain graph $\mathcal{K}$ having nine binary variables, whose structure is given in Figure 6.1. The directed edges represent the causal influence. The moral graph of the chain graph $\mathcal{K}^m$ is obtained by adding the pair of edges $\{$*Lung cancer?*, *Tuberculosis?*$\}$ and $\{$*Bronchitis?*, *Either... ?*$\}$. This moral graph is not triangulated, but adding the edge $\{$*Lung cancer?*, *Bronchitis?*$\}$ makes it so; Figure 6.2 shows an *elimination tree* derived from the moralized and triangulated graph of Figure 6.1, which we shall use to illustrate local propagation.

6.5.3 *Numerical specification*

With the exception of the pair CD, all chain components consist of single nodes. Hence, each such node requires a specification consisting of the conditional probability distribution of the node, given its parents (if any). We shall take the values given on page 21 for the ASIA example, reproduced in Table 6.2. Note how probabilities of 0 and 1 are used to specify the logical *Either... ?* node.

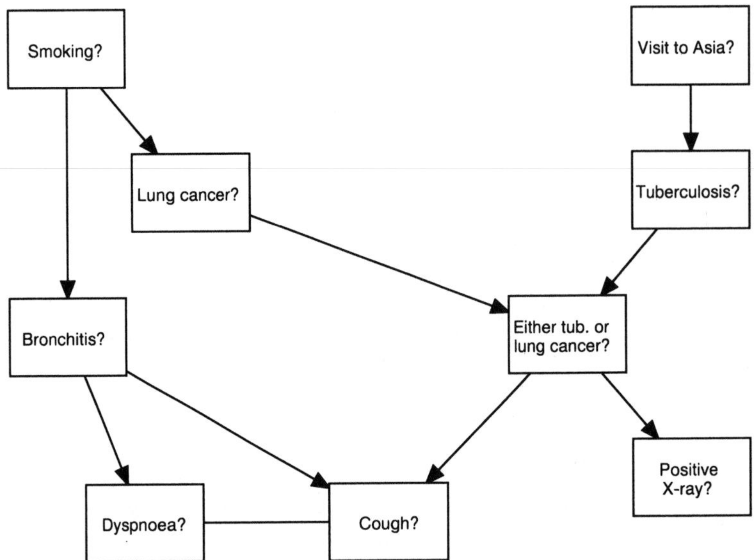

FIGURE 6.1. Graph of the CH-ASIA example.

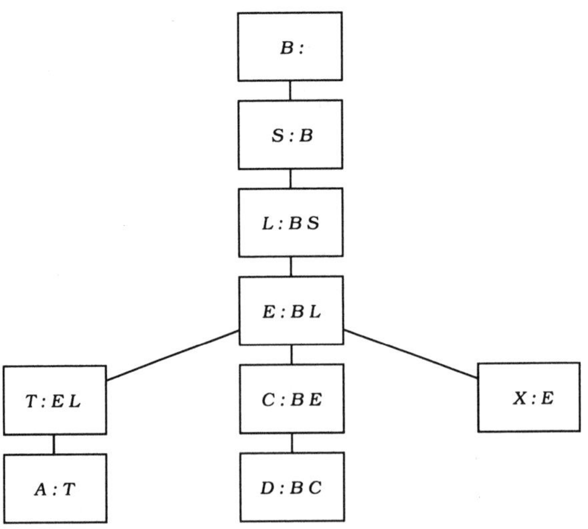

FIGURE 6.2. Elimination tree for local propagation. Collect flows flow upwards, and distribute flows flow downwards. Elimination set B is the root. Note that removing the top two elimination sets yields a junction tree of cliques.

TABLE 6.2. Initial potentials for chain-component singletons. A=yes is denoted by a, A=no by $\bar{a}$ etc.

a	0.01
$\bar{a}$	0.99

	t	$\bar{t}$
a	0.05	0.95
$\bar{a}$	0.01	0.99

	x	$\bar{x}$
e	0.98	0.02
$\bar{e}$	0.05	0.95

	e	$\bar{e}$
$l\ t$	1	0
$\bar{l}\ t$	1	0
$l\ \bar{t}$	1	0
$\bar{l}\ \bar{t}$	0	1

s	0.5
$\bar{s}$	0.5

	b	$\bar{b}$
s	0.6	0.4
$\bar{s}$	0.3	0.7

	l	$\bar{l}$
s	0.1	0.9
$\bar{s}$	0.01	0.99

There remains the conditional probability $P(C, D \mid B, E)$ to specify. From the graphical structure this factorizes as:

$$P(C, D \mid B, E) = \phi(C, D, B)\, \phi(C, B, E)\, Z^{-1}(B, E), \tag{6.35}$$

The last term is a normalizing constant. It is not specified separately, but calculated from the other factors.

The elicitation of chain graph potentials is more complex than for simple conditional probabilities because the extra factorization has to be taken into account. One way to proceed is by specifying interaction terms of the conditional probabilities, in analogy with procedures in log-linear models and Markov random fields (Darroch et al. 1980). We begin with the main effects.

The *main effects* are the odds for presence of the feature, given that neighbours in the same chain component are in a chosen reference state; in our example we everywhere select the reference state "no." Hence, the main effect for C is the odds for cough (c), given that there is no dyspnoea ($\bar{d}$). The main effect for C is allowed to depend on the states of both parents B and E, whereas the main effect for D is allowed to depend only on B, since there is no edge between D and E.

Denoting the binary states "no" by 0 and "yes" by 1, we set our reference state to be $C = 0, D = 0$ for each level of B, E. We introduce

$$\phi(B, E) = P(C = 0, D = 0 \mid B, E). \tag{6.36}$$

(However, we do not calculate $\phi(B, E)$ as yet.) Then the odds for cough C, given $D = 0$, B and E, are used to define $\phi_{C \mid B,E}(1)$, where

$$\begin{aligned} \phi_{C \mid B,E}(i) &:= \frac{P(C = i, D = 0 \mid B, E)}{P(C = 0, D = 0 \mid B, E)} \\ &= \frac{P(C = i, D = 0 \mid B, E)}{\phi(B, E)}. \end{aligned} \tag{6.37}$$

This implies in particular that

$$\phi_{C\,|\,B,E}(0) = P(C=0, D=0\,|\,B,E)/\phi(B,E) = 1.$$

Similarly we use the odds for D, given $C = 0, E$ and B, to set $\phi_{D\,|\,B}(1)$, where

$$\begin{aligned}\phi_{D\,|\,B}(j) &:= \frac{P(C=0, D=j\,|\,B,E)}{P(C=0, D=0\,|\,B,E)} \\ &= \frac{P(C=0, D=j\,|\,B,E)}{\phi(B,E)},\end{aligned} \tag{6.38}$$

from which we have $\phi_{D\,|\,B}(0) = 1$.

The interaction is the ratio between the odds for presence of dyspnoea given presence of cough and the corresponding odds given absence of cough. From (6.35), this interaction is only allowed to depend on B, the common parent of C and D. This odds ratio is used to define $\phi_{C,D\,|\,B}(1,1)$ where

$$\phi_{C,D\,|\,B}(i,j) := \frac{P(C=i, D=j\,|\,B,E)\,P(C=0, D=0\,|\,B,E)}{P(C=i, D=0\,|\,B,E)\,P(C=0, D=j\,|\,B,E)}, \tag{6.39}$$

and thus $\phi_{C,D\,|\,B}(i,j) = 1$ if $(i,j) \neq (1,1)$.

These potentials specify the conditional probability $P(C,D\,|\,B,E)$:

$$P(C,D\,|\,B,E) = \phi_{C\,|\,B,E}(C)\,\phi_{D\,|\,B}(D)\,\phi_{C,D\,|\,B}(C,D)\,\phi(B,E) \tag{6.40}$$

from which $\phi(B,E)$ can be evaluated:

$$\phi(B,E)^{-1} = \sum_{C,D} \phi_{C\,|\,B,E}(C)\,\phi_{D\,|\,B}(D)\,\phi_{C,D\,|\,B}(C,D). \tag{6.41}$$

This normalization can be found by direct calculation, but it can also be found by propagation on the subgraph induced by $\{B,C,D,E\}$ of the triangulated moral graph, initialized with potentials $\phi_{C\,|\,B,E}$, $\phi_{D\,|\,B}$ and $\phi_{C,D\,|\,B}(C,D)$.

The odds and odds-ratios for the CD chain component were taken as in Table 6.3. These define the factors for the chain-component potentials given in Table 6.4; the normalized conditional probability table is given in Table 6.5.

6.5.4 Initialization

We work with the elimination tree of Figure 6.2, an expanded version of the junction tree of cliques, to illustrate propagation; for brevity we refer to the elimination sets of the elimination tree as cliques. The elimination tree can be initialized in a number of ways. Table 6.6 shows one assignment

TABLE 6.3. Odds and odds-ratio for the CD chain component.

B	Main effects $\phi_{C \mid BE}$		$\phi_{D \mid B}$	Odds-ratio $\phi_{CD \mid B}$
E	e	$\bar{e}$		
b	5	2	4	4
$\bar{b}$	3	1	1	2

TABLE 6.4. Chain component potentials $\phi(C, D, B) = \phi_{CD \mid B}(C, D)\, \phi_{D \mid B}(D)$ and $\phi(C, B, E) = \phi_{C \mid BE}(C)$ defined from the odds and odds-ratios of Table 6.3, together with the calculated normalization constant $\phi(B, E)$.

$\phi(C, D, B)$	cd	$\bar{c}d$	$c\bar{d}$	$\bar{c}\bar{d}$
b	16	4	1	1
$\bar{b}$	2	1	1	1

$\phi(C, B, E)$	c	$\bar{c}$
be	5	1
$\bar{b}e$	3	1
$b\bar{e}$	2	1
$\bar{b}\bar{e}$	1	1

$\phi(B, E)$	
be	1/90
$\bar{b}e$	1/11
$b\bar{e}$	1/39
$\bar{b}\bar{e}$	1/5

TABLE 6.5. Normalized conditional probability table $P(C, D \mid BE)$ obtained by multiplying and normalizing potentials of Table 6.4.

	cd	$\bar{c}d$	$c\bar{d}$	$\bar{c}\bar{d}$
be	0.8889	0.0444	0.0556	0.0111
$\bar{b}e$	0.5455	0.0909	0.2727	0.0909
$b\bar{e}$	0.8205	0.1026	0.0513	0.0256
$\bar{b}\bar{e}$	0.4	0.2	0.2	0.2

of potentials to initialize the cliques. Note that not all cliques have a non-trivial potential assigned to them (clique B has none in this case). Note also that the factors of the chain component CD are not all assigned to the same clique, though the normalization $\phi(B, E)$ could have been put into the CBE clique. The initial numerical values of the clique potentials are shown in Table 6.7. The separator potentials are all unity at this stage.

TABLE 6.6. Assignment of potentials for initializing the junction tree.

Clique	Potentials
B	1
SB	$P(B \mid S)P(S)$
LBS	$P(L \mid S)$
EBL	$\phi(EB)$
TEL	$P(E \mid LT)$
CBE	$\phi(CBE)$
XE	$P(X \mid E)$
AT	$P(T \mid A)P(A)$
DBC	$\phi(CDB)$

If we had been working with a junction tree of cliques by removing the elimination sets B and SB, then the clique LBS would have been initialized with the potential $P(B \mid S)P(S)P(L \mid S)$.

6.5.5 *Propagation without evidence*

The full set of clique potentials after calling Collect Evidence (to B) and then Distribute Evidence is shown in Table 6.8. These operations generate the marginal representation, so by further sum-marginalizations within the cliques we may obtain the marginal distribution of each variable: these are given in Table 6.9.

6.5.6 *Propagation with evidence*

We now consider entering evidence. Suppose that we enter evidence $\mathcal{E} = (a \ \& \ s \ \& \ \bar{x})$. Then the initial clique potentials AT, XE, and SB are modified by the evidence $\mathcal{E}$ to take the following values:

$a \ t$	0.0005
$\bar{a} \ t$	0
$a \ \bar{t}$	0.0095
$\bar{a} \ \bar{t}$	0

$x \ e$	0
$\bar{x} \ e$	0.02
$x \ \bar{e}$	0
$\bar{x} \ \bar{e}$	0.95

$s \ b$	0.3
$\bar{s} \ b$	0
$s \ \bar{b}$	0.2
$\bar{s} \ \bar{b}$	0

TABLE 6.7. Initialization of junction tree of Figure 6.2, using potentials of Table 6.2 and Table 6.4.

b	1
$\bar{b}$	1

$s\ b$	0.3000
$\bar{s}\ b$	0.1500
$s\ \bar{b}$	0.2000
$\bar{s}\ \bar{b}$	0.3500

$l\ b\ s$	0.1000
$\bar{l}\ b\ s$	0.9000
$l\ \bar{b}\ s$	0.1000
$\bar{l}\ \bar{b}\ s$	0.9000
$l\ b\ \bar{s}$	0.0100
$\bar{l}\ b\ \bar{s}$	0.9900
$l\ \bar{b}\ \bar{s}$	0.0100
$\bar{l}\ \bar{b}\ \bar{s}$	0.9900

$e\ b\ l$	0.0111
$\bar{e}\ b\ l$	0.0256
$e\ \bar{b}\ l$	0.0909
$\bar{e}\ \bar{b}\ l$	0.2000
$e\ b\ \bar{l}$	0.0111
$\bar{e}\ b\ \bar{l}$	0.0256
$e\ \bar{b}\ \bar{l}$	0.0909
$\bar{e}\ \bar{b}\ \bar{l}$	0.2000

$t\ e\ l$	1
$\bar{t}\ e\ l$	1
$t\ \bar{e}\ l$	0
$\bar{t}\ \bar{e}\ l$	0
$t\ e\ \bar{l}$	1
$\bar{t}\ e\ \bar{l}$	0
$t\ \bar{e}\ \bar{l}$	0
$\bar{t}\ \bar{e}\ \bar{l}$	1

$a\ t$	0.0005
$\bar{a}\ t$	0.0099
$a\ \bar{t}$	0.0095
$\bar{a}\ \bar{t}$	0.9801

$c\ b\ e$	5
$\bar{c}\ b\ e$	1
$c\ \bar{b}\ e$	3
$\bar{c}\ \bar{b}\ e$	1
$c\ b\ \bar{e}$	2
$\bar{c}\ b\ \bar{e}$	1
$c\ \bar{b}\ \bar{e}$	1
$\bar{c}\ \bar{b}\ \bar{e}$	1

$d\ b\ c$	16
$\bar{d}\ b\ c$	1
$d\ \bar{b}\ c$	2
$\bar{d}\ \bar{b}\ c$	1
$d\ b\ \bar{c}$	4
$\bar{d}\ b\ \bar{c}$	1
$d\ \bar{b}\ \bar{c}$	1
$\bar{d}\ \bar{b}\ \bar{c}$	1

$x\ e$	0.9800
$\bar{x}\ e$	0.0200
$x\ \bar{e}$	0.0500
$\bar{x}\ \bar{e}$	0.9500

TABLE 6.8. Clique potentials after a call to COLLECT EVIDENCE to B (left) and then a subsequent call to DISTRIBUTE EVIDENCE (right). Note that after DISTRIBUTE EVIDENCE we have the marginal representation of the prior on the junction tree.

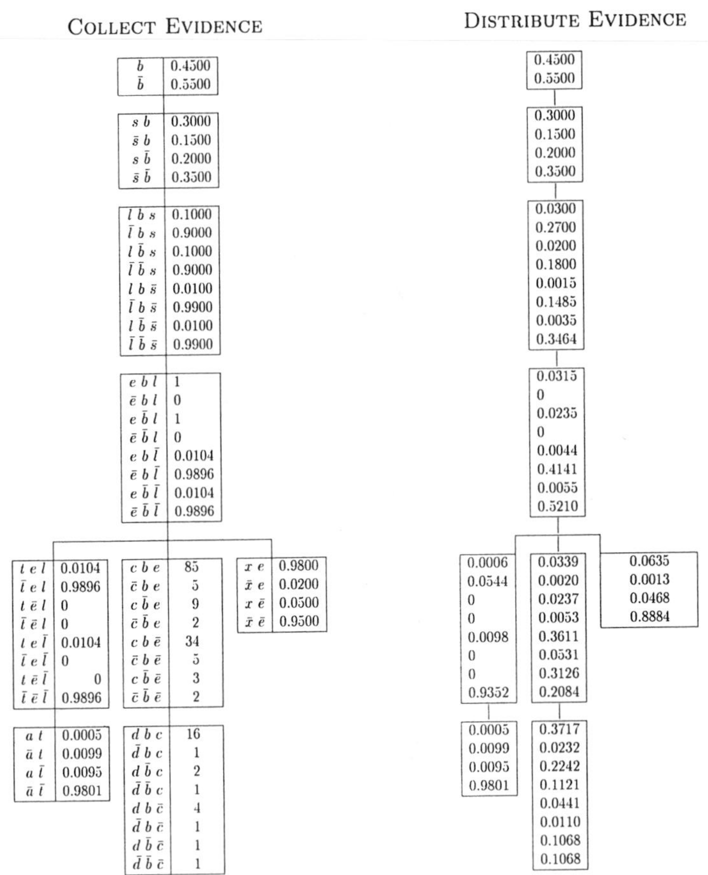

COLLECT EVIDENCE

b	0.4500
$\bar{b}$	0.5500

$s\ b$	0.3000
$\bar{s}\ b$	0.1500
$s\ \bar{b}$	0.2000
$\bar{s}\ \bar{b}$	0.3500

$l\ b\ s$	0.1000
$\bar{l}\ b\ s$	0.9000
$l\ \bar{b}\ s$	0.1000
$\bar{l}\ \bar{b}\ s$	0.9000
$l\ b\ \bar{s}$	0.0100
$\bar{l}\ b\ \bar{s}$	0.9900
$l\ \bar{b}\ \bar{s}$	0.0100
$\bar{l}\ \bar{b}\ \bar{s}$	0.9900

$e\ b\ l$	1
$\bar{e}\ b\ l$	0
$e\ \bar{b}\ l$	1
$\bar{e}\ \bar{b}\ l$	0
$e\ b\ \bar{l}$	0.0104
$\bar{e}\ b\ \bar{l}$	0.9896
$e\ \bar{b}\ \bar{l}$	0.0104
$\bar{e}\ \bar{b}\ \bar{l}$	0.9896

$t\ e\ l$	0.0104
$\bar{t}\ e\ l$	0.9896
$t\ \bar{e}\ l$	0
$\bar{t}\ \bar{e}\ l$	0
$t\ e\ \bar{l}$	0.0104
$\bar{t}\ e\ \bar{l}$	0
$t\ \bar{e}\ \bar{l}$	0
$\bar{t}\ \bar{e}\ \bar{l}$	0.9896

$c\ b\ e$	85
$\bar{c}\ b\ e$	5
$c\ \bar{b}\ e$	9
$\bar{c}\ \bar{b}\ e$	2
$c\ b\ \bar{e}$	34
$\bar{c}\ b\ \bar{e}$	5
$c\ \bar{b}\ \bar{e}$	3
$\bar{c}\ \bar{b}\ \bar{e}$	2

$x\ e$	0.9800
$\bar{x}\ e$	0.0200
$x\ \bar{e}$	0.0500
$\bar{x}\ \bar{e}$	0.9500

$a\ t$	0.0005
$\bar{a}\ t$	0.0099
$a\ \bar{t}$	0.0095
$\bar{a}\ \bar{t}$	0.9801

$d\ b\ c$	16
$\bar{d}\ b\ c$	1
$d\ \bar{b}\ c$	2
$\bar{d}\ \bar{b}\ c$	1
$d\ b\ \bar{c}$	4
$\bar{d}\ b\ \bar{c}$	1
$d\ \bar{b}\ \bar{c}$	1
$\bar{d}\ \bar{b}\ \bar{c}$	1

DISTRIBUTE EVIDENCE

0.4500
0.5500

0.3000
0.1500
0.2000
0.3500

0.0300
0.2700
0.0200
0.1800
0.0015
0.1485
0.0035
0.3464

0.0315
0
0.0235
0
0.0044
0.4141
0.0055
0.5210

0.0006
0.0544
0
0
0.0098
0
0
0.9352

0.0339
0.0020
0.0237
0.0053
0.3611
0.0531
0.3126
0.2084

0.0635
0.0013
0.0468
0.8884

0.0005
0.0099
0.0095
0.9801

0.3717
0.0232
0.2242
0.1121
0.0441
0.0110
0.1068
0.1068

TABLE 6.9. Marginal probabilities for the variables in CH-ASIA.

a	0.01	d	0.7468	t	0.0104
$\bar{a}$	0.99	$\bar{d}$	0.2532	$\bar{t}$	0.9896
c	0.7312	x	0.1103	e	0.0648
$\bar{c}$	0.2688	$\bar{x}$	0.8897	$\bar{e}$	0.9352
l	0.055	s	0.5	b	0.45
$\bar{l}$	0.945	$\bar{s}$	0.5	$\bar{b}$	0.55

Note that the evidence for $S = s$ could have been entered into the LBS clique instead. Also, we could have introduced these modifications into any representation of the prior, for example its marginal representation with clique potentials as on the right of Table 6.8. Table 6.10 shows the clique potentials after propagating this evidence. The left-hand tree shows the potentials after sum-propagation, and it shows for each clique the joint distribution of the variables and the evidence. The right-hand tree shows the potentials after propagating with fast retraction.

TABLE 6.10. Marginal and out-marginal clique potentials (multiplied by 1000) after propagating the evidence $\mathcal{E} = (a \;\&\; s \;\&\; \bar{x})$.

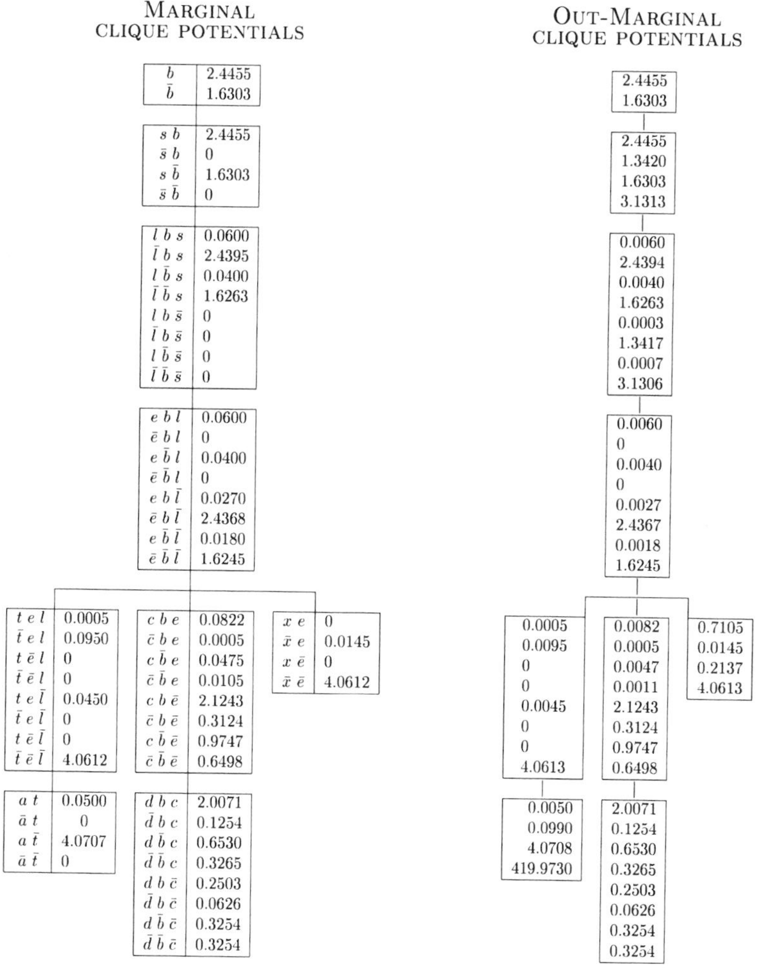

	MARGINAL CLIQUE POTENTIALS	OUT-MARGINAL CLIQUE POTENTIALS
b	2.4455	2.4455
$\bar{b}$	1.6303	1.6303

	MARGINAL CLIQUE POTENTIALS	OUT-MARGINAL CLIQUE POTENTIALS
$s\ b$	2.4455	2.4455
$\bar{s}\ b$	0	1.3420
$s\ \bar{b}$	1.6303	1.6303
$\bar{s}\ \bar{b}$	0	3.1313

	MARGINAL CLIQUE POTENTIALS	OUT-MARGINAL CLIQUE POTENTIALS
$l\ b\ s$	0.0600	0.0060
$\bar{l}\ b\ s$	2.4395	2.4394
$l\ \bar{b}\ s$	0.0400	0.0040
$\bar{l}\ \bar{b}\ s$	1.6263	1.6263
$l\ b\ \bar{s}$	0	0.0003
$\bar{l}\ b\ \bar{s}$	0	1.3417
$l\ \bar{b}\ \bar{s}$	0	0.0007
$\bar{l}\ \bar{b}\ \bar{s}$	0	3.1306

	MARGINAL CLIQUE POTENTIALS	OUT-MARGINAL CLIQUE POTENTIALS
$e\ b\ l$	0.0600	0.0060
$\bar{e}\ b\ l$	0	0
$e\ \bar{b}\ l$	0.0400	0.0040
$\bar{e}\ \bar{b}\ l$	0	0
$e\ b\ \bar{l}$	0.0270	0.0027
$\bar{e}\ b\ \bar{l}$	2.4368	2.4367
$e\ \bar{b}\ \bar{l}$	0.0180	0.0018
$\bar{e}\ \bar{b}\ \bar{l}$	1.6245	1.6245

	MARGINAL CLIQUE POTENTIALS	OUT-MARGINAL CLIQUE POTENTIALS
$t\ e\ l$	0.0005	0.0005
$\bar{t}\ e\ l$	0.0950	0.0095
$t\ \bar{e}\ l$	0	0
$\bar{t}\ \bar{e}\ l$	0	0
$t\ e\ \bar{l}$	0.0450	0.0045
$\bar{t}\ e\ \bar{l}$	0	0
$t\ \bar{e}\ \bar{l}$	0	0
$\bar{t}\ \bar{e}\ \bar{l}$	4.0612	4.0613

	MARGINAL CLIQUE POTENTIALS	OUT-MARGINAL CLIQUE POTENTIALS
$c\ b\ e$	0.0822	0.0082
$\bar{c}\ b\ e$	0.0005	0.0005
$c\ \bar{b}\ e$	0.0475	0.0047
$\bar{c}\ \bar{b}\ e$	0.0105	0.0011
$c\ b\ \bar{e}$	2.1243	2.1243
$\bar{c}\ b\ \bar{e}$	0.3124	0.3124
$c\ \bar{b}\ \bar{e}$	0.9747	0.9747
$\bar{c}\ \bar{b}\ \bar{e}$	0.6498	0.6498

	MARGINAL CLIQUE POTENTIALS	OUT-MARGINAL CLIQUE POTENTIALS
$x\ e$	0	0.7105
$\bar{x}\ e$	0.0145	0.0145
$x\ \bar{e}$	0	0.2137
$\bar{x}\ \bar{e}$	4.0612	4.0613

	MARGINAL CLIQUE POTENTIALS	OUT-MARGINAL CLIQUE POTENTIALS
$a\ t$	0.0500	0.0050
$\bar{a}\ t$	0	0.0990
$a\ \bar{t}$	4.0707	4.0708
$\bar{a}\ \bar{t}$	0	419.9730

	MARGINAL CLIQUE POTENTIALS	OUT-MARGINAL CLIQUE POTENTIALS
$d\ b\ c$	2.0071	2.0071
$\bar{d}\ b\ c$	0.1254	0.1254
$d\ \bar{b}\ c$	0.6530	0.6530
$\bar{d}\ \bar{b}\ c$	0.3265	0.3265
$d\ b\ \bar{c}$	0.2503	0.2503
$\bar{d}\ b\ \bar{c}$	0.0626	0.0626
$d\ \bar{b}\ \bar{c}$	0.3254	0.3254
$\bar{d}\ \bar{b}\ \bar{c}$	0.3254	0.3254

The potentials can then be normalized by dividing by the sum of values in the cliques. For ordinary sum-marginalization these sums are identical in all cliques and equal to the prior probability of the evidence (0.0041), whereas for out-marginalization each clique has its own normalizing constant, the prior probability of the evidence on the variables not in the clique. The resulting normalized potentials are displayed in Table 6.11 with the out-margins of individual variables shown in Table 6.12.

TABLE 6.11. Normalized marginal and out-marginal clique potentials after propagating the evidence $\mathcal{E} = (a \;\&\; s \;\&\; \bar{x})$.

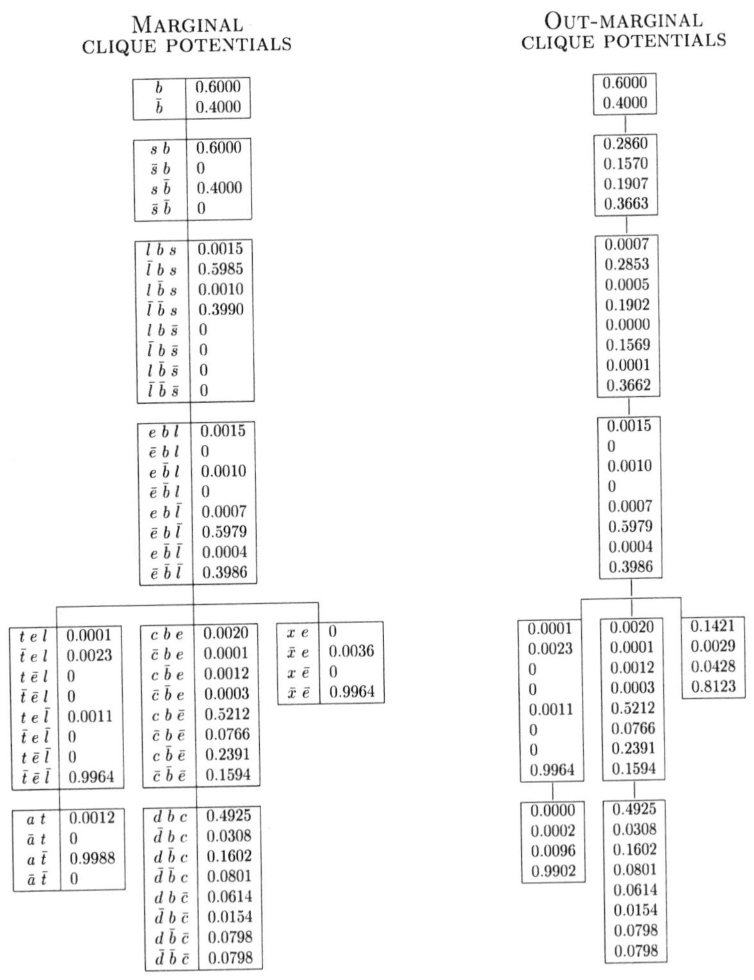

MARGINAL CLIQUE POTENTIALS / OUT-MARGINAL CLIQUE POTENTIALS

	Marginal	Out-marginal
b	0.6000	0.6000
$\bar{b}$	0.4000	0.4000

	Marginal	Out-marginal
$s\ b$	0.6000	0.2860
$\bar{s}\ b$	0	0.1570
$s\ \bar{b}$	0.4000	0.1907
$\bar{s}\ \bar{b}$	0	0.3663

	Marginal	Out-marginal
$l\ b\ s$	0.0015	0.0007
$\bar{l}\ b\ s$	0.5985	0.2853
$l\ \bar{b}\ s$	0.0010	0.0005
$\bar{l}\ \bar{b}\ s$	0.3990	0.1902
$l\ b\ \bar{s}$	0	0.0000
$\bar{l}\ b\ \bar{s}$	0	0.1569
$l\ \bar{b}\ \bar{s}$	0	0.0001
$\bar{l}\ \bar{b}\ \bar{s}$	0	0.3662

	Marginal	Out-marginal
$e\ b\ l$	0.0015	0.0015
$\bar{e}\ b\ l$	0	0
$e\ \bar{b}\ l$	0.0010	0.0010
$\bar{e}\ \bar{b}\ l$	0	0
$e\ b\ \bar{l}$	0.0007	0.0007
$\bar{e}\ b\ \bar{l}$	0.5979	0.5979
$e\ \bar{b}\ \bar{l}$	0.0004	0.0004
$\bar{e}\ \bar{b}\ \bar{l}$	0.3986	0.3986

	Marginal	Out-marginal
$t\ e\ l$	0.0001	0.0001
$\bar{t}\ e\ l$	0.0023	0.0023
$t\ \bar{e}\ l$	0	0
$\bar{t}\ \bar{e}\ l$	0	0
$t\ e\ \bar{l}$	0.0011	0.0011
$\bar{t}\ e\ \bar{l}$	0	0
$t\ \bar{e}\ \bar{l}$	0	0
$\bar{t}\ \bar{e}\ \bar{l}$	0.9964	0.9964

	Marginal	Out-marginal
$c\ b\ e$	0.0020	0.0020
$\bar{c}\ b\ e$	0.0001	0.0001
$c\ \bar{b}\ e$	0.0012	0.0012
$\bar{c}\ \bar{b}\ e$	0.0003	0.0003
$c\ b\ \bar{e}$	0.5212	0.5212
$\bar{c}\ b\ \bar{e}$	0.0766	0.0766
$c\ \bar{b}\ \bar{e}$	0.2391	0.2391
$\bar{c}\ \bar{b}\ \bar{e}$	0.1594	0.1594

	Marginal	Out-marginal
$x\ e$	0	0.1421
$\bar{x}\ e$	0.0036	0.0029
$x\ \bar{e}$	0	0.0428
$\bar{x}\ \bar{e}$	0.9964	0.8123

	Marginal	Out-marginal
$a\ t$	0.0012	0.0000
$\bar{a}\ t$	0	0.0002
$a\ \bar{t}$	0.9988	0.0096
$\bar{a}\ \bar{t}$	0	0.9902

	Marginal	Out-marginal
$d\ b\ c$	0.4925	0.4925
$\bar{d}\ b\ c$	0.0308	0.0308
$d\ \bar{b}\ c$	0.1602	0.1602
$\bar{d}\ \bar{b}\ c$	0.0801	0.0801
$d\ b\ \bar{c}$	0.0614	0.0614
$\bar{d}\ b\ \bar{c}$	0.0154	0.0154
$d\ \bar{b}\ \bar{c}$	0.0798	0.0798
$\bar{d}\ \bar{b}\ \bar{c}$	0.0798	0.0798

TABLE 6.12. Out-margins based on evidence $\mathcal{E} = (a \ \& \ s \ \& \ \bar{x})$. The asterisks mark states fixed by the evidence.

$a*$	0.0096	d	0.7939	t	0.0012
$\bar{a}$	0.9904	$\bar{d}$	0.2061	$\bar{t}$	0.9988
c	0.7635	x	0.1849	e	0.0036
$\bar{c}$	0.2365	$\bar{x}*$	0.8151	$\bar{e}$	0.9964
l	0.0025	$s*$	0.4767	b	0.6000
$\bar{l}$	0.9975	$\bar{s}$	0.5233	$\bar{b}$	0.4000

6.5.7 *Max-propagation*

In Table 6.13 we show clique potentials after max-propagations both with and without the evidence $\mathcal{E} = (a \ \& \ s \ \& \ \bar{x})$, but without normalizing. Note that the maximum value of each tree appears in each of the respective cliques.

From the left-hand table we can read off the most probable configuration: it is $(b, s, \bar{l}, \bar{e}, \bar{t}, c, \bar{x}, d, \bar{a})$ having probability 0.2063. From the right-hand table, we see that, with the evidence $\mathcal{E} = (a \ \& \ s \ \& \ \bar{x})$, the most probable configuration is now $x^1 = (b, s, \bar{l}, \bar{e}, \bar{t}, c, \bar{x}, d, a)$, having prior probability of 0.002.

To find the second-most likely configuration, x^2, under this evidence, we may use the method of Section 6.4.4 taking the ordered residual sets as $(B, S, L, E, T, C, X, D, A)$. (Note that for an elimination tree the residual sets are singletons.) Then we have to consider the following nine virtual evidences:

$$
\begin{aligned}
\mathcal{E}_0^1 &:= \mathcal{E} \ \& \ B \neq b \\
\mathcal{E}_1^1 &:= \mathcal{E} \ \& \ B = b \ \& \ S \neq s \\
\mathcal{E}_2^1 &:= \mathcal{E} \ \& \ B = b \ \& \ S = s \ \ \& \ L \neq \bar{l} \\
\mathcal{E}_3^1 &:= \mathcal{E} \ \& \ B = b \ \& \ S = s \ \ \& \ L = \bar{l} \ \ \& \ E \neq \bar{e} \\
\mathcal{E}_4^1 &:= \mathcal{E} \ \& \ B = b \ \& \ \ \ldots \ \ \& \ E = \bar{e} \ \ \& \ T \neq \bar{t} \\
\mathcal{E}_5^1 &:= \mathcal{E} \ \& \ B = b \ \& \ \ \ldots \ \ \& \ T = \bar{t} \ \ \& \ C \neq c \\
\mathcal{E}_6^1 &:= \mathcal{E} \ \& \ B = b \ \& \ \ \ldots \ \ \& \ C = c \ \ \& \ X \neq \bar{x} \\
\mathcal{E}_7^1 &:= \mathcal{E} \ \& \ B = b \ \& \ \ \ldots \ \ \& \ X = \bar{x} \ \& \ D \neq d \\
\mathcal{E}_8^1 &:= \mathcal{E} \ \& \ B = b \ \& \ \ \ldots \ \ \& \ D = d \ \ \& \ A \neq a.
\end{aligned}
$$

However, because all variables are binary and because of the prior evidence $\mathcal{E} = (a \ \& \ s \ \& \ \bar{x})$, some of these virtual evidences, specifically $\mathcal{E}_1^1$, $\mathcal{E}_6^1$, and

$\mathcal{E}_8^1$, are impossible because they exclude both states of a variable. Of the six remaining evidences $\mathcal{E}_0^1$ has the greatest max-normalization, hence we deduce that the second most likely configuration subject to the evidence $\mathcal{E} = (a \ \& \ s \ \& \ \bar{x})$ has $B = \bar{b}$. A call to FIND MAX-CONFIGURATION using evidence $\mathcal{E}_0^1$ yields $x^2 = (\bar{b}, s, \bar{l}, \bar{e}, \bar{t}, c, \bar{x}, d, a)$, which has prior probability 0.0006.

TABLE 6.13. Max-marginal representation, after max-propagating without (left) and with (right) the evidence $\mathcal{E} = (a \ \& \ s \ \& \ \bar{x})$. (The values on the right have been multiplied by 10^4.)

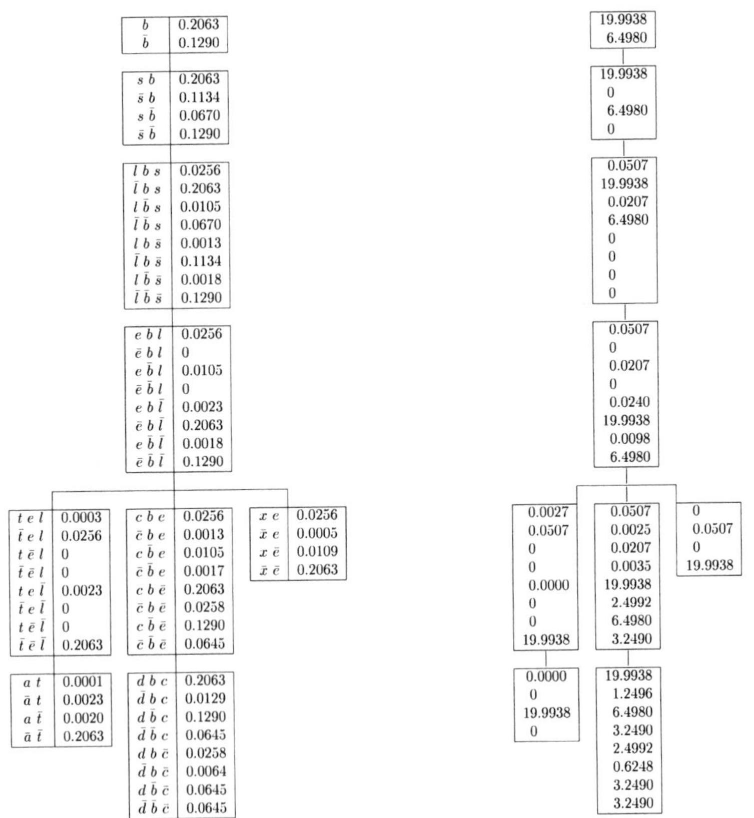

b	0.2063	19.9938
$\bar{b}$	0.1290	6.4980

$s\ b$	0.2063	19.9938
$\bar{s}\ b$	0.1134	0
$s\ \bar{b}$	0.0670	6.4980
$\bar{s}\ \bar{b}$	0.1290	0

$l\ b\ s$	0.0256	0.0507
$\bar{l}\ b\ s$	0.2063	19.9938
$l\ \bar{b}\ s$	0.0105	0.0207
$\bar{l}\ \bar{b}\ s$	0.0670	6.4980
$l\ b\ \bar{s}$	0.0013	0
$\bar{l}\ b\ \bar{s}$	0.1134	0
$l\ \bar{b}\ \bar{s}$	0.0018	0
$\bar{l}\ \bar{b}\ \bar{s}$	0.1290	0

$e\ b\ l$	0.0256	0.0507
$\bar{e}\ b\ l$	0	0
$e\ \bar{b}\ l$	0.0105	0.0207
$\bar{e}\ \bar{b}\ l$	0	0
$e\ b\ \bar{l}$	0.0023	0.0240
$\bar{e}\ b\ \bar{l}$	0.2063	19.9938
$e\ \bar{b}\ \bar{l}$	0.0018	0.0098
$\bar{e}\ \bar{b}\ \bar{l}$	0.1290	6.4980

$t\ e\ l$	0.0003	0.0027
$\bar{t}\ e\ l$	0.0256	0.0507
$t\ \bar{e}\ l$	0	0
$\bar{t}\ \bar{e}\ l$	0	0
$t\ e\ \bar{l}$	0.0023	0.0000
$\bar{t}\ e\ \bar{l}$	0	0
$t\ \bar{e}\ \bar{l}$	0	0
$\bar{t}\ \bar{e}\ \bar{l}$	0.2063	19.9938

$c\ b\ e$	0.0256	0.0507
$\bar{c}\ b\ e$	0.0013	0.0025
$c\ \bar{b}\ e$	0.0105	0.0207
$\bar{c}\ \bar{b}\ e$	0.0017	0.0035
$c\ b\ \bar{e}$	0.2063	19.9938
$\bar{c}\ b\ \bar{e}$	0.0258	2.4992
$c\ \bar{b}\ \bar{e}$	0.1290	6.4980
$\bar{c}\ \bar{b}\ \bar{e}$	0.0645	3.2490

$x\ e$	0.0256	0
$\bar{x}\ e$	0.0005	0.0507
$x\ \bar{e}$	0.0109	0
$\bar{x}\ \bar{e}$	0.2063	19.9938

$a\ t$	0.0001	0.0000
$\bar{a}\ t$	0.0023	0
$a\ \bar{t}$	0.0020	19.9938
$\bar{a}\ \bar{t}$	0.2063	0

$d\ b\ c$	0.2063	19.9938
$\bar{d}\ b\ c$	0.0129	1.2496
$d\ \bar{b}\ c$	0.1290	6.4980
$\bar{d}\ \bar{b}\ c$	0.0645	3.2490
$d\ b\ \bar{c}$	0.0258	2.4992
$\bar{d}\ b\ \bar{c}$	0.0064	0.6248
$d\ \bar{b}\ \bar{c}$	0.0645	3.2490
$\bar{d}\ \bar{b}\ \bar{c}$	0.0645	3.2490

6.6 Dealing with large cliques

The feasibility of the propagation algorithms described above depends on the sizes of the state spaces of the cliques obtained after triangulation.

With good triangulation algorithms, remarkably large and dense networks can be handled, but there comes a point when computational limits are exceeded.

One common context in which this happens is when there is regularity in the graphical structure, with many cycles, such as when nodes form a lattice (e.g., in image analysis) or repeated multiply connected large blocks (e.g., in complex temporal models). Various analytical approximations are possible: current attention focuses on variational methods and mean field approximations (Saul et al. 1996). One can also use simulation schemes which are derivatives of those explored in image processing (Geman and Geman 1984). This general statistical computational technique was originally suggested by Pearl (1987) for use in expert systems. We deal with this topic separately in Appendix B and Section 9.8. Here we describe some other techniques which have been developed to extend the range of applications of probabilistic networks.

6.6.1 Truncating small numbers

One method of improving the efficiency of the calculations is to make some approximation to the representation of the joint probability on the junction tree, and to use exact propagation methods on the result. For some large applications it happens that many entries in the clique potentials are zero. This can happen when there are logical constraints in the conditional probability tables of the probabilistic network model; for example, in the CH-ASIA example, the *Either...?* node is a logical node in which half of the conditional probability entries are zero. When several such constraints operate they can create many zeros in the clique and separator tables. This has been exploited in the HUGIN system (Andersen et al. 1989) by a procedure called *compression*, in which an efficient representation of the clique tables is used so that these zeros do not need to be stored explicitly. The savings can be large: a reduction in storage space of 93 percent was achieved in one of the MUNIN networks (Jensen and Andersen 1990). Such compression does not affect the accuracy of the inference process, as it introduces no approximations.

Further savings in space can be achieved by choosing some threshold value ϵ, and setting to zero all entries in the clique and separator potentials having values less than ϵ. The resulting tables can then be compressed, as above, leading to significant reductions in storage space requirements. Essentially this truncation means that individual configurations that are relatively unlikely are ignored. While ignoring one such configuration may not have much effect, ignoring many could have an important cumulative effect. However, one cannot simply use a single global value as a threshold, since what appears to be a small entry on initialization might become large after propagation, and vice versa. Furthermore if a clique table is very large, then all the entries will tend to be smaller than for a small clique. Thus, one should adjust the cut-off in each clique in some appropriate manner.

Observe that setting a clique element to zero corresponds to a finding that excludes that state. Hence, propagating that as evidence one can obtain from the normalization constant the probability of that finding. Setting several entries to zero and propagating will yield, via the normalization, the probability of those exclusion findings. Jensen and Andersen (1990) use this to construct an iterative method of selecting elements of a clique table to set to zero. First one fixes ϵ, and then one sets a threshold δ, having initial value ϵ. Then one repeatedly halves δ until the sum of those elements that are less than δ of the clique table is no greater than ϵ; when this occurs those elements are set to zero, which ensures that the fraction $1 - \epsilon$ of the probability mass of the clique table is retained, which is a measure of the local error. The global error can be found by propagating: it is $1-\mu$ where μ is the normalization constant. If a case is observed that has been excluded by this approximation, then a zero normalization constant will result from propagation; this will occur with probability $1-\mu$ (according to the model). Usually the approximation will be good, resulting in significant savings in computational requirements. See Jensen and Andersen (1990) for further details and analysis.

6.6.2 Splitting cliques

Another approach to the problem of one or more unmanageably large cliques is to take an axe to them, splitting them into smaller cliques by removing edges in the triangulated graph and approximating the probability specification by introducing the corresponding additional independence. Such a procedure would lead to slightly incorrect inferences. However, it can lead to large reductions in complexity: even removing a single edge might make many other edges introduced by moralization and fill-in redundant and, hence, removable also. Such an approach has been examined by Kjærulff (1994), who presents an approximation scheme for removing edges, providing local and overall bounds on the errors induced by edge removals, using the Kullback–Leibler divergence(Kullback and Leibler 1951) to measure the effect of removing edges. He shows that in some, but not all, cases this approach can be significantly better than eliminating small entries in the clique tables. He suggests a hybrid approach in which first edges having little effect on the joint distribution are removed to yield a new chordal graph and junction tree, and then truncation of small entries as described by Jensen and Andersen (1990) is performed.

6.7 Current research directions and further reading

Current research concerning propagation in discrete probabilistic networks is concerned with studying and improving the efficiency of one or more variants of the propagation algorithms (Kjærulff 1998; Shenoy 1997; Madsen and Jensen 1998), or exploiting cutset conditioning to trade time for space (Shachter et al. 1994).

Another trend is the development of variants of the algorithm that perform slightly different or additional tasks (Jensen 1995), or that work in a somewhat modified way (Bloemeke and Valtorta 1998) or approximately (Kjærulff 1993; Jensen et al. 1995; Saul et al. 1996; Curds 1997; Kearns and Saul 1998).

There has also been some activity concerned with algorithms to solve rather different tasks, possibly at a more abstract level, including belief function propagation, constraint satisfaction, discrete optimization, and decision analysis (Shenoy and Shafer 1990; Shenoy 1991, 1992; Lauritzen and Jensen 1997).

The much-used Viterbi algorithm for hidden Markov models (Viterbi 1967) is a special instance of max-propagation as described here; likewise, other decoding algorithms such as the BCJR algorithm (Bahl et al. 1974) are instances of propagation algorithms (see Frey (1998) for further details). Recent experiments show exceptional performance of so-called *turbo-codes* which are decoded by algorithms based on approximate probability propagation (McEliece et al. 1998).

Algorithms for solving related propagation tasks go at least as far back as 1880 (see Section 7.11). Algorithms very close indeed to those described here were derived by Cannings et al. (1976, 1978) in the context of human genetic pedigrees. Pearl (1982), Kim and Pearl (1983), and Pearl (1986b, 1988) demonstrated the feasibility of local computation in probabilistic networks, inspiring further work by Shachter (1986), Shenoy and Shafer (1986), Lauritzen and Spiegelhalter (1988), and others, thereby paving the way for the exploitation of probability-based models as parts of realistic systems for planning and decision support. The technique was subsequently improved by Jensen et al. (1990b), utilizing the propagation of simple flows through the junction-tree, each such flow involving only two adjacent vertices and the associated edge, that is pairs of cliques joined by a separator. Dawid (1992a) described a generalization of this algorithm and showed how the basic idea can be extended to solve numerous other tasks. Dechter (1996) proposed a general algorithm, *bucket elimination*, which is closely related to propagation in the elimination tree.

An alternative version of the algorithm was given by Shafer and Shenoy (1988) and by Shenoy and Shafer (1990); see Shafer (1996) for a detailed comparison of some of the variants.

7
Gaussian and Mixed Discrete-Gaussian Networks

The previous chapter described inference procedures for probabilistic networks of discrete random variables. For many purposes one would wish to be able to handle models in which some or all random variables can take values in a continuous range. In general, exact inference in such networks is not possible when arbitrary conditional densities are employed. One way out of this dilemma is to discretize the value ranges of the continuous random variables, and so approximate the model by one in which all variables are discrete. The quality of such an approximation will depend upon the discretization employed: in principle one can increase the accuracy of the approximation by increasing the number of 'bins' in discretizing a continuous range, but this will result in the state space of the junction tree of cliques becoming unmanageably large. Thus, there will inevitably be a trade-off between improving the approximation of the discretization process and keeping the computational complexity within feasible bounds.

However, there is a very important and analytically tractable model involving continuous random variables — the *multivariate Gaussian distribution.* The nature of this analytic tractability allows one to develop exact evidence propagation for probabilistic networks of continuous variables where the network models a multivariate Gaussian distribution. A generalization to mixed models was developed by Lauritzen (1992), in which both discrete and continuous variables can be present and for which the conditional distribution of the continuous (or quantitative) variables given the discrete (or qualitative) variables is restricted to be multivariate Gaussian. Under this condition this scheme, presented here, permits the local computation on the junction tree of exact probabilities, means, and variances for mixed qualitative and quantitative variables. The distributional theory of graphical models with both quantitative and qualitative variables is fundamental to the computational methods. A brief account of the basic elements of this is contained in Section 7.1.

The asymmetry between the quantitative and qualitative variables leads to another formal element of the computational structure: the decomposition theory of *marked graphs* which are graphs with two types of vertices, here corresponding to the discrete and continuous variables. This leads to a *strongly rooted junction tree* and sets some additional limitations for the specification and propagation structure. The necessary concepts are explained in Section 7.3.

7.1 CG distributions

Chapter 6 showed how to exploit the local structure in the specification of a discrete probability model for fast and efficient computation. This chapter extends that computational scheme to networks where some vertices represent entities that are measured on a quantitative and some on a qualitative (discrete) scale. This extension allows more flexible and faithful modelling and speeds up computation as well. To handle this more general case, the properties of conditional Gaussian (CG) distributions introduced by Lauritzen and Wermuth (1984, 1989) are exploited. We shall briefly review some standard notation, but refer the reader to Lauritzen (1996) for further details and derivations of formulae.

The set of variables V is partitioned as $V = \Delta \cup \Gamma$ into variables of *discrete* (Δ) and *continuous* (Γ) type. A typical element of the joint state space is denoted as in one of the possibilities below:

$$x = (x_\alpha)_{\alpha \in V} = (i, y) = \big((i_\delta)_{\delta \in \Delta}, (y_\gamma)_{\gamma \in \Gamma}\big),$$

where i_δ are qualitative and y_γ are real-valued. A particular combination $i = (i_\delta)_{\delta \in \Delta}$ is referred to as a *cell* and the set of cells is denoted by $\mathcal{I}$. The joint distribution of the variables is supposed to have a density f with

$$f(x) = f(i, y) = \chi(i) \exp\{g(i) + h(i)'y - y'K(i)y/2\}, \qquad (7.1)$$

where $\chi(i) \in \{0, 1\}$ indicates whether f is positive at i and A' denotes the transpose of a vector or matrix A. We then say that X follows a *CG distribution* which is equivalent to the statement that the continuous variables follow a multivariate Gaussian distribution given the discrete, i.e.,

$$\mathcal{L}(X_\Gamma \mid X_\Delta = i) = \mathcal{N}_{|\Gamma|}(\xi(i), \Sigma(i)) \quad \text{whenever} \quad p(i) = P\{X_\Delta = i\} > 0,$$

where $X_A = (X_\alpha)_{\alpha \in A}$ and so on, $|\Gamma|$ denotes the cardinality of Γ, and

$$\xi(i) = K(i)^{-1}h(i), \quad \Sigma(i) = K(i)^{-1}, \qquad (7.2)$$

the latter being positive definite. The triple (g, h, K) — only defined for $\chi(i) > 0$ — constitutes the *canonical characteristics* of the distribution, and $\{p, \xi, \Sigma\}$ are the *moment characteristics*. In the case where we have only one kind of variable the undefined components are denoted by zeros, i.e., $(g, 0, 0)$ or $(0, h, K)$.

Note that there is a slight difference between the notation used here and that of Lauritzen (1996), in that we allow $p(i)$ to be equal to 0 for some entries i. Also, strictly speaking, χ belongs to the characteristics of the distribution, but we assume this to be implicitly represented through the domain where the other components are well-defined.

An important element of the ability to perform local computation on the junction tree for discrete models is the generalization of probability distributions by potentials. A similar generalization for the mixed case proves very fruitful here. Thus, we extend the notion of a CG distribution to that of a *CG potential*, which is any function ϕ of the form

$$\phi(x) = \phi(i, y) = \chi(i) \exp\{g(i) + h(i)'y - y'K(i)y/2\}, \qquad (7.3)$$

where $K(i)$ is only assumed to be a symmetric matrix. Thus, ϕ is not necessarily a density. We still use the triple (g, h, K) as canonical characteristics for the potential ϕ.

A basic difference is that the moment characteristics for a CG potential are only well-defined when K is positive definite for all i with $\chi(i) > 0$. Then Σ and ξ are given as in (7.2), whereas by integrating y from the joint distribution (7.1) we obtain

$$p(i) \propto \{\det \Sigma(i)\}^{\frac{1}{2}} \exp\{g(i) + h(i)'\Sigma(i)h(i)/2\}. \qquad (7.4)$$

Conversely, if the moment characteristics $\{p, \xi, \Sigma\}$ are given, we can calculate the canonical characteristics as $K(i) = \Sigma(i)^{-1}$, $h(i) = K(i)\xi(i)$, and

$$g(i) = \log p(i) + \{\log \det K(i) - |\Gamma| \log(2\pi) - \xi(i)'K(i)\xi(i)\}/2.$$

7.2 Basic operations on CG potentials

The basis for the computational scheme relies partly on a set of fundamental operations on the CG potentials, partly on a message-passing scheme in the junction tree. The latter is to be described in Section 7.5. Here we describe the elements of the local computations which generalize those for discrete models.

Definition 7.1 [EXTENSION]
Let ϕ be a CG potential defined on $\mathcal{U} = (\mathcal{I} \times \mathcal{Y})$. The potential is extended to $\bar{\phi}$ defined on $\mathcal{W} = (\mathcal{I} \times \mathcal{J}) \times (\mathcal{Y} \times \mathcal{Z})$ by setting $\bar{\phi}(i, j, y, z) = \phi(i, y)$. □

If (g, h, K) are the characteristics of the CG potential ϕ, then the extension essentially amounts to adjoining zeros to the characteristics so as to give them the desired dimensions; the corresponding characteristics are:

$$\bar{g}(i,j) = g(i), \quad \bar{h}(i,j) = \begin{pmatrix} h(i) \\ 0 \end{pmatrix}, \quad \bar{K}(i,j) = \begin{pmatrix} K(i) & 0 \\ 0 & 0 \end{pmatrix}.$$

When it is clear from the context, we do not distinguish between a potential and its extensions.

Definition 7.2 [RESTRICTION]
Let ϕ be a CG potential defined on $\mathcal{W} = (\mathcal{I} \times \mathcal{J}) \times (\mathcal{Y} \times \mathcal{Z})$, and let $(j, z) \in (\mathcal{J} \times \mathcal{Z})$. The restriction $\phi^{(j,z)}$ of ϕ to $\mathcal{I} \times \mathcal{Y}$ is defined as

$$\phi^{(j,z)}(i, y) = \phi(i, j, y, z).$$

□

Thus, in terms of the canonical characteristic, this operation just corresponds to 'picking out' the relevant parts.

Definition 7.3 [MULTIPLICATION]
Multiplication is defined in the obvious way, i.e., for ϕ and ψ defined on $\mathcal{U}$ and $\mathcal{W}$ respectively we define $\phi\psi$ on $\mathcal{U} \cup \mathcal{W}$ as

$$(\phi\psi)(x) = \phi(x)\psi(x)$$

where ϕ and ψ on the right-hand side first have been extended to $\mathcal{U} \cup \mathcal{W}$.
□

Expressed in terms of the canonical characteristics, multiplication becomes simple addition of components:

$$(g_1, h_1, K_1) \times (g_2, h_2, K_2) = (g_1 + g_2, h_1 + h_2, K_1 + K_2).$$

Definition 7.4 [DIVISION]
Division is likewise defined in the obvious way, but special care has to be taken when dividing by zero. Thus, for $x = (i, y)$, we let

$$(\phi/\psi)(x) = \begin{cases} \phi(x)/\psi(x), & \text{if } \psi(x) \neq 0 \\ 0, & \text{otherwise.} \end{cases}$$

□

Thus, when both potentials have positive characteristics, division becomes simple subtraction expressed in terms of the canonical characteristics:

$$(g_1, h_1, K_1)/(g_2, h_2, K_2) = (g_1 - g_2, h_1 - h_2, K_1 - K_2).$$

An essential difference between the pure discrete case and the situation here is due to fact that adding two CG potentials in general will result in a mixture of CG potentials — a function of a different algebraic structure. Hence, there will be some complications requiring attention.

We distinguish several cases. First we discuss *marginals over continuous variables*. In this case we simply integrate. Let

$$y = \begin{pmatrix} y_1 \\ y_2 \end{pmatrix}, \quad h = \begin{pmatrix} h_1 \\ h_2 \end{pmatrix}, \quad K = \begin{pmatrix} K_{11} & K_{12} \\ K_{21} & K_{22} \end{pmatrix},$$

with y_1 having dimension p and y_2 dimension q. We then have the following lemma.

Lemma 7.5 *The integral* $\int \phi(i, y_1, y_2)\, dy_1$ *is finite if and only if* K_{11} *is positive definite. It is then equal to a CG potential* $\tilde{\phi}$ *with canonical characteristics given as*

$$\begin{aligned}
\tilde{g}(i) &= g(i) + \left\{p \log(2\pi) - \log\det K_{11}(i) + h_1(i)' K_{11}(i)^{-1} h_1(i)\right\}/2, \\
\tilde{h}(i) &= h_2(i) - K_{21}(i) K_{11}(i)^{-1} h_1(i), \\
\tilde{K}(i) &= K_{22}(i) - K_{21}(i) K_{11}(i)^{-1} K_{12}(i).
\end{aligned}$$

Proof. Let

$$\mu(i) = -K_{11}(i)^{-1} K_{12}(i) y_2 + K_{11}(i)^{-1} h_1(i).$$

Then we find by direct calculation that

$$\begin{aligned}
\phi(i, y) &= \exp\left\{-\left(y_1 - \mu(i)\right)' K_{11}(i) \left(y_1 - \mu(i)\right)/2\right\} \\
&\quad \times \exp\left\{y_2'\left(h_2(i) - K_{21}(i) K_{11}(i)^{-1} h_1(i)\right)\right\} \\
&\quad \times \exp\left\{-y_2'\left(K_{22}(i) - K_{21}(i) K_{11}(i)^{-1} K_{12}(i)\right) y_2/2\right\} \\
&\quad \times \exp\left\{g(i) + h_1(i)' K_{11}(i)^{-1} h_1(i)/2\right\}.
\end{aligned}$$

Now y_1 only appears in the first factor. This can be integrated by letting $z = y_1 - \mu(i)$ and recalling that if $z \in \mathcal{R}^p$ and K is positive definite, then

$$\int e^{-z' K z/2}\, dz = (2\pi)^{\frac{p}{2}} (\det K)^{-\frac{1}{2}}.$$

The result follows. □

When calculating *marginals over discrete variables* we distinguish two cases. First, if h and K do not depend on j, i.e., $h(i, j) \equiv h(i)$ and $K(i, j) \equiv K(i)$, we define the marginal $\tilde{\phi}$ of ϕ over j the direct way:

$$\begin{aligned}
\tilde{\phi}(i, y) &= \sum_j \phi(i, j, y) = \sum_j \chi(i, j) \exp\{g(i, j) + h(i)' y - y' K(i) y/2\} \\
&= \exp\{h(i)' y - y' K(i) y/2\} \sum_j \chi(i, j) \exp g(i, j),
\end{aligned}$$

which leads to the following canonical characteristics for the marginal:

$$\tilde{g}(i) = \log \sum_{j:\chi(i,j)=1} \exp g(i,j), \quad \tilde{h}(i) = h(i), \quad \tilde{K}(i) = K(i).$$

If either of h or K depends on j, the marginalization process is more subtle since simple addition of CG potentials will not result in a CG potential. The procedure we shall then use is only well-defined for $K(i,j)$ positive definite and is best described in terms of the moment characteristics $\{p, \xi, \Sigma\}$. The marginal $\tilde{\phi}$ is defined as the potential with moment characteristics $\{\tilde{p}, \tilde{\xi}, \tilde{\Sigma}\}$ where

$$\tilde{p}(i) = \sum_j p(i,j), \quad \tilde{\xi}(i) = \sum_j \xi(i,j)p(i,j)/\tilde{p}(i),$$

and

$$\tilde{\Sigma}(i) = \sum_j \Sigma(i,j)p(i,j)/\tilde{p}(i) + \sum_j (\xi(i,j) - \tilde{\xi}(i))'(\xi(i,j) - \tilde{\xi}(i))p(i,j)/\tilde{p}(i).$$

The 'marginalized' density will then have the correct moments, i.e.,

$$P(I = i) = \tilde{p}(i), \quad \mathrm{E}(Y \mid I = i) = \tilde{\xi}(i), \quad \mathrm{var}(Y \mid I = i) = \tilde{\Sigma}(i),$$

where expectations are taken with respect to the CG distribution determined by ϕ. This is a direct consequence of the familiar relations:

$$\begin{aligned} \mathrm{E}(Y \mid I = i) &= \mathrm{E}\left\{\mathrm{E}(Y \mid (I,J)) \mid I = i\right\} \\ \mathrm{var}(Y \mid I = i) &= \mathrm{E}\left\{\mathrm{var}(Y \mid (I,J)) \mid I = i\right\} + \mathrm{var}\left\{\mathrm{E}(Y \mid (I,J)) \mid I = i\right\}. \end{aligned}$$

Having enumerated the various possibilities, we now collect them together in the following definition:

Definition 7.6 [MARGINALIZATION]
When *marginalizing over both continuous and discrete variables* we first marginalize over the continuous variables and then over the discrete. If in the second of these stages we have (h, K) independent of j, we say that we have a *strong marginalization.* In the other case we must use the marginalization process just described, and we speak of a *weak marginalization.* In both cases we use the symbol $\sum_{W \setminus U} \phi_W$ for the marginalized potential, where U denotes the set of variables marginalized *to*, and $W \setminus U$ the set of variables marginalized *over.* □

We leave it for the reader to verify that the weak marginalization satisfies the standard composition rule such that when $U \subset V \subset W$, then

$$\sum_{V \setminus U} \left(\sum_{W \setminus V} \phi_W \right) = \sum_{W \setminus U} \phi_W. \tag{7.5}$$

However, only the strong marginalizations behave well when products are involved. In general, we have:

$$\sum_{W\setminus V} (\phi_W \phi_V) \neq \phi_V \left(\sum_{W\setminus V} \phi_W \right). \tag{7.6}$$

This forces us to establish correctness of the local propagation scheme directly by extending the theory of Chapter 6. Note that another consequence is that the axioms of Shafer and Shenoy (1990) or Lauritzen and Jensen (1997) are not fulfilled.

In the special case of strong marginalizations, equality holds in (7.6). This follows by elementary calculations since strong marginalizations are just ordinary integrations.

7.3 Marked graphs and their junction trees

The non-closure of CG distributions under marginalization means that, to exploit their attractive analytic properties in application to probabilistic networks, restrictions on the type of models that can be considered have to be imposed. To handle these restrictions we have to extend the graph-theoretic terminology of Chapter 4. As before, we work with a chain graph $\mathcal{G} = (V, E)$, where V is a finite set of vertices and E the set of edges. In particular we need to work with graphs where the vertices are *marked* in the sense that they are partitioned into two groups, Δ and Γ. We use the term *marked graph* for a graph of this type. The vertices in the set Δ will represent qualitative (discrete) variables and those in Γ quantitative (continuous) variables. Therefore, we say that the vertices in Δ are *discrete* and those in Γ are *continuous.*

7.3.1 Decomposition of marked graphs

The basic trick enabling the computational task to be performed locally is the decomposition of a suitably modified network into partly independent components formed by the cliques of that graph. However, the inherent asymmetry between discrete and continuous variables in the CG distributions implies that one also needs to take proper account of the behaviour of the markings of the graph. We refer the interested reader to Leimer (1989) or Lauritzen (1996) for a detailed graph-theoretic study of the problems, as well as all proofs. Here we introduce formally the notion of a decomposition.

Definition 7.7 [Strong Decomposition] A triple (A, B, C) of disjoint subsets of the vertex set V of an undirected, marked graph $\mathcal{G}$ is said to form a *strong decomposition* of $\mathcal{G}$ if $V = A \cup B \cup C$ and the three conditions below all hold:

(i) C separates A from B;

(ii) C is a complete subset of V;

(iii) $C \subseteq \Delta$ or $B \subseteq \Gamma$. □

When this is the case we say that (A, B, C) *decomposes* $\mathcal{G}$ into the *components* $\mathcal{G}_{A \cup C}$ and $\mathcal{G}_{B \cup C}$.

If only (i) and (ii) hold, we say that (A, B, C) forms a *weak decomposition.* Thus, weak decompositions ignore the markings of the graph, and were introduced in Section 4.2, Definition 4.2. In the pure cases, (iii) holds automatically and all weak decompositions are also decompositions. Note that what we have chosen in this chapter to call a decomposition (without a qualifier) is what Leimer (1989) calls a strong decomposition. If the sets A and B in (A, B, C) are both non-empty, we say that the decomposition is *proper.* Figure 7.1 illustrates the notions of strong and weak decompositions.

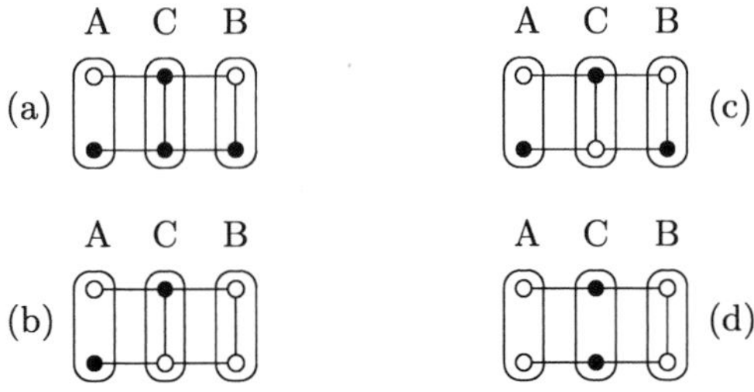

FIGURE 7.1. Illustration of the notions of strong and weak decompositions: nodes represented by dots are discrete, by circles are continuous. In (a) we see a decomposition with $C \subseteq \Delta$ and in (b) with $B \subset \Gamma$. In (c) the decomposition is only weak because neither of these two conditions is fulfilled. In (d) we do not have a decomposition because the separator C is not complete.

A decomposable graph is one that can be successively decomposed into its cliques. Again we choose to state this formally through a recursive definition as follows.

Definition 7.8 [DECOMPOSABLE]

An undirected, marked graph is said to be *decomposable* if it is complete, or if there exists a proper decomposition (A, B, C) into decomposable subgraphs $\mathcal{G}_{A \cup C}$ and $\mathcal{G}_{B \cup C}$. □

Note that the definition makes sense because both subgraphs $\mathcal{G}_{A \cup C}$ and $\mathcal{G}_{B \cup C}$ must have fewer vertices than the original graph $\mathcal{G}$.

Decomposable unmarked graphs are characterized by being triangulated, i.e., not having cycles of length greater than three. Decomposable marked graphs are characterized further by not having a particular type of path. More precisely, we have

Proposition 7.9 *An undirected marked graph is decomposable if and only if it is triangulated and does not contain any path* $(\delta_1 = \alpha_0, \dots, \alpha_n = \delta_2)$ *between two discrete vertices passing through only continuous vertices, with the discrete vertices not being neighbours.*

We have illustrated a typical forbidden path in Figure 7.2.

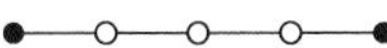

FIGURE 7.2. A path which is forbidden in a decomposable graph.

7.3.2 Junction trees with strong roots

The construction of the computational structure for discrete models begins with a chain graph, forms its moral graph, adds links to make it triangulated and forms a junction tree of the cliques of the triangulated graph. This procedure needs to be generalized and modified for the mixed case.

We first form the moral graph as described in Chapter 4. Then we need to add further links in such a way as to obtain a decomposable marked graph. We can do this using a modification of the one-step look ahead Algorithm 4.13, in which one first eliminates the continuous variables and then the discrete variables.

Algorithm 7.10 [ONE-STEP LOOK AHEAD STRONG TRIANGULATION]

- Start with all vertices unnumbered, set counter $i := k$.
- While there are still some unnumbered vertices:
  - Select an unnumbered vertex $v \in \Gamma$ to optimize the criterion $c(v)$. If there are no such vertices, select an unnumbered vertex $v \in \Delta$ to optimize the criterion $c(v)$.
  - Label it with the number i, i.e., let $v_i := v$.
  - Form the set C_i consisting of v_i and its unnumbered neighbours.
  - Fill in edges where none exist between all pairs of vertices in C_i.
  - Eliminate v_i and decrement i by 1. □

Equivalently, if the initial criterion function (to be minimized, say) is $c(\cdot)$, we can introduce a new criterion function $c^*(\cdot)$ such that $c^*(v) = c(v)$ for $v \in \Gamma$, while $c^*(v) = c(v)+K$ for $v \in \Delta$, choosing K so that $\min_{v\in\Delta} c^*(v) > \max_{v\in\Gamma} c^*(v)$; and then apply Algorithm 4.13 to c^*.

Lemma 7.11 *On completion of Algorithm 7.10 one has a strongly triangulated graph $\mathcal{G}$ with a perfect numbering $(v_1, \ldots, v_k)$ of its nodes, and an associated sequence of elimination sets $(C_1, \ldots, C_k)$ satisfying the running intersection property.*

Proof. Since the algorithm can be regarded as a specialization of Algorithm 4.13, the graph is triangulated and the numbering is perfect. We have only to show that $\mathcal{G}$ does not contain a forbidden path. Suppose that there were such a forbidden path, which we can then choose to be as short as possible: say $\delta_1 = \alpha_0 \sim \ldots \sim \alpha_n = \delta_2$, between two discrete nodes δ_1 and δ_2 which are not connected, and which otherwise passes through only continuous nodes. Let γ be the continuous node on this path that was eliminated first, and denote its neighbours in the path by u and v. Since the forbidden path is minimal there can be no edge connecting u and v. But on eliminating γ such a fill-in edge would have been introduced. Hence, there can be no such path, so that $\mathcal{G}$ is strongly triangulated, i.e., strongly decomposable. □

The next step is the construction of the junction tree from the decomposable graph. This can be done in a number of ways. For example, one can use the following strong version of the maximum cardinality search algorithm, always breaking ties in favour of discrete vertices:

Algorithm 7.12 [STRONG MAXIMUM CARDINALITY SEARCH]

- Set **Output**:= '$\mathcal{G}$ is decomposable'.
- Set counter $i := 1$.
- Set $L = \emptyset$.
- For all $v \in V$, set $c(v) := 0$.
- While $L \neq V$:
  - Set $U := V \setminus L$.
  - Select any vertex v maximizing $c(v)$ over $v \in U$, always breaking ties in favour of discrete vertices, and call it v_i.
  - If $\Pi_{v_i} := \mathrm{ne}(v_i) \cap L$ is not complete in $\mathcal{G}$, set **Output**:= '$\mathcal{G}$ is not decomposable'.
  - Otherwise, set $c(w) = c(w) + 1$ for each vertex $w \in \mathrm{ne}(v_i) \cap U$.
  - Set $L = L \cup \{v_i\}$.
  - Increment i by 1.
- Report **Output**. □

Leimer (1989) showed that this algorithm correctly checks whether a given marked graph is decomposable, and that, if it is, its cliques can be found by using Algorithm 7.12 in combination with Algorithm 4.11. Alternatively, they can be found by removing redundant sets from the elimination sequence produced by Algorithm 7.10.

The junction tree or elimination tree again forms the basis of the computational structure. One proceeds as in Section 4.4, but now the asymmetry between continuous and discrete variables makes a further condition necessary for the propagation scheme to work properly.

Definition 7.13 [Strong Root]

A node R on a junction tree or an elimination tree is a *strong root* if any pair A, B of neighbours on the tree with A closer to R than B satisfies

$$(B \setminus A) \subseteq \Gamma \text{ or } (B \cap A) \subseteq \Delta. \tag{7.7}$$

□

For a junction tree of cliques, condition (7.7) is equivalent to $(A \setminus B, B \setminus A, A \cap B)$ forming a strong decomposition of $\mathcal{G}_{A \cup B}$. It requires that, when a separator between two neighbouring cliques is not purely discrete, the clique further away from the root should have only continuous vertices outside the separator.

The statement *iii′* of Theorem 2′ of Leimer (1989) ensures that the cliques of a decomposable marked graph can be organized in a junction tree with at least one strong root. In fact, one can apply Algorithm 7.10 using the numbered elimination sets to construct such a rooted elimination tree, with C_1 as the strong root. The algorithm also produces a sequence of sets $(C_1, C_2, \ldots, C_k)$ that satisfies the *strong running intersection property*, meaning that for all $1 < j \leq k$ there is an $i < j$ such that we have $C_j \cap (C_1 \cup \cdots \cup C_{j-1}) \subseteq C_i$, and additionally,

$$C_j \setminus (C_1 \cup \cdots \cup C_{j-1}) \subseteq \Gamma \quad \text{or} \quad C_j \cap (C_1 \cup \cdots \cup C_{j-1}) \subseteq \Delta.$$

As before, redundant sets can be removed using Lemma 4.16, which continues to hold for the strong running intersection property (see Lemma 2.13 of Lauritzen (1996)). A junction tree of cliques with a strong root can now be constructed using Algorithm 4.8. We assume henceforth that this has been done.

7.4 Model specification

As in the discrete case, the qualitative part of the model is initially specified by an acyclic graph, be it chain graph or one of its specializations, such as the one in Figure 7.3.

The graph specifies the basic dependencies among the variables by assuming that the joint distribution of these has the chain graph Markov property with respect to the graph.

Chain components can have either discrete or continuous or both types of variables. However, to exploit the properties of CG potentials we require that the graph satisfies the following constraint: if a chain component has discrete variables then the component must not have any continuous parents. That is, either all variables in a chain component are continuous, or all parents of a chain component are discrete.

With these constraints on the chain graph $\mathcal{K}$ with chain components K, each conditional density will factorize into CG densities as in the discrete case, as

$$f(x) = \prod_{k \in K} f_k(x_k \mid x_{\mathrm{pa}(k)}) \tag{7.8}$$

$$\propto \prod_{k \in K} Z^{-1}\left(x_{\mathrm{pa}(k)}\right) \prod_{A \in \mathcal{A}_k} \psi_A(x_A) \tag{7.9}$$

where in (7.9) $\mathcal{A}_k$ denotes the set of maximal subsets of $k \cup \mathrm{pa}(k)$ that are complete in the moralized version $\mathcal{K}^m$ of $\mathcal{K}$ and contain at least one child variable in k.

The essential reason for the constraints can be readily comprehended from this factorization. In order for the product of terms to be a CG density, each chain component normalization term Z^{-1} must have the form of a CG potential. However, the normalization is found by marginalization, over the child variables of the product of the remaining terms. If the constraints were violated such a marginal would in general be a mixture of CG potentials, and the required division normalization would yield an entity outside the CG potential family.

However, while theoretically correct when it comes to a prior specification of CG potentials, such a general discussion is at the time of writing somewhat academic. Recall from Section 6.5.3 that the prior specification of chain component potentials for chain graphs of discrete models required discussion of odds-ratios, etc. The analogous extension for CG potentials will be just as technically demanding. Henceforth we shall restrict discussion to models specified by DAGs. For this special case, the constraints simplify to the condition that in the graph *no continuous vertices have discrete children.* This assumption can be relaxed, but then we have to use approximate methods in the specification phase, see Lauritzen (1992) for some suggestions; and also Gammerman et al. (1995).

Thus, we specify, for each discrete variable A, the conditional distribution at A given the states at its parent vertices (which are then all discrete). For each continuous variable A we assume the conditional distribution of the response variable Y associated with A to be of the type

$$\mathcal{L}\big(Y \mid \mathrm{pa}(A)\big) = \mathcal{N}\big(\alpha(i) + \beta(i)'z, \gamma(i)\big).$$

Here pa(A) is a short notation for the combination of discrete and continuous states (i, z) of the variables that are parents of A. In this formula $\gamma(i) > 0$, $\alpha(i)$ is a real number, and $\beta(i)$ is a vector of the same dimension as the continuous part z of the parent variables.

Note that we assume that the mean depends linearly on continuous parent variables and that the variance does not depend on the continuous part of the parent variables. The linear function as well as the variance is allowed to depend on the discrete part of the parent variables.

The conditional density then corresponds to a CG potential ϕ_A defined on the combination (i, z, y) of parent variables (i, z) and response variable y with canonical characteristics (g_A, h_A, K_A), where

$$g_A(i) = -\frac{\alpha(i)^2}{2\gamma(i)} - [\log\{2\pi\gamma(i)\}]/2, \qquad (7.10)$$

$$h_A(i) = \frac{\alpha(i)}{\gamma(i)}\begin{pmatrix} 1 \\ -\beta(i) \end{pmatrix}, \qquad (7.11)$$

$$K_A(i) = \frac{1}{\gamma(i)}\begin{pmatrix} 1 & -\beta(i)' \\ -\beta(i) & \beta(i)\beta(i)' \end{pmatrix}. \qquad (7.12)$$

This follows from direct calculation using the expression for the normal density. We simply write

$$\phi(i, z, y) = \{2\pi\gamma(i)\}^{-1/2} \exp\left[-\{y - \alpha(i) - \beta(i)'z\}^2/\{2\gamma(i)\}\right],$$

resolve the parentheses, take logarithms, and identify terms. Note that $K_A(i)$ has rank 1 and is therefore typically not positive definite.

7.5 Operating in the junction tree

When the model has been specified, the handling of incoming evidence and calculation of specific probabilities is done in the junction tree representation using the elementary operations described in Section 7.2.

Essentially the junction tree representation of the cliques in the strongly triangulated moralized graph captures the computationally interesting aspects of the product structure in the joint density of the variables involved. Then the computations can be performed locally within the cliques and between the cliques that are neighbours in the junction tree.

Hence, we assume that a junction tree with strong root has been established on the basis of the original graph such as discussed in Section 7.3. The collection of belief universes in the junction tree is denoted by $\mathcal{C}$ to indicate that it is the set of cliques in a strongly decomposable graph. The collection of separators is denoted by $\mathcal{S}$, where this may involve multiple copies of the same separator set. Both the belief universes and the separators can have belief potentials ϕ_W attached to them and these are all assumed to be CG potentials defined on the corresponding spaces of variables. The *joint system belief* ϕ_U associated with the given attachment of potentials is then defined as

$$\phi_U = \frac{\prod_{V \in \mathcal{C}} \phi_V}{\prod_{S \in \mathcal{S}} \phi_S}, \tag{7.13}$$

and is assumed to be integrable and proportional to the joint density of all the variables. Since all potentials involved are CG potentials, the joint density and system belief will be a CG potential itself.

We always assume that the tree is *supportive* meaning that for any universe V with neighbouring separator S we have $\phi_S(x) = 0 \Rightarrow \phi_V(x) = 0$. This enables us to deal correctly with cases where some states are ruled out as impossible by having potentials equal to zero.

7.5.1 *Initializing the junction tree*

As a first step, the junction tree with strong root has to be initialized according to the model specification. This is done by generalizing the discrete case as follows.

First we assign each vertex A in the original graph to a universe U in the tree. This has to be done in such a way that $\mathrm{fa}(A) \subseteq U$ but is otherwise arbitrary (where we recall $\mathrm{fa}(A) := \{A\} \cup \mathrm{pa}(A)$). This ensures that the universe is large enough so that the CG potential ϕ_A obtained from the conditional density of X_A given $X_{\mathrm{pa}(A)}$ can be extended to U.

For each universe U we then let ϕ_U be the product of all the (extensions of) potentials ϕ_A for vertices assigned to it. On the separators we let $\phi_S \equiv 1$, i.e., the potential with canonical characteristic $(0, 0, 0)$. This is also the potential on universes with no vertices assigned to them.

7.5.2 *Charges*

As in the discrete case, a collection of CG potentials $\Phi = \{\phi_A, A \in \mathcal{C} \cup \mathcal{S}\}$ will be termed a (CG) *charge* on $\mathcal{T}$.

For any charge, the expression on the right-hand side of (7.13) defines its *contraction.* Whenever (7.13) holds, we may call Φ a *representation* of f. The representation constructed in Section 7.5.1 is termed the *initial representation* and its potentials the *initial potentials.* The computational algorithm to be introduced works by transforming one representation for the joint system belief f to another in a series of simple steps, starting from the initial (or any other convenient) representation of f, modified by the incorporation of evidence described below in Section 7.5.3, and finishing with the *marginal representation* of f in which, for each $A \in \mathcal{C} \cup \mathcal{S}$, ϕ_A is equal to $\sum_{U \setminus A} f$, the (weak) marginal to A of f, given the evidence.

7.5.3 Entering evidence

Incoming evidence is envisaged, or restricted, to be of the type that asserts that certain states of particular discrete variables or combinations of these are impossible, and that certain continuous variables have specified values.

To be able to handle evidence propagation on the junction tree, each item of evidence must be concerned with groups of variables that are members of the same universe in the junction tree. Thus, an *item of evidence* is one of the following:

- A function $\chi_W(i_W) \in \{0, 1\}$, where W is a set of discrete variables that is a subset of some universe U in the junction tree.
- A statement that $Y_\gamma = y^*_\gamma$ for a particular continuous variable γ.

The first type of evidence — that we shall term *discrete* evidence — is entered simply by multiplying χ_W onto the potential ϕ_U.

If the second type of evidence — *continuous* evidence — is entered, the potentials have to be modified in all universes U containing γ. If we have observed that $Y_\gamma = y^*_\gamma$, we have to modify the potentials to become those where y_γ becomes fixed at the value y^*_γ. If the potential ϕ has canonical characteristics (g, h, K) with

$$h(i) = \begin{pmatrix} h_1(i) \\ h_\gamma(i) \end{pmatrix}, \quad K(i) = \begin{pmatrix} K_{11}(i) & K_{1\gamma}(i) \\ K_{\gamma 1}(i) & K_{\gamma\gamma}(i) \end{pmatrix},$$

the transformed potentials ϕ^* will have canonical characteristics (g^*, h^*, K^*) given by

$$K^*(i) = K_{11}(i), \tag{7.14}$$

$$h^*(i) = h_1(i) - y^*_\gamma K_{\gamma 1}(i), \tag{7.15}$$

$$g^*(i) = g(i) + h_\gamma(i) y^*_\gamma - K_{\gamma\gamma}(i)(y^*_\gamma)^2/2, \tag{7.16}$$

derived from standard results on conditioning on one of the variables in a multivariate distribution. Note that such evidence will involve a dimensional reduction in the h vectors and K matrices, and for this reason a continuous item of evidence has to be entered to all universes and separators, of which γ is a member.

When evidence has been entered, it holds that the contraction of the potential will be proportional to the conditional density, given that the states for which χ_W is equal to zero, are impossible, and $y_\gamma = y^*_\gamma$, i.e., it represents the conditional belief, given the evidence.

7.5.4 Flow of information between adjacent cliques

As in the discrete case, the fundamental idea in the algorithm is that of a universe passing a message, or flow, to a neighbour in the junction tree.

So consider a tree of belief universes with collection $\mathcal{C}$ and separators $\mathcal{S}$. Let $U \in \mathcal{C}$ and let W be a neighbour of U with separator S. A *flow* from a source U to the sink W consists of the following operations on the belief potentials (cf. Section 6.3.4):

$$\begin{aligned} \phi_S^* &= \sum_{U \setminus S} \phi_U, \\ \phi_W^* &= \phi_W(\phi_S^*/\phi_S), \end{aligned}$$

and all other potentials are unchanged. In other words, just as in the discrete case, the potential of the source U is marginalized to the separator, and the update ratio between the new and old separator potentials is passed on and multiplied onto the potential at the sink W.

There is, however, an important difference here as the marginal may not always be well-defined: if the marginal is not a strong marginal then ϕ_U must be a CG density for the weak marginal to be defined, and even for strong marginals to be defined there are integrability restrictions (cf. Lemma 7.5).

Note that we have

$$\frac{\phi_U^* \phi_W^*}{\phi_S^*} = \frac{\phi_U(\phi_W(\phi_S^*/\phi_S))}{\phi_S^*} = \frac{\phi_U \phi_W}{\phi_S} \tag{7.17}$$

whence the joint system belief is invariant under the flow process.

In some cases the universes U and W will after a flow 'contain the same information' on common variables. We say that U and W are *consistent* if $\sum_{U \setminus S} \phi_U \propto \phi_S \propto \sum_{W \setminus S} \phi_W$, and a tree of belief universes is said to be *locally consistent* if all mutual neighbours in the tree are consistent. We then have the following result.

Lemma 7.14 *If ϕ_S is the strong marginal of ϕ_W, then U and W are consistent after a flow from U to W. In fact, $\sum_{U \setminus W} \phi_U^* = \phi_S^* = \sum_{W \setminus U} \phi_W^*$.*

Proof. Since the marginalization over $W \setminus U$ is strong it is composed of integrations and summations only. Hence, we find that:

$$\sum_{W \setminus U} \phi_W^* = \sum_{W \setminus U} \phi_W(\phi_S^*/\phi_S) = (\phi_S^*/\phi_S) \sum_{W \setminus U} \phi_W = \phi_S^*.$$

The other equality is trivial. □

We emphasize that the corresponding result is false when ϕ_S is only the weak marginal of ϕ_W, see (7.6). The necessity of using junction trees with strong roots to obtain exact propagation is a consequence of this fact. In the situation described in Lemma 7.14 we say that W has *calibrated* to U.

7.5.5 Two-phase propagation

Because of the special asymmetry and the fact that marginals are not always well-defined, there is not the same liberty for scheduling the flows. Fortunately two-phase propagation is possible as in the discrete case.

Collecting evidence

Definition 7.15 [Collect Evidence]
Each $U \in \mathcal{C}$ is given the action Collect Evidence. When Collect Evidence in U is called from a neighbour W, then U calls Collect Evidence in all its other neighbours, and when they have finished their Collect Evidence, a flow is sent from them toward U. □

We note that since Collect Evidence is composed of flows, after Collect Evidence, the joint system belief is unchanged.

The idea is now to evoke Collect Evidence from a strong root R in the junction tree. A wave of activation of neighbours will move through the tree and a wave of flows towards the root will take place. When the wave terminates, the root R will have absorbed the information available from all parts of the tree.

If Collect Evidence is evoked from a strong root R, and W and W^* are neighbours with separator S such that W is closer in the tree to R than W^*, then the Collect Evidence from R has caused W to absorb from W^*. Thus, after Collect Evidence, the belief potential for S is the marginal of W^* with respect to S. Since the root is strong, the marginal will be strong. This can be exploited for a second distribute flow through the tree originating form the strong root and described below.

Distributing evidence

After Collect Evidence the root R has absorbed all information available. Next it must pass this information on to the remaining universes in the tree, formalized as the operation Distribute Evidence.

Definition 7.16 [Distribute Evidence]
Each $U \in \mathcal{C}$ is given the action Distribute Evidence. When Distribute Evidence is called in U from a neighbour W then W passes a message to U and calls Distribute Evidence in all its other neighbours. □

The activation of Distribute Evidence from the root R will create an outward wave of flows that will stop when it has reached the leaves of the tree. Again the joint system belief remains unchanged under Distribute Evidence. But it even holds that when Distribute Evidence has terminated, the resulting tree of belief universes will be locally consistent.

This follows since after Collect Evidence all separator potentials will be strong marginals of potentials further away from the strong root. When

DISTRIBUTE EVIDENCE is subsequently performed, Lemma 7.14 ensures that all absorptions are calibrations, and the tree is locally consistent with a potential representation which has the following property.

Theorem 7.17 *Let $\mathcal{T}$ be a locally consistent junction tree of belief universes with a strong root R and collection $\mathcal{C}$. Let ϕ_V be the joint system belief for $\mathcal{T}$ and let $U \in \mathcal{C}$. Then*

$$\sum_{V \setminus U} \phi_V \propto \phi_U. \tag{7.18}$$

That is, each local potential is the correct (weak) marginal of the joint system belief.

Proof. Let n denote the number of universes in the collection $\mathcal{C}$. We first realize that it is enough to consider the case $n = 2$. If $n = 1$ the statement obviously holds. If $n > 2$ we can find a leaf L in the tree and use the case $n = 2$ on the junction tree with strong root $R' = \cup_{U \in \mathcal{C} \setminus \{L\}} U$ and one leaf L. By induction the case gets reduced to $n = 2$.

So assume that $V = R \cup L$ where R is a strong root and let $S = R \cap L$ be the separator. The marginal to R is a strong marginal and we find by Lemma 7.14 that

$$\sum_{L \setminus R} \phi_V = \sum_{L \setminus R} \frac{\phi_L \phi_R}{\phi_S} = \frac{\phi_R}{\phi_S} \sum_{L \setminus R} \phi_L = \phi_R.$$

If S contains only discrete variables the marginal to L is also strong and the same calculation applies.

Else, $L \setminus R$ contains only continuous variables. Then $L \setminus S \subseteq \Gamma$, i.e., only continuous vertices are in the external part of the leaf. Denote the states of the variables in S by (i, y) and those in $L \setminus S$ by z. Since ϕ_S is the weak marginal of ϕ_V, the moments $p(i)$, $\mathrm{E}(Y \mid I = i)$, and $\mathrm{var}(Y \mid I = i)$ are correct when calculated according to ϕ_S or, since these are identical, also according to ϕ_U. That the remaining moments now are correct follows since

$$\begin{aligned} \mathrm{E}(Z \mid I = i) = \mathrm{E}\{\mathrm{E}(Z \mid Y, I = i)\} &= \mathrm{E}\{A(i) + B(i)Y \mid I = i\} \\ &= A(i) + B(i)\mathrm{E}(Y \mid I = i), \end{aligned}$$

where $A(i)$ and $B(i)$ are determined from ϕ_L / ϕ_S alone. Similarly

$$\begin{aligned} \mathrm{E}(Z'Y \mid I = i) &= \mathrm{E}\{\mathrm{E}(Z'Y \mid Y, I = i)\} \\ &= \mathrm{E}\{(A(i) + B(i)Y)'Y \mid I = i\} \\ &= A(i)'\mathrm{E}(Y \mid I = i) + \mathrm{E}(Y'B(i)Y \mid I = i) \\ &= A(i)'\mathrm{E}(Y \mid I = i) + \mathrm{E}(Y \mid I = i)'B(i)\mathrm{E}(Y \mid I = i) \\ &\quad + \mathrm{tr}\{B(i)\mathrm{var}(Y \mid I = i)\}. \end{aligned}$$

Finally,

$$\begin{aligned}\operatorname{var}(Z \mid I = i) &= \mathrm{E}\{\operatorname{var}(Z \mid Y, I = i)\} + \operatorname{var}\{\mathrm{E}(Z \mid Y, I = i)\} \\ &= C(i) + B(i)'C(i)B(i),\end{aligned}$$

where also the conditional covariance $C(i)$ is determined from ϕ_L/ϕ_S alone, whence the moments are correct. □

In summary, after entering evidence the junction tree can be made consistent by evoking COLLECT EVIDENCE and then DISTRIBUTE EVIDENCE from a strong root. The weak marginal of the belief at a vertex α can subsequently be obtained from any universe (or even separator) containing α by further weak marginalization.

In particular this gives the correct updated probabilities of the states of any discrete variable and the correct updated mean and variance of any continuous variable.

If the full marginal density is required for a continuous variable, further computations are needed which typically involve all discrete variables on the path between a strong root and a universe containing the variable in question.

We emphasize that despite the use of weak marginalizations, the tree still contains a fully correct representation of the joint system belief, given the evidence. More precisely it holds that

$$f(x) = \frac{\prod_{C \in \mathcal{C}} f_{[C]}(x_C)}{\prod_{S \in \mathcal{S}} f_{[S]}(x_S)},$$

where $f_{[A]}(x_A)$ is the (weak) marginal to A of f. This follows from combining the flow invariance as expressed in (7.17) with Theorem 7.17. (See also Lauritzen (1996), page 188.)

Thus, no information is lost under propagation of evidence and the system remains ready for a correct, exact updating of beliefs when more evidence is obtained.

7.6 A simple Gaussian example

Before considering the more complicated example of the next section, we shall illustrate the types of calculations involved using a simple purely continuous model. Thus, consider the simple three-node model,

$$\boxed{X} \longrightarrow \boxed{Y} \longrightarrow \boxed{Z}$$

specified with initial conditional and unconditional distributions as follows:

$$\begin{aligned}\mathcal{L}(X) &= \mathcal{N}(0, 1), \\ \mathcal{L}(Y \mid X) &= \mathcal{N}(x, 1), \\ \mathcal{L}(Z \mid Y) &= \mathcal{N}(y, 1).\end{aligned}$$

The cliques for this tree are $\boxed{XY}$ and $\boxed{YZ}$. Now any clique can be taken as a root for the pure case. After initializing and propagating, the clique potentials are

$$\phi(x,y) \propto \exp\left(-\frac{1}{2}\begin{pmatrix}x & y\end{pmatrix}\begin{pmatrix}2 & -1\\ -1 & 1\end{pmatrix}\begin{pmatrix}x\\ y\end{pmatrix}\right)$$

and

$$\phi(y,z) \propto \exp\left(-\frac{1}{2}\begin{pmatrix}y & z\end{pmatrix}\begin{pmatrix}1.5 & -1\\ -1 & 1\end{pmatrix}\begin{pmatrix}y\\ z\end{pmatrix}\right),$$

with separator $\phi(y) \propto \exp(-y^2/4)$. Now if we enter evidence $y^* = 1.5$, say, then the potentials reduce to

$$\phi^*(x) = \phi(x, 1.5) \propto \exp(1.5x - x^2)$$

and

$$\phi^*(z) = \phi(1.5, z) \propto \exp(1.5z - 0.5z^2),$$

because in this example Y makes up the separator between the two cliques. The marginal distributions are then:

$$\begin{aligned} \pounds^*(X) &= \pounds(X \mid Y = 1.5) &&= \mathcal{N}(0.75, 0.5),\\ \pounds^*(Z) &= \pounds(Z \mid Y = 1.5) &&= \mathcal{N}(1.5, 1). \end{aligned}$$

Alternatively, suppose we take $\boxed{XY}$ as the root clique, and enter evidence that $Z = 1.5$. Then the message from $\boxed{YZ}$ to $\boxed{XY}$ is given by the update-ratio $\phi^*(y)/\phi(y)$ where $\phi^*(y) \propto \exp(1.5y - 0.75y^2)$ so that after propagation the clique potential on $\boxed{XY}$ is of the form

$$\phi(x,y) \propto \exp\left(\begin{pmatrix}x & y\end{pmatrix}\begin{pmatrix}0\\ 1.5\end{pmatrix} - \frac{1}{2}\begin{pmatrix}x & y\end{pmatrix}\begin{pmatrix}2 & -1\\ -1 & 2\end{pmatrix}\begin{pmatrix}x\\ y\end{pmatrix}\right),$$

with these new marginal distributions:

$$\begin{aligned} \pounds^*(X) &= \pounds(X \mid Z = 1.5) &&= \mathcal{N}(1/2, 2/3),\\ \pounds^*(Y) &= \pounds(Y \mid Z = 1.5) &&= \mathcal{N}(1, 2/3). \end{aligned}$$

7.7 Example: WASTE

A fictitious but simple example taken from Lauritzen (1992) is used to illustrate the general theory for conditional Gaussian models. We shall refer to this as the WASTE example.

7.7.1 *Structural specification*

The illustration is concerned with the control of the emission of heavy metals from a waste incinerator:

> The emission from a waste incinerator differs because of compositional differences in incoming waste. Another important factor is the waste burning regime which can be monitored by measuring the concentration of CO_2 in the emission. The filter efficiency depends on the technical state of the electrofilter and the amount and composition of waste. The emission of heavy metal depends both on the concentration of metal in the incoming waste and the emission of dust particulates in general. The emission of dust is monitored through measuring the penetrability of light.

Here we have ignored the obvious time aspect of the monitoring problem and concentrated on a single point in time, for the sake of simplicity. The essence of the description is represented in the DAG of Figure 7.3.

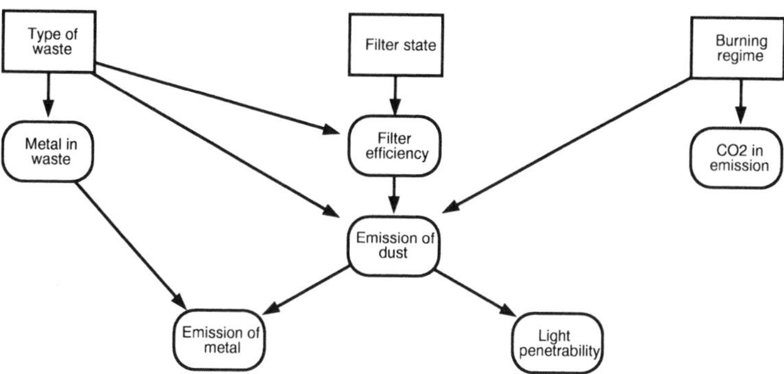

FIGURE 7.3. WASTE incinerator example; variables will also be denoted by W (Type of waste), F (Filter state), B (Burning regime), M_{in} (Metal in waste), E (Filter efficiency), C (CO_2 emission), D (Emission of dust), M_{out} (Emission of metal), and L (Light penetrability).

The described network could in principle be used for several purposes. Typically the emission of dust and heavy metal, the filter efficiency, as well as the actual concentration of heavy metal in the waste would normally not be directly available. The filter state might or might not be known as is also the case for the type of waste.

From the measurements and knowledge available at any time, the emission of heavy metal can be predicted, in particular the mean and standard deviation of the predictive distribution for that emission is of interest. Diagnostic probabilities for stability of the burning regime and/or the state of the filter could be required. Finally the network can be used for management purposes, in that the influence of filter efficiency, burning regime, etc., on objective variables, such as the emission of heavy metals, can be computed.

7.7.2 Numerical specification

For our example we need to specify the conditional distributions, which we do as follows

Burning regime. This variable is discrete and denoted by B. We let

$$P(B = \mathit{stable}) = 0.85 = 1 - P(B = \mathit{unstable}).$$

Filter state. This is discrete and denoted by F. We let

$$P(F = \mathit{intact}) = 0.95 = 1 - P(F = \mathit{defective}).$$

Type of waste. This is discrete and denoted by W. We let

$$P(W = \mathit{industrial}) = 2/7 = 1 - P(W = \mathit{household}).$$

Filter efficiency. This is represented on a logarithmic scale and denoted by E. We assume the relation $\mathrm{dust}_{\mathrm{out}} = \mathrm{dust}_{\mathrm{in}} \times \rho$ and get on the logarithmic scale that $\log \mathrm{dust}_{\mathrm{out}} = \log \mathrm{dust}_{\mathrm{in}} + \log \rho$. We let $E = \log \rho$, admitting that filter inefficiency might better describe the variable E. We then specify

$$\begin{aligned}
\pounds(E \mid \mathit{intact},\ \mathit{household}) &= \mathcal{N}(-3.2, 0.00002) \\
\pounds(E \mid \mathit{defective},\ \mathit{household}) &= \mathcal{N}(-0.5, 0.0001) \\
\pounds(E \mid \mathit{intact},\ \mathit{industrial}) &= \mathcal{N}(-3.9, 0.00002) \\
\pounds(E \mid \mathit{defective},\ \mathit{industrial}) &= \mathcal{N}(-0.4, 0.0001)
\end{aligned}$$

This corresponds to filter efficiencies $1 - \rho$ of about 96 percent, 39 percent, 98 percent and 33 percent respectively. For example, when the filter is defective and household waste is burnt, the filter removes a fraction of $1 - \exp(-0.5) = 0.39$ of the dust.

Emission of dust. This is represented on a logarithmic scale as a variable D. We let

$$\begin{aligned}
\pounds(D \mid \mathit{stable}, \mathit{industrial}, e) &= \mathcal{N}(6.5 + e, 0.03) \\
\pounds(D \mid \mathit{stable}, \mathit{household}, e) &= \mathcal{N}(6.0 + e, 0.04) \\
\pounds(D \mid \mathit{unstable}, \mathit{industrial}, e) &= \mathcal{N}(7.5 + e, 0.1) \\
\pounds(D \mid \mathit{unstable}, \mathit{household}, e) &= \mathcal{N}(7.0 + e, 0.1).
\end{aligned}$$

Thus, on a day when household waste is burned under a stable regime and the filter works perfectly, the typical concentration will be $\exp(6.0 - 3.2) = 16.4\text{mg/Nm}^3$ where 1Nm^3 is the amount of air in 1m^3 at standardized conditions for air temperature and pressure. Similarly, if the filter is defective on a day with industrial waste and the burning regime is unstable, we will typically see an output concentration of dust on $\exp(7.5 - 0.4) = 1212\text{mg/Nm}^3$.

Concentration of CO_2. This is represented on a logarithmic scale as a variable C. We let

$$\mathcal{L}(C \mid \mathit{stable}) = \mathcal{N}(-2, 0.1), \quad \mathcal{L}(C \mid \mathit{unstable}) = \mathcal{N}(-1, 0.3).$$

Thus, the concentration of CO_2 is typically around 14 percent under a stable regime and 37 percent when the burning process is unstable.

Penetrability of light. Represented on a logarithmic scale as a variable L. We let

$$\mathcal{L}(L \mid D = d) = \mathcal{N}(3 - d/2, 0.25).$$

This corresponds to the penetrability being roughly inversely proportional to the square root of the concentration of dust.

Metal in waste. The concentration of heavy metal in the waste is represented as a continuous variable M_{in} on a logarithmic scale. We let

$$\begin{aligned} \mathcal{L}(M_{\text{in}} \mid \mathit{industrial}) &= \mathcal{N}(0.5, 0.01) \\ \mathcal{L}(M_{\text{in}} \mid \mathit{household}) &= \mathcal{N}(-0.5, 0.005). \end{aligned}$$

The precise interpretation is unit dependent, but the main point is that industrial waste tends to contain heavy metal in concentrations that are about three times as high as in household waste. Also the variability of the metal concentrations is higher in industrial waste.

Emission of metal. A continuous variable M_{out} on a logarithmic scale. We let

$$\mathcal{L}(M_{\text{out}} \mid d, M_{\text{in}}) = \mathcal{N}(d + M_{\text{in}}, 0.002).$$

Thus, we simply assume that the concentration of emitted metal is about the same in the dust emitted as in the original waste.

7.7.3 Strong triangulation

The first part in forming a strongly triangulated graph from the model is to moralize the graph $\mathcal{G}$. Then we add the links to make it strongly decomposable. The result of these modifications is shown in Figure 7.4. Note that the link between B and F is necessary to remove the forbidden path (B, E, F) and make the graph decomposable, whereas it is (weakly) triangulated even without this. In this particular example the discrete variables end up forming a complete subset, but this is not always the case.

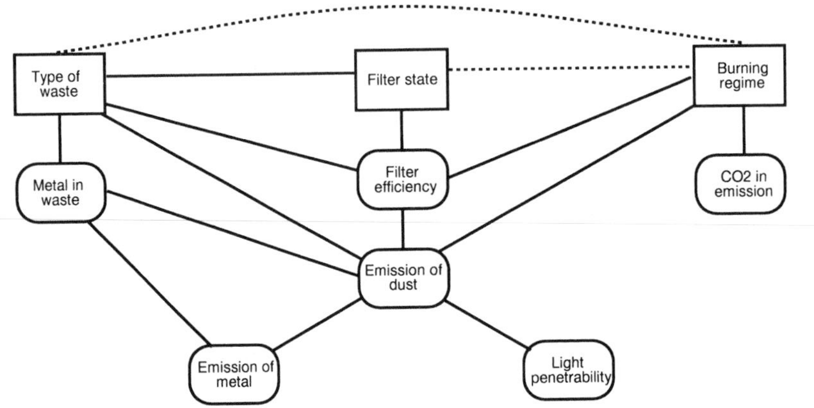

FIGURE 7.4. Moralized and strongly triangulated version of the WASTE graph. Dotted lines indicate fill-in edges required to make the moralized graph strongly triangulated

7.7.4 *Forming the junction tree*

Figure 7.5 displays a junction tree of cliques for our example, with strong root $\{W, E, B, F\}$. Thus, for example, $\{W, M_{\text{in}}, D\}$ has only the continuous variable M_{in} beyond the separator $\{W, D\}$.

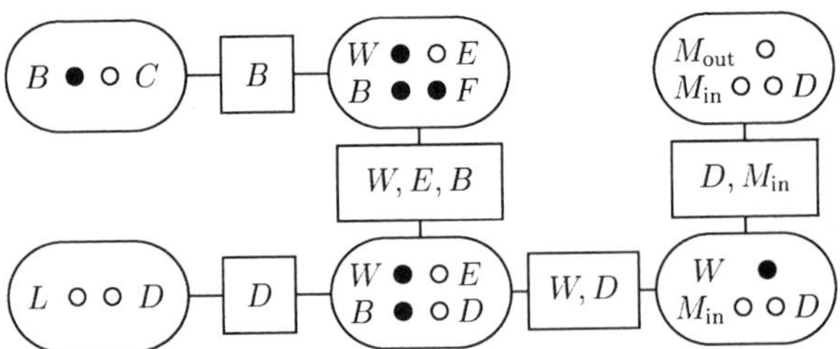

FIGURE 7.5. Junction tree of cliques for the WASTE example, with separators on the edges displayed as rectangular nodes. Clique $\{W, E, B, F\}$ is the strong root.

7.7.5 *Initializing the junction tree*

In our basic example, there are several possibilities for initializing the junction tree in Figure 7.5. An assignment of vertices to universes could be: B

and C to $\{B, C\}$, F, W, and E to $\{W, E, B, F\}$, D to $\{W, E, B, D\}$, L to $\{L, D\}$, M_{in} to $\{W, M_{\text{in}}, D\}$, and M_{out} to $\{M_{\text{out}}, M_{\text{in}}, D\}$.

The potentials in the belief universes are then obtained as follows: The assignment of B to the universe $\{B, C\}$ gives for the value *stable* a potential with characteristics $(\log 0.85, 0, 0) = (-0.16252, 0, 0)$. The assignment of C gives for the same value a potential where, from (7.10),

$$g_{\{C\}}(\textit{stable}) = -\frac{(-2)^2}{2 \times 0.1} - \{\log(2\pi \times 0.1)\}/2 = -20 + 0.23235 = -19.76765,$$

and from (7.11) and (7.12) we get $h_{\{C\}}(\textit{stable}) = -2/0.1 = 20$, and $k_{\{C\}}(\textit{stable}) = (1)/0.1 = 10$. Adding these numbers and rounding off leads to the potentials

$$g_{\{B,C\}}(\textit{stable}) = -19.930, \ h_{\{B,C\}}(\textit{stable}) = -20, \ k_{\{B,C\}}(\textit{stable}) = 10.$$

Analogously for the unstable case we find

$$\begin{aligned} g_{\{B,C\}}(\textit{unstable}) &= -3.881, \\ h_{\{B,C\}}(\textit{unstable}) &= -3.333, \\ k_{\{B,C\}}(\textit{unstable}) &= 3.333. \end{aligned}$$

Only the variable L was assigned to the universe $\{L, D\}$, and this variable has no discrete parents. Hence, only $h_{\{L,D\}}$ and $K_{\{L,D\}}$ are needed. From (7.11) we find

$$h_{\{L,D\}} = \frac{3}{0.25} \begin{pmatrix} 1 \\ -0.5 \end{pmatrix} = \begin{pmatrix} 12 \\ 6 \end{pmatrix}$$

and from (7.12) that

$$K_{\{L,D\}} = \frac{1}{0.25} \begin{pmatrix} 1 & 0.5 \\ 0.5 & 0.25 \end{pmatrix} = \begin{pmatrix} 4 & 2 \\ 2 & 1 \end{pmatrix}.$$

Similar calculations have to be performed for the remaining belief universes.

7.7.6 *Entering evidence*

In the example we might know that the waste burned was of industrial type, and enter this information as the function χ_W with $\chi_W(\textit{industrial}) = 1$ and $\chi_W(\textit{household}) = 0$. Similarly we might have measured the light penetration to be 1.1 and the concentration of CO_2 to -0.9, both on the logarithmic scale applied when specifying the conditional distributions. The latter translates to a concentration of 41 percent CO_2 in the emission. Then using (7.16) the potentials from the initialization are modified to become, for example

$$g^*_{\{B\}}(\textit{stable}) = -19.930 + 18 - 4.050 = -5.980 \qquad (7.19)$$

and

$$g^*_{\{B\}}(unstable) = -3.881 + 3 - 1.350 = -2.231 \tag{7.20}$$

as well as

$$h^*_{\{D\}} = 6 - 1.1 \times 2 = 3.8, \quad K^*_{\{D\}} = 1.$$

TABLE 7.1. Marginal probabilities, means and standard deviations before and after entering of evidence.

		initially	after evidence
Burning regime	*stable*	0.85	0.0122
	unstable	0.15	0.9878
Filter state	*intact*	0.95	0.9995
	defective	0.05	0.0005
Type of waste	*industrial*	0.29	1
	household	0.71	0
Filter efficiency	mean	-3.25	-3.90
	standard deviation	0.709	0.076
Emission of dust	mean	3.04	3.61
	standard deviation	0.770	0.326
Concentration of CO_2	mean	-1.85	-0.9
	standard deviation	0.507	0
Penetrability of light	mean	1.48	1.1
	standard deviation	0.631	0
Metal in waste	mean	-0.214	0.5
	standard deviation	0.459	0.1
Emission of metal	mean	2.83	4.11
	standard deviation	0.860	0.344

In the example we have displayed the initial and updated marginal probabilities, the means, and the standard deviations in Table 7.1. As was to be expected from measuring a CO_2 emission of 41 percent, there is strong evidence for an unstable burning regime. On the other hand, the filter must be intact to explain the penetrability. Combining these with the fact that industrial waste is burned, the expected emission of heavy metal has been increased by a factor of $\exp(1.3) \approx 3.7$.

7.8 Complexity considerations

We shall briefly discuss some of the issues involving feasibility of the computations. The most complex operation is the weak marginalization over a given clique. If the clique contains discrete variables $\delta \in d$ with state spaces of cardinality n_δ and q continuous variables, the computational complexity of marginalization is of the order of magnitude $q^3 \prod_{\delta \in d} n_\delta$ whereas the storage requirements are about $q^2 \prod_{\delta \in d} n_\delta$. This is because matrix inversion of a $q \times q$ matrix takes about q^3 operations and about q^2 space, and this has to be performed for every cell in the table of discrete configurations. These quantities should be compared with $2^q \prod_{\delta \in d} n_\delta$, which is the complexity, both of computation and storage, when the q variables are discretized as to binary variables. Thus, when q is large, dramatic savings are possible.

However, it should be remembered that extra links may have to be added to make the graph strongly triangulated instead of just triangulated. These extra links may in particular cases increase clique size so much that the savings are thereby lost.

7.9 Numerical instability problems

Although the theory presented above is algebraically correct, when it comes to computer implementation there are problems arising from rounding the finite precision of computer arithmetic, which manifest themselves when transforming between the canonical and moment characteristics of a CG potential; the exponent in (7.4) will typically take on a very wide range of values.

To illustrate why this could lead to problems, consider the following example due to Frank Jensen (personal communication). It simply consists of a discrete binary variable A with one continuous child Y, with probabilities given by

$$\begin{aligned} P(A=1) &= 0.5 \\ \pounds(Y \mid A=0) &= \mathcal{N}(0, 10^{-5}) \\ \pounds(Y \mid A=1) &= \mathcal{N}(1, 10^{-6}). \end{aligned}$$

Thus, the conditional densities of Y are sharply peaked, and the marginal density has a highly peaked bimodal distribution. In terms of CG potentials we obtain, assuming six significant digits and recalling that the vector h and matrix K are one-dimensional for this example,

$$\begin{aligned} \phi_A &= ((-0.693147, -, -), (-0.693147, -, -)) \\ \phi_{Y \mid A} &= ((4.83752, 0, 10^5), (-499994, 10^6, 10^6)) \\ \phi_{YA} &= ((4.14437, 0, 10^5), (-499995, 10^6, 10^6)). \end{aligned}$$

Marginalizing Y out of this product we obtain

$$\phi_A^* = ((-0.69315, -, -)(-1, -, -)),$$

which should have been equal to the original potential for A.

Modern computers use a higher precision, so to obtain the same disastrous results would require either increasing the difference in the mean and/or decreasing the variance to observe rounding errors. However, for large networks it may not be clear whether or not such serious rounding errors are taking place, thus one needs to be aware of the problem in applications.

7.9.1 *Exact marginal densities*

In the theory presented, the full marginal densities are not calculated; only their means and variances are. This comes from the use of the weak marginalization operation, which can be seen as a way of replacing a mixture of multivariate Gaussians by a single multivariate Gaussian by matching the expectation and variance. However, it would be more informative to have the full marginal density.

Curds (1997) has attacked this problem using computer algebra. This permits one to work directly with mixtures without approximations using strong marginalizations only. Such a procedure also automatically overcomes the numerical instability problems. Computer algebra takes care of the exact analysis, even though this is expensive in terms of computational resources. In this way one can obtain exact analytic expressions for the marginal of any node using propagation on the junction tree, and compare this with the results obtained using the above theory. His results clearly indicate some multimodality for the marginals of the continuous variables. In particular the node M_i is bimodal with a mean value which is very unlikely to be observed, while the marginal density for the E node is formed from four normal distributions each having small variances. Hence, in this case the moments of normal mixture distributions may thus be of limited value.

7.10 Current research directions

The most important line of research is associated with taking care of the two problems mentioned above: numerical instability and finding the full mixture distribution. Work is currently in progress, but it basically holds that the full mixture distribution can be obtained without numerical problems by representing the CG potentials and the messages in a slightly more sophisticated form. This gives a procedure that is numerically stable, fast, gives full mixture densities, and permits linear deterministic relationships between the continuous variables. The procedure is implemented in the commercial program HUGIN which has been used to calculate the numbers in Table 7.1.

7.11 Background references and further reading

A very early variant of the basic propagation algorithm was derived and used by Thiele (1880), who derived the simplest special case of what is now known as the Kalman filter and smoother (Kalman and Bucy 1961); see Lauritzen (1981). This algorithm is essentially an instance of the propagation algorithm as described here; see Normand and Tritchler (1992) for the pure continuous case. In effect one is then using local computation to solve sparse systems of linear equations, which also was an early instance of the algorithm (Parter 1961; Rose 1970). Shachter and Kenley (1989) introduced Gaussian influence diagrams to model networks of Gaussian random variables for decision problems and they were later extended to the mixed-discrete Gaussian case (Poland 1994); see also the next chapter.

Interesting instances of mixed discrete-Gaussian networks appear in polygenic models for genetic pedigrees (Sham 1998), where major genes correspond to discrete variables and polygenic effects to continuous. However, this is unfortunately an instance where the demand for a strong junction tree forces all the discrete variables to be in the same clique so that exact propagation becomes infeasible.

8

Discrete Multistage Decision Networks

Discrete multistage decision problems under uncertainty involve random variables, decision variables, and a loss or utility function. A decision variable has a number of states called *decisions* or *choices*; making a decision is modelled by selecting one choice from the range of possible choices of the decision variable. Decisions are taken in a specified order and each has consequences either by its contribution to an overall loss or utility, or by revealing the state of one or more random variables, or both. For example, a decision might be to choose to perform one from among a set of possible tests, as a result of which certain results or diagnoses become possible. However, tests usually have costs associated with them, typically financial and/or in terms of time, pain, morbidity, and so on. The objective is to find for each decision variable the optimal choice to make given what is known at the time the decision is to be taken. The optimality criterion by which decisions are judged is that of *maximizing overall expected utility*, or the equivalent criterion of *minimizing overall expected loss* (Savage 1954; Raiffa and Schlaifer 1961; Lindley 1985). Without loss of generality we use utilities throughout; by changing the sign, we may obtain an equivalent loss function representation.

Here we describe an extension of the probabilistic networks of Chapter 6 to describe such problems and solve them by local computation. Essentially identical extensions were proposed by Cowell (1994) and, independently, by Jensen et al. (1994), the latter taking influence diagrams (Howard and Matheson 1984; Shachter 1986) as their starting point, though both approaches have much in common with the valuation network formulation advocated earlier by Shenoy (1992). We shall describe a variation of the formulation proposed by Cowell (1994), where the concept of an influence diagram is extended to *decision networks* which encompass chain graph structures. The solution process requires a rooted elimination tree structure, similar in spirit to the case of local propagation in the conditional Gaussian (CG) networks of Chapter 7. In addition, the potentials on the elimination tree now consist of pairs, one part representing the probabilistic part, the other the utility part. The solution process is facilitated by defining a special algebra for manipulating such potentials.

Throughout this chapter we consider the simplest case in which both the decisions and random variables are discrete.

8.1 The nature of multistage decision problems

Let R denote a set of random variables, where each $r \in R$ takes values from a finite set $\mathcal{I}_r$ of states. Thus, a configuration of R takes values in the product space $\mathcal{I}_R = \times_{r \in R} \mathcal{I}_r$. Similarly, let D denote a set of decision variables, each $d \in D$ taking values from a finite set Δ_d of decisions, so that a configuration of D takes values in the product space $\Delta_D = \times_{d \in D} \Delta_d$, and a joint configuration of $V = (R, D)$ takes values in $\mathcal{X}_V = \mathcal{I}_R \times \Delta_D$. We will denote a typical element of $\mathcal{X}_V$ by $x = (i, \delta)$ with $i \in \mathcal{I}_R$ and $\delta \in \Delta_D$.

In a multistage decision problem, the decision variables are ordered as $(d_1, d_2, \ldots, d_n)$, where d_1 is the decision to be taken first, d_2 second, ..., and d_n the decision to be taken last. Complementing this ordering of the decision variables is a partition of the random variables into an ordered collection $(R_1, R_2, \ldots, R_{n+1})$ of sets such that variables in the set $\bigcup_{j \leq m} R_j$ are known to the decision maker when the choice for the decision variable d_m is made, and those in the set $\bigcup_{j>m} R_j$ are unknown. Consequent to taking the decision for d_m the states of the random variables R_{m+1} are revealed. These two sequences may be combined to form $(R_1, d_1, R_2, d_2, \ldots, R_n, d_n, R_{n+1})$, an ordered set which we will call a *decision sequence*. The above characterizes the temporal structure of a multistage decision problem. To specify the problem fully we require the following numerical inputs: a set of conditional probability tables of the form

$$\{P(R_1), P(R_2 \mid R_1, d_1), \ldots, P(R_{n+1} \mid R_1, d_1, \ldots, R_n, d_n)\}, \tag{8.1}$$

where $P(R_1)$ is used as a shorthand notation for the table of values $\{p(i_{R_1}) : i_{R_1} \in \mathcal{I}_{R_1}\}$ etc., and a real-valued utility function dependent upon both the random variables and decisions; let us denote this by

$$U_n(R_1, d_1, \ldots, R_n, d_n, R_{n+1}), \tag{8.2}$$

where again the expression in (8.2) is a shorthand notation for the collection of values $U_n(i_{R_1}, \delta_{d_1}, \ldots, i_{R_n}, \delta_{d_n}, i_{R_{n+1}})$. If a set R_m is empty it could be deleted, the decision variables d_{m-1} and d_m combined into one, and the decision sequence relabelled accordingly.

Generally the probabilities in (8.1) required for solving a decision problem are not given in this form when the problem is specified or elicited. Usually the specified probabilities have to be manipulated using the laws of probability in order to derive the above expressions. In the formulation of decision trees (Raiffa and Schlaifer 1961) this is called *pre-processing*, and is performed before the optimal decisions are found. In influence diagrams the required probabilities are found during the solution process by arc-reversals and barren node eliminations, the standard processes of solution via influence diagrams (Shachter 1986).

The solution of the multistage decision problem requires, for each decision variable d_m a *tabulation of choices*, which we shall refer to as a *decision mapping*, or more briefly *mapping*, of what optimal action $\hat{d}_m$ to take given the (possibly sub-optimal) choices actually taken earlier for the set of decision variables $D^{m-1} := \bigcup_{j=1}^{m-1} d_j$, and the states of the revealed random variables $R^m := \bigcup_{j=1}^{m} R_j$. A specification of these mappings for all decision variables represents an explicit solution to the problem, i.e., an ordered collection of tabulations, S, of the optimal decision to make at each stage:

$$\begin{aligned} S &:= (\hat{d}_1(R_1), \hat{d}_2(R_1, d_1, R_2), \ldots, \hat{d}_n(R_1, d_1, \ldots, d_{n-1}, R_n)) \\ &= (\hat{d}_1(R^1), \hat{d}_2(R^2, D^1), \ldots, \hat{d}_n(R^n, D^{n-1})), \end{aligned}$$

where again $\hat{d}_1(R_1)$ is used as a shorthand notation for the table of values $\{\hat{d}_1(i_{R_1}) : i_{R_1} \in \mathcal{I}_{R_1}\}$ etc. The optimal strategy, $\hat{S}$, in which all decisions are taken optimally, can be then found from the above tabulation, as follows:

$$\hat{S} := (\hat{d}_1(R^1), \hat{d}_2(R^2, \hat{d}_1(R^1)), \ldots).$$

Note that the collection S contains the mappings for optimal decisions when earlier — possibly sub-optimal — decisions have been taken. Although the mapping for each decision variable d_m depends only upon previous decision and random variables which would be known when the decision for d_m is taken, finding the optimal decision $\hat{d}_m$ depends upon the requirement that all future decisions will be made optimally too. This implies that consideration must be taken into account of the potential effects of taking a particular decision for d_m on all subsequent decision and random variables. It is this that makes the general decision problem NP-hard (Cooper 1990).

8.2 Solving the decision problem

Mathematically the routine for finding the maximum expected utility is simple to perform, by evaluating the alternating sequence of summations and maximizations given by

$$\sum_{R_1} p(R_1) \max_{d_1} \sum_{R_2} p(R_2 \mid R_1, d_1) \times$$
$$\max_{d_2} \sum_{R_3} p(R_3 \mid R_1, d_1, R_2, d_2) \times$$
$$\vdots$$
$$\max_{d_n} \sum_{R_{n+1}} p(R_{n+1} \mid R_1, \dots, d_n)\, U_n(R_1, d_1, \dots, R_{n+1}). \tag{8.3}$$

In (8.3) and similar expressions, $\sum_R p(R)$ is a compact substitute for the more complex $\sum_{i_R \in \mathcal{I}_R} p(i_R)$.

We may evaluate (8.3) by working backwards; detailed proofs and justification of the solution method can be found in Raiffa and Schlaifer (1961). Finding the decision mapping for each decision variable is done in two steps, first by forming an expectation and then by maximizing the result. Thus, one begins by evaluating the following conditional expectation of the utility function

$$\overline{U}_n(R_1, d_1, \dots, R_n, d_n) := \sum_{R_{n+1}} p(R_{n+1} \mid R_1, d_1, \dots, R_n, d_n)\, U_n(R_1, d_1, \dots, R_n, d_n, R_{n+1}). \tag{8.4}$$

This function $\overline{U}_n$ is then maximized over d_n for all possible choices of the previous decision variables $d_1, \dots, d_{n-1}$ and states of the random variables in the sets $R_1, \dots, R_n$, to produce the mapping $\hat{d}_n(R_1, d_1, \dots, d_{n-1}, R_n)$. If two or more alternative choices for d_n yield the same maximum for the conditional expectation of the utility, then any of the choices can be taken as the optimal. Substituting $\hat{d}_n$ into the function $\overline{U}_n$ defines a new utility function U_{n-1} which is independent of d_n and R_{n+1}:

$$U_{n-1}(R_1, d_1, \dots, d_{n-1}, R_n) := \overline{U}_n(R_1, d_1, \dots, d_{n-1}, R_n, \hat{d}_n(R_1, d_1, \dots, d_{n-1}, R_n)). \tag{8.5}$$

Thus, the original problem in n decision variables, (8.3), is reduced to a decision problem in $n-1$ decisions with the utility function U_{n-1}. By recursively working backwards one solves for the mappings for the decision variables $d_{n-1}, d_{n-2}, \dots, d_1$ in turn. This 'backward induction' solution method is mathematically equivalent to the procedure of tracing back along the forks of a decision tree used in solving decision trees.

Now each probability term in (8.3) can be taken through the summation and maximization operations to its right, hence, we can rewrite (8.3) in the compact form

$$\sum_{R_1} \max_{d_1} \sum_{R_2} \max_{d_2} \sum_{R_3} \cdots \max_{d_n} \sum_{R_{n+1}} p(R\!:\!D)\, U_n(R, D), \tag{8.6}$$

in which the maximization and summation operations *cannot* be interchanged, and

$$p(R:D) := p(R_{n+1} \mid R_1, \ldots, d_n) \cdots p(R_2 \mid R_1, d_1) p(R_1). \qquad (8.7)$$

The quantity $p(R : D)$ must be interpreted with care. If the values of the decision variables D are set externally, $p(R:D)$ gives the induced joint distribution of the set of random variables R. However, when each d_m is allowed to depend on previous variables, as will be the case, for example, for the optimal decision $\hat{d}_m$, then $p(R:D)$ will *not*, in general, be equal to the conditional distribution $p(R \mid D)$ of R given D. For example, when $n = 1$ the conditional distribution $p(R_1 \mid d_1)$ of R_1 given d_1 will then be affected by the built-in dependence of d_1 on R_1, and thus may differ from the marginal distribution of R_1 under $p(R:D)$, namely $p(R_1)$, which is not so affected. In consequence, $p(R_2 \mid d_1) = \sum_{R_1} p(R_2 \mid R_1, d_1)\, p(R_1 \mid d_1)$ will generally differ from $\sum_{R_1} p(R_2, R_1 : d_1) = \sum_{R_1} p(R_2 \mid R_1, d_1)\, p(R_1)$. Thus, we cannot find an arbitrary marginal $p(R_m \mid D)$ by summing unwanted variables out of $p(R:D)$. However, this operation *is* valid for eliminating R_{n+1} from (8.7), and this is all we shall require for solving a decision problem. For general j we have

$$\sum_{R_{j+1}} p(R^{j+1} : D^j) = \sum_{R_{j+1}} p(R_{j+1} \mid R^j, D^j)\, p(R^j : D^{j-1}) = p(R^j : D^{j-1}). \qquad (8.8)$$

Thus, for any j,

$$\sum_{R_{j+1}, \ldots R_{n+1}} p(R:D) = p(R^j : D^{j-1}), \qquad (8.9)$$

so that the left-hand side of (8.9) cannot depend on $d_j, \ldots, d_n$. This is an important consistency constraint on the function $p(R:D)$. Fortunately for the formulation of decision problems given in Section 8.4 there is a simple graphical test of consistency which we present in Section 8.4.2. With the above caveats in mind, we shall nevertheless refer to $P(R:D)$ as a 'joint distribution' over R.

8.3 Decision potentials

We shall show in Section 8.4.6 that we can represent the backward induction solution method by a *collect* operation on an elimination tree, in which the product of the probability and utility is implicitly represented by a pair of functions, generalizing the potentials of Chapter 6. To this end we introduce the following definition.

Definition 8.1 [DECISION POTENTIAL]
Let $U = (R', D')$, $(R' \subseteq R, D' \subseteq D)$, be a subset of the set of random and decision variables $V = (R, D)$. We have $\mathcal{X}_U = \mathcal{I}_{R'} \times \Delta_{D'}$. A *decision potential* on $U = (R', D')$, or on $\mathcal{X}_U$, is a function $\phi : \mathcal{X}_U \to \mathbb{R}_{\geq 0} \times \mathbb{R}$ or, equivalently, an ordered pair of real-valued functions (γ, μ), with $\gamma \geq 0$. We call the first part, γ, the *probability part*, and the second part, μ, the *utility part*. The *domain* of ϕ is $\mathcal{X}_U$, or just U for short. □

Note that we do not require $\sum_U \gamma(U) = 1$. We call the probability part *vacuous* if it is identically equal to unity, and we call the utility part vacuous if it is identically equal to zero. A decision potential is called vacuous if both its probability and utility parts are vacuous.

To solve the decision problem in terms of decision potentials (henceforth in this chapter just called potentials), we define an algebra with which to manipulate them that will facilitate the evaluation of the ordered operations of (8.3) or (8.6). Thus, we introduce variations of familiar operations and definitions of the previous chapters, beginning with

Definition 8.2 [EXTENSION]
Let ϕ be a potential on $\mathcal{I} \times \Delta$. Then the extension of ϕ to a potential $\phi^\dagger$ on $(\mathcal{I} \times \mathcal{J}) \times (\Delta \times \tilde{\Delta})$ is defined by

$$\phi^\dagger(i, j, \delta, \tilde{\delta}) = \phi(i, \delta). \tag{8.10}$$

□

In particular, an extension of a vacuous potential is again vacuous.

Definition 8.3 [RESTRICTION]
Let ϕ have domain $(\mathcal{I} \times \mathcal{J}) \times (\Delta \times \tilde{\Delta})$, and let $(j, \tilde{\delta}) \in (\mathcal{J} \times \tilde{\Delta})$. Then the restriction $\phi^{(j,\tilde{\delta})}$ of ϕ, with domain $\mathcal{I} \times \Delta$, is defined by

$$\phi^{(j,\tilde{\delta})}(i, \delta) = \phi(i, j, \delta, \tilde{\delta}).$$

□

Definition 8.4 [MULTIPLICATION]
Let $\phi_1 = (\gamma_1, \mu_1)$ have domain U_1 and $\phi_2 = (\gamma_2, \mu_2)$ have domain U_2. Then their product $\phi_1\phi_2$ has domain $U_1 \cup U_2$ and is given by

$$(\phi_1\phi_2)(x) := (\gamma_1(x)\gamma_2(x), \mu_1(x) + \mu_2(x)),$$

where we let $\mathcal{X} := \mathcal{X}_{U_1 \cup U_2}$, and either or both of ϕ_1, ϕ_2 is extended if necessary to have domain $U_1 \cup U_2$ before combining. □

Clearly multiplication of potentials is associative and commutative. Also, the product of two vacuous potentials is again vacuous.

Division is defined similarly to multiplication, with care taken to handle divisions by zero.

Definition 8.5 [DIVISION]
Let $\phi_U = (\gamma_U, \mu_U)$ have domain U and $\phi_W = (\gamma_W, \mu_W)$ have domain $W \subseteq U$. Then the division of ϕ_U by ϕ_W has domain U, and is given by

$$(\phi_U/\phi_W)(x_U) := (\gamma_U(x_U)/\gamma_W(x_U), \mu_U(x_U) - \mu_W(x_U)),$$

where $\gamma_U(x_U)/\gamma_W(x_U) := 0$ if $\gamma_W(x_U) = 0$, and ϕ_W is extended if necessary to have domain U before the division is performed. □

In particular, the division of one vacuous potential by another vacuous potential is again vacuous.

Definition 8.6 [SUM-MARGINALIZATION]
Let $W \subseteq U \subseteq V$, and let $\phi_U = (\gamma_U, \mu_U)$ be a potential on U. Then the *sum-marginalization* of the potential ϕ_U to W, denoted by $\mathbf{S}_{U \setminus W}\, \phi_U$, is the potential on W defined by

$$\mathop{\mathbf{S}}_{U \setminus W} \phi_U := \left(\sum_{U \setminus W} \gamma_U, \frac{\sum_{U \setminus W} \gamma_U \mu_U}{\sum_{U \setminus W} \gamma_U} \right), \tag{8.11}$$

where $\sum_{U \setminus W}$ denotes the standard sum-marginalization of Definition 6.5. □

Thus, the probability part is marginalized in the usual way as in Chapter 6, whereas the utility part is marginalized by taking weighted averages using the probability part as weights, corresponding to computation of expected utilities.

The propagation scheme in this chapter uses yet another marginalization operation, associated with optimizing the decisions, defined as follows.

Definition 8.7 [MAX-MARGINALIZATION]
Let $W \subseteq U \subseteq V$, and let $\phi_U = (\gamma_U, \mu_U)$ be a decision potential on U. Then the *max-marginalization* of the potential ϕ_U to W, denoted by $\mathbf{M}_{U \setminus W} \phi_U$, is defined by

$$\mathop{\mathbf{M}}_{U \setminus W} \phi_U := \left(\gamma_U^{z^*}, \frac{(\gamma_U \mu_U)^{z^*}}{\gamma_U^{z^*}} \right), \tag{8.12}$$

where z^* maximizes $\gamma_U \mu_U$ for each $w \in \mathcal{X}_W$, i.e., $z^* = \operatorname{argmax}_{U \setminus W} (\gamma_U \mu_U)^z$. □

The reader may readily verify that both marginalizations satisfy the usual property: for $U \subseteq W \subseteq V$,

$$\mathop{\mathbf{S}}_{W \setminus U} \left(\mathop{\mathbf{S}}_{V \setminus W} \phi_V \right) = \mathop{\mathbf{S}}_{V \setminus U} \phi_V, \quad \mathop{\mathbf{M}}_{W \setminus U} \left(\mathop{\mathbf{M}}_{V \setminus W} \phi_V \right) = \mathop{\mathbf{M}}_{V \setminus U} \phi_V,$$

whereas, in general,

$$\mathop{\mathbf{S}}_{W\setminus U}\left(\mathop{\mathbf{M}}_{V\setminus W}\phi_V\right)\neq\mathop{\mathbf{M}}_{W\setminus U}\left(\mathop{\mathbf{S}}_{V\setminus W}\phi_V\right).$$

We further have the following two simple lemmas.

Lemma 8.8 *Let $\phi_U=(\gamma_U,\mu_U)$ be a potential on U and $\phi_W=(\gamma_W,\mu_W)$ a potential on W, and let their product be $\phi_V=\phi_U\,\phi_W$ (with domain $V:=U\cup W$). Then $\mathbf{S}_{V\setminus U}\,\phi_V=\phi_U\,\mathbf{S}_{V\setminus U}\,\phi_W$.*

Proof. We have

$$\begin{aligned}
\mathop{\mathbf{S}}_{V\setminus U}\phi_V &= \mathop{\mathbf{S}}_{V\setminus U}\phi_U\,\phi_W\\
&= \mathop{\mathbf{S}}_{V\setminus U}(\gamma_U\,\gamma_W,\mu_U+\mu_W)\\
&= \left(\sum_{V\setminus U}\gamma_U\,\gamma_W,\ \frac{\sum_{V\setminus U}(\gamma_U\,\gamma_W(\mu_U+\mu_W))}{\sum_{V\setminus U}\gamma_U\,\gamma_W}\right)\\
&= \left(\gamma_U\sum_{V\setminus U}\gamma_W,\ \frac{\gamma_U\,\mu_U\sum_{V\setminus U}\gamma_W+\gamma_U\sum_{V\setminus U}\gamma_W\,\mu_W}{\gamma_U\sum_{V\setminus U}\gamma_W}\right)\\
&= \left(\gamma_U\sum_{V\setminus U}\gamma_W,\ \mu_U+\frac{\sum_{V\setminus U}\gamma_W\,\mu_W}{\sum_{V\setminus U}\gamma_W}\right)\\
&= (\gamma_U,\mu_U)\left(\sum_{V\setminus U}\gamma_W,\ \frac{\sum_{V\setminus U}\gamma_W\mu_W}{\sum_{V\setminus U}\gamma_W}\right)\\
&= (\gamma_U,\mu_U)\mathop{\mathbf{S}}_{V\setminus U}(\gamma_W,\mu_W)=\phi_U\mathop{\mathbf{S}}_{V\setminus U}\phi_W,
\end{aligned}$$

which was to be proved. □

Comment: The reader will have noticed in the above proof a cancellation in the numerator and denominator of the term γ_U. This does not present a problem if γ_U is zero, because γ_U is still a factor of ϕ_U, and Definition 8.5 defines $1/0=0$.

Lemma 8.9 *Let U, V, and W be pairwise disjoint sets. Let $\phi_{V,U}=(\gamma_{V,U},\mu_{V,U})$ and $\phi_{U,W}=(\gamma_{U,W},\mu_{U,W})$ be decision potentials on $V\cup U$ and $U\cup W$ respectively. Furthermore, let the probability part $\gamma_{U,W}$ be independent of $w\in\mathcal{X}_W$ for fixed $u\in\mathcal{X}_U$, that is, $\gamma_{U,W}(u,w)=\gamma_{U,W}(u,w')$ for all $w\neq w'\in\mathcal{X}_W$ and $u\in\mathcal{X}_U$. Then*

$$\mathop{\mathbf{M}}_{W}(\phi_{V,U}\,\phi_{U,W})=\phi_{V,U}\mathop{\mathbf{M}}_{W}\phi_{U,W}. \tag{8.13}$$

Proof. For each $u \in \mathcal{X}_U$, let $\overline{w}(u) \in \mathcal{X}_W$ maximize $\mu_{U,W}(u,w)$ over w, and define $\overline{\mu}_{U,W}(u,w) := \mu_{U,W}(u,\overline{w}(u))$. Since by assumption $\gamma_{U,W}$ is a function of u alone, we have

$$\max_W \{\gamma_{V,U}\,\gamma_{U,W}(\mu_{V,U}+\mu_{U,W})\} = \gamma_{V,U}\,\gamma_{U,W}(\mu_{V,U}+\overline{\mu}_{U,W}),$$

achieved at $w = \overline{w}(u)$. Thus, we get

$$\begin{aligned}
\mathop{\mathbf{M}}_W(\phi_{V,U}\,\phi_{U,W}) &= \mathop{\mathbf{M}}_W(\gamma_{V,U}\,\gamma_{U,W},\ \mu_{V,U}+\mu_{U,W}) \\
&= (\gamma_{V,U}\,\gamma_{U,W},\ \mu_{V,U}+\overline{\mu}_{U,W}) \\
&= (\gamma_{V,U},\mu_{V,U})(\gamma_{U,W},\overline{\mu}_{U,W}) \\
&= (\gamma_{V,U},\ \mu_{V,U})\mathop{\mathbf{M}}_W(\gamma_{U,W},\ \mu_{U,W}) \\
&= \phi_{V,U}\mathop{\mathbf{M}}_W\phi_{U,W},
\end{aligned}$$

which was to be proved. □

In the application of these operations to decision networks, sum-marginalization will eliminate random variables, while max-marginalization will eliminate a single decision variable. In the latter case the function $\overline{w}$ will then correspond to a decision mapping.

If there are no decision variables or utilities in the problem then the potentials and their operations reduce to the probability potentials defined for probabilistic networks, with a vacuous utility part tagging along.

8.4 Network specification and solution

The extension of probabilistic networks to handle multistage decision problems requires different types of nodes to represent the random variables, the decision variables, and the utility. Additionally, the numerical specification of the probabilities and utilities is required. A further wrinkle is that because of the non-commutativity of the summations and maximizations in (8.3), in order to implement the backward induction method by local computation on the tree, additional constraints in forming the elimination tree are necessary and the propagation will be performed in a rooted tree, as was the case in Chapter 7.

8.4.1 Structural and numerical specification

Let $(R_1, d_1, R_2, d_2, \ldots, R_n, d_n, R_{n+1})$ be a decision sequence for the random variables $R = (R_1, \ldots, R_{n+1})$ and decision variables $D = (d_1, \ldots, d_n)$ of a multistage decision problem, with $V = (R \cup D)$. Then we define a *decision network* constructed over V to consist of a set of nodes organized into a chain graph $\mathcal{K}$ having the following structure: Each node is of one of three possible types: *random variable* $\in R$, *decision variable* $\in D$, or *utility* $\in \mathcal{U}$. There are constraints on the structure that the chain graph can take, specifically,

1. Decision nodes have neither parents nor neighbours.
2. Utility nodes have neither children nor neighbours.

We shall assume that the decision network $\mathcal{K}$ is connected, for otherwise we have independent decision problems which can be treated separately. From the two constraints it follows that the chain components consist of isolated decision nodes, isolated utility nodes, and groups of random variable nodes.

The network $\mathcal{K}$ represents the conditional independence properties of the joint distribution of the random variables given the decisions, $P(R : D)$, in the sense that $P(R : D)$ is assumed to factorize appropriately over the chain graph $\mathcal{K}_{R \cup D}$, together with the functional decomposition into additive components of the total utility, $U(R, D)$. The directed edges from a decision node represent the direct influences that taking the decision can have, either in terms of its contribution to the total utility, or, for links going into random variables, in affecting their outcomes. Now the decision sequence captures the time-order in which random variables are revealed and decisions are taken, and because decisions cannot affect the past it follows that if a decision sequence is to be *causally consistent* with the decision network, there can be no links from any decision d to any random variable occurring before d in the decision sequence. As we shall show in Section 8.4.2, every random variable descendant of a decision d must appear later than d in the decision sequence for the network $\mathcal{K}$ and decision sequence to be consistent.

Readers familiar with the theory of influence diagrams may be puzzled by the first constraint imposed upon the decision nodes. In the standard representation using influence diagrams developed for DAGs (Howard and Matheson 1984), decision nodes generally do have parent nodes. The directed arcs going into decision nodes are called *information arcs*; the minimal ancestral set of nodes to a particular decision node d (excluding d itself) represents the information that will be known to the decision maker when the decision d will be taken. Typically this will be the set of all previous decisions taken and observed random variables (the *no forgetting principle*, see Shachter (1986)). Thus, incidentally, the set of directed edges which need to be included as information arcs in an influence diagram is not necessarily unique for a particular problem, though some minimal non-empty set is generally required. The causal-consistency condition is then equivalent to decisions not being able to affect the past, hence that the influence diagram is acyclic. From a *modelling* viewpoint and the solution process described below, the decision sequence can be regarded as basic, and the information arcs of an influence diagram are not required. Nevertheless, to highlight the importance of the decision sequence in the solution process described below, we shall omit information arcs in our examples. Readers uncomfortable with this may imagine them to be present.

Having defined the graphical structure and decision sequence, numerical inputs are required to complete the specification of the multistage decision problem. These take the following form:

- To each chain component $k \in \mathcal{K}$ consisting of random variables is attached a set of conditional distributions $\pounds(X_k \mid X_{\mathrm{pa}(k)})$. If $\mathrm{pa}(k) = \emptyset$ these reduce to unconditional probabilities. These distributions factorize further if there is more than one random variable in the chain component as also described in Chapter 6. Denote these chain components by K_R. The joint probability represented by $\mathcal{K}$ is then given by

$$p(i_R : \delta_D) = \prod_{k \in K_R} p(i_k \mid x_{\mathrm{pa}(k)}). \tag{8.14}$$

- To each chain component $\mu \in \mathcal{K}$ consisting of a single utility node μ is attached a real-valued function depending upon the parents, which we denote by ψ_μ. Denote these chain components by $K_{\mathcal{U}}$. The total utility of the decision network, assumed additive over the utility nodes, is

$$U(i_R, \delta_D) = \sum_{\mu \in K_{\mathcal{U}}} \psi_\mu(x_{\mathrm{pa}(\mu)}). \tag{8.15}$$

- There are no functions or tables associated with the chain components comprising the isolated decision nodes.

8.4.2 Causal consistency lemma

As remarked in Section 8.2 not all joint distributions (8.14) will be consistent with a given decision sequence, i.e., obey the constraints of (8.9), to form a well-posed decision problem. Fortunately, there is a simple graphical criterion to check for consistency, given by the following lemma.

Lemma 8.10 $(R_1, d_1, R_2, d_2, \ldots, R_n, d_n, R_{n+1})$ *is a consistent decision sequence for a decision network having structure* $\mathcal{K}$ *if, in* $\mathcal{K}$ *with the utility nodes removed, for each decision node* d_m *descendants of* d_m *all occur later in the decision sequence,* i.e.,

$$\mathrm{de}(d_m) \subseteq \bigcup_{j=m+1}^{n+1} R_j.$$

Proof. Let $\mathcal{K}^\dagger$ denote the decision network $\mathcal{K}$ with utility nodes removed. We need to show that any conditional distribution $p(R : D)$ factorizing according to $\mathcal{K}^\dagger$ obeys the consistency requirements of (8.9), in the form

$$p(R^{j+1} : D) = p(R^{j+1} : D^j), \quad j = 1, \ldots, n+1. \tag{8.16}$$

Now recall that if one has a probability distribution that factorizes according to a chain graph $\mathcal{G}$, then marginalizing over the variables of some chain component C that does not have any children results in a marginal distribution which factorizes on the graph obtained from $\mathcal{G}$ by removing C and its incident edges.

Note that for any decision node d_m the descendants $\mathrm{de}(d_m)$ in $\mathcal{K}^\dagger$ will all be random variables. It follows that for any decision node d_m the marginal distribution $p(R \setminus \mathrm{de}(d_m) : D)$ factorizes over the chain graph $\mathcal{K}^\dagger$ with the nodes $\mathrm{de}(d_m)$ removed; denote this graph by $\mathcal{K}^\dagger_m$. (To see this, simply organize the random variables in the set $\mathrm{de}(d_m)$ into their well-ordered chain components, and then marginalize out these variables in an order reverse to the well-ordering.) But in the graph $\mathcal{K}^\dagger_m$ the decision node is isolated. Thus, we have that $p(R \setminus \mathrm{de}(d_m) : D)$ is independent of d_m.

Thus, if $(R_1, d_1, R_2, d_2, \ldots, R_n, d_n, R_{n+1})$ is a decision sequence and if for every decision d_m its descendants $\mathrm{de}(d_m)$ occur later in the decision sequence, then for $j \leq m$ the marginal distribution $p(R^j : D)$ is independent of d_m. Hence, (8.16) holds. □

A simple interpretation of the consistency requirement is this: selecting a choice for a particular decision variable will affect the (distribution of the) random variables that are descendants of the decision; thus, in the decision network directed edges from decisions to random variables represent causal influence. When a decision is taken it cannot influence the past; hence, the random variable descendants of a decision should come later in the decision sequence, because the decision sequence represents the time (partial-) ordering of the decision and random variables of the decision problem.

8.4.3 Making the elimination tree

To construct the inference engine for the decision problem one first moralizes the network as usual, using the procedure in Section 4.1. Note that if one started with an influence diagram representation, that is, one included information arcs directed into the decisions, then such arcs *must* be deleted before moralizing the graph. This is the approach of Jensen et al. (1994). After the moralization process is completed, one then deletes the utility nodes. The utility nodes can induce moral edges, hence, it is important first to eliminate them after the moralization.

The moralization procedure ensures that each multiplicative factor of the joint probability and each additive factor of the total utility can be associated with the decision potential of at least one elimination set of the elimination tree derived by triangulating the moral graph.

Denote the resulting moral graph with the utility nodes removed by $(\mathcal{K}^\dagger)^m$. The next stage is to form a rooted elimination tree. This uses a modification of the triangulation algorithms of Chapter 4 and Chapter 7, which takes into account the order constraints given by the decision sequence $(R_1, d_1, \ldots, R_n, d_n, R_{n+1})$:

Algorithm 8.11 [ONE-STEP LOOK AHEAD TRIANGULATION]

- Start with all vertices unnumbered, set counter $i := |R \cup D|$ and counter $j := n + 1$.
- While there are still unnumbered vertices:
  - Select an unnumbered vertex $v \in R_j$ to optimize the criterion $c(v)$. If there are no such vertices, select $v = d_{j-1}$ and decrement j by 1.
  - Label the selected vertex with the number i, i.e., let $v_i := v$.
  - Form the set C_i consisting of v_i and its unnumbered neighbours.
  - Fill in edges where none exist between all pairs of vertices in C_i.
  - Eliminate v_i and decrement i by 1. □

Next we use Algorithm 4.14 to create an elimination tree $\mathcal{T}$, with an elimination set C_v for each node v of $(\mathcal{K}^\dagger)^m$.

For convenience we add a final empty set C_0, which is joined to all other singleton elimination sets. We designate C_0 as root of the tree. The separator of C_v from nodes nearer the root is $\text{sep}(v) := C_v \setminus \{v\}$; the adjacent elimination set nearer the root than C_v will be denoted by $e(v)$. We denote the collection of elimination sets by $\mathcal{C}$ and the collection of separators by $\mathcal{S}$.

8.4.4 Initializing the elimination tree

To initialize the tree $\mathcal{T}$ one first associates to each elimination set $C \in \mathcal{C}$ and each separator $S \in \mathcal{S}$ a vacuous decision potential. One then takes each factor in the joint probability $P(R\!:\!D)$ occurring in (8.14) (recall that for general chain graphs the conditional probability distribution of a chain component given its parents may itself factorize; see Section 5.4) and *multiplies* it into the probability part of the potential of any single elimination set containing its set of variables. When this is done, one then takes each additive term ψ_μ of the total utility in (8.15) and *adds* it to the utility part of the potential of any single one of the elimination sets containing $\text{pa}(\mu)$ (extending ψ_μ as necessary before addition). Clearly, after these initializations have been performed, we have on the tree $\mathcal{T}$ the following generalized potential representation of the joint probability and utility:

$$\phi_V = \frac{\prod_{C \in \mathcal{C}} \phi_C}{\prod_{S \in \mathcal{S}} \phi_S} = (P(R\!:\!D), U(R, D)). \tag{8.17}$$

The separator potentials $\{\phi_S\}$ are all vacuous at this point.

8.4.5 *Message passing in the elimination tree*

We now have a representation of the joint distribution and total utility in terms of decision potentials on the elimination tree. We use this representation to solve the decision problem (8.3) by performing a suitable collect operation from the root C_0.

Definition 8.12 [ABSORPTION]
Let C be an elimination set of the elimination tree $\mathcal{T}$ and let $C_{v_1}, \ldots, C_{v_m}$ be neighbours of C, with separators $S_1, \ldots, S_m$ respectively, which are further away than C from the root C_0. Thus, $C_{v_m} \setminus S_m = \{v_m\}$. The elimination set C is said to *absorb* from $C_{v_1}, \ldots, C_{v_m}$ if the decision potentials ϕ_{S_m} and ϕ_C are changed to $\phi^*_{S_m}$ and ϕ^*_C, where

$$\phi^*_{S_m} = \begin{cases} \mathbf{S}_{v_m} \phi_{C_{v_m}}, & \text{if } v_m \text{ is a random node} \\ \mathbf{M}_{v_m} \phi_{C_{v_m}}, & \text{if } v_m \text{ is a decision node} \end{cases}$$

and

$$\phi^*_C = \phi_C(\phi^*_{S_1}/\phi_{S_1}) \cdots (\phi^*_{S_m}/\phi_{S_m}). \tag{8.18}$$

□

In our propagation scheme the separator potentials in the denominators above will all be vacuous, so the divisions in (8.18) are in fact redundant and are only indicated to maintain analogy with previous chapters. Note also that the absorption can be implemented to take place by sending repeated single messages from $C_{v_1}, \ldots, C_{v_m}$ to C, in an arbitrary order.

Absorption leaves the overall decision potential (8.17) invariant. Based on the notion of absorption we can now construct the basic propagation algorithm, using the following definition.

Definition 8.13 [COLLECT EVIDENCE]
Each $C \in \mathcal{C}$ is given the action COLLECT EVIDENCE. When COLLECT EVIDENCE in C is called from a neighbour W, then C calls COLLECT EVIDENCE in all its other neighbours, and when they have finished their COLLECT EVIDENCE, C absorbs from them. □

As we prove in detail in Section 8.4.6, a call of COLLECT EVIDENCE from the root C_0 leads to a sequence of marginalizations, either summations over random variables or maximization over decisions, which emulate the backward sequence of operations in (8.3). In summary this is true because: (i) the elimination tree has the running intersection property; (ii) its construction encodes the decision sequence; (iii) Lemma 8.8 and Lemma 8.9 apply; (iv) in particular when maximizing over a decision node the probability part of the elimination potential is independent of the decision (cf. (8.9)) so that Lemma 8.9 applies, so that one indeed maximizes over the expected utility at each stage.

The result is that after a call of COLLECT EVIDENCE the potential representation (8.17) has maintained its overall invariance. More importantly, if C is an elimination set in which its elimination node v is a decision variable d, then the optimal choice for d conditional on each configuration of the remaining variables (which are in the neighbouring separator pointing toward the root) can be identified from the entries in the utility part for the elimination set potential, viz. any entry that maximizes this entry.

8.4.6 Proof of elimination tree solution

Recall that the maximum expected utility is found by evaluating (8.6), i.e.,

$$\sum_{R_1} \max_{d_1} \sum_{R_2} \max_{d_2} \sum_{R_3} \cdots \max_{d_n} \sum_{R_{n+1}} p(R:D)\, U_n(R,D),$$

in which the operations of summation and maximization cannot be interchanged. If $(r_1^1, \ldots, r_1^{k_1}, d_1, r_2^1, \ldots, r_2^{k_2}, d_2, \ldots, r_{n+1}^{k_{n+1}})$ is an elimination ordering consistent with the decision sequence, this can be expressed as

$$\sum_{r_1^1} \cdots \sum_{r_1^{k_1}} \max_{d_1} \sum_{r_2^1} \cdots \sum_{r_2^{k_2}} \max_{d_2} \sum_{r_3^1} \cdots \\ \sum_{r_3^{k_3}} \cdots \max_{d_n} \sum_{r_{n+1}^1} \cdots \sum_{r_{n+1}^{k_{n+1}}} p(R:D)\, U_n(R,D).$$

We now introduce some more notation: let

$$P_i^j := \sum_{r_i^{j+1}} \cdots \sum_{r_i^{k_i}} \cdots \sum_{r_{n+1}^{k_{n+1}}} P(R:D),$$

and

$$V_i^j := \sum_{r_i^{j+1}} \cdots \sum_{r_i^{k_i}} \max_{d_i} \cdots \sum_{r_{n+1}^{k_{n+1}}} P(R:D)\, U_n(R,D).$$

Thus, V_i^j is an intermediate utility expression which arises in the solution of the problem and P_i^j is the associated intermediate probability; in particular, $P_1^0 = 1$, and V_0^1 is the maximum expected utility. Recall that, because of the consistency condition, P_i^j depends only on $d_1, \ldots, d_{i-1}$ for $j > 0$, and further P_i^0 is independent of d_{i-1}.

Let $(C_0, C_1^1, \ldots, C_1^{k_1}, C_{d_1}, C_2^1, \ldots, C_2^{k_2}, C_{d_2}, \ldots, C_{n+1}^{k_{n+1}})$ denote the elimination sets corresponding to the elimination ordering, and define the following collections

$$T_i^j = \{C_0, C_1^1, \dots, C_1^{k_1}, C_{d_1}, C_2^1, \dots, C_2^{k_2}, \dots, C_{d_{i-1}}, C_i^1, \dots, C_i^j\},$$

for $j > 0$, and

$$T_i^0 = \{C_0, C_1^1, \dots, C_1^{k_1}, C_{d_1}, C_2^1, \dots, C_2^{k_2}, \dots, C_i^{k_{i-1}}, C_{d_{i-1}}\}.$$

Let $\Phi[T_i^j]$ denote the product of the decision potentials on the elimination sets in T_i^j. Then we can augment the elimination tree solution algorithm with various assertions. Proof of every assertion proves the correctness of the algorithm. We assume for simplicity that collect operations and associated absorptions are implemented by passing a sequence of single messages in the elimination tree following the elimination ordering.

Algorithm 8.14 [ELIMINATION TREE SOLUTION]

for $i := n+1$ step -1 until 1 do
  {
    Assertion 1: $\Phi[T_i^{k_i}] = (P_i^{k_i}, V_i^{k_i}/P_i^{k_i})$.
    for $j := k_i$ step -1 until 1 do
      {
        Assertion 2: $\Phi[T_i^j] = (P_i^j, V_i^j/P_i^j)$.
        Equate $\phi_{sep(r_i^j)}$ to sum-marginal of the potential on C_i^j:
            $\phi_{sep(r_i^j)} := \mathbf{S}_{r_i^j}\, \phi_{C_i^j}$.
        Assertion 3: $\Phi[T_i^{j-1}]\phi_{sep(r_i^j)} = (P_i^{j-1}, V_i^{j-1}/P_i^{j-1})$;
        Multiply $\phi_{sep(r_i^j)}$ into the elimination set to
        which C_i^j is connected toward the root:
            $\phi_{e(r_i^j)} := \phi_{e(r_i^j)}\phi_{sep(r_i^j)}$;
        Assertion 4: $\Phi[T_i^{j-1}] = (P_i^{j-1}, V_i^{j-1}/P_i^{j-1})$.
      }
    Assertion 5: $\Phi[T_i^0] = (P_i^0, V_i^0/P_i^0)$.
    if $i > 1$ then
      {
        Equate $\phi_{sep(d_{i-1})}$ to the max-marginal of the potential on $C_{d_{i-1}}$:
            $\phi_{sep(d_{i-1})} := \mathbf{M}_{d_{i-1}}\, \phi_{C_{d_{i-1}}}$,
        and store the corresponding decision mapping;
        Assertion 6: $\Phi[T_{i-1}^{k_{i-1}}]\phi_{sep(d_{i-1})} = (P_{i-1}^{k_{i-1}}, V_{i-1}^{k_{i-1}}/P_{i-1}^{k_{i-1}})$.
        Multiply $\phi_{sep(d_{i-1})}$ into the elimination set to
        which $C_{d_{i-1}}$ is connected toward the root:
            $\phi_{e(d_{i-1})} := \phi_{e(d_{i-1})}\phi_{sep(d_{i-1})}$;
        Assertion 7: $\Phi[T_{i-1}^{k_{i-1}}] = (P_{i-1}^{k_{i-1}}, V_{i-1}^{k_{i-1}}/P_{i-1}^{k_{i-1}})$.
      }
  }

Assertion 8: $\Phi[T_0^1] = (P_0^1, V_0^1)$. □

Proof. We show by double induction on i and j that each assertion is true.

- Assertion 1 is true when $i = n+1$, for which $P_{n+1}^{k_{n+1}} = P(R\colon D)$ and $V_{n+1}^{k_{n+1}} = P(R\colon D)U_n(R, D)$. Assume it is true for i. Showing that Assertion 7 then follows provides the inductive proof of this assumption.

- Assertion 2 is true when $j = k_i$, as this simply restates Assertion 1. Assume it is true for j. Showing Assertion 4 then follows provides the inductive proof of this assumption.

- Assertion 3: $\Phi[T_i^j] = \Phi[T_i^{j-1}]\phi_{C_i^j}$ by definition, and by Assertion 2 $\Phi[T_i^j] = (P_i^j, V_i^j/P_i^j)$. By the definition of sum-marginalization of decision potentials,

$$\mathop{\mathbf{S}}_{r_i^j} \Phi[T_i^j] = \left(\sum_{r_i^j} P_i^j, \frac{\sum_{r_i^j} P_i^j (V_i^j/P_i^j)}{\sum_{r_i^j} P_i^j} \right) = (P_i^{j-1}, V_i^{j-1}/P_i^{j-1}).$$

However, by Lemma 8.8,

$$\begin{aligned} \mathop{\mathbf{S}}_{r_i^j} \Phi[T_i^j] &= \mathop{\mathbf{S}}_{r_i^j} \Phi[T_i^{j-1}]\phi_{C_i^j} \\ &= \Phi[T_i^{j-1}] \mathop{\mathbf{S}}_{r_i^j} \phi_{C_i^j}. \end{aligned}$$

- Assertion 4 follows simply from Assertion 3, noting that $e(r_i^j) \in T_i^{j-1}$. Note also that Assertion 4 is Assertion 2 with the j index decremented by 1. Hence, by induction both assertions hold true for all j values (under the assumption of Assertion 1 being true for i.)

- Assertion 5 is simply Assertion 4 at the end of the j loop, and so is also true.

- Assertion 6: $\Phi[T_i^0] = \Phi[T_{i-1}^{k_{i-1}}]\phi_{C_{d_{i-1}}}$ by definition, and $\Phi[T_i^0] = (P_i^0, V_i^0/P_i^0)$ by Assertion 5. By the definition of max-marginalization of decision potentials,

$$\mathop{\mathbf{M}}_{d_{i-1}} \Phi[T_i^0] = (P_{i-1}^{k_{i-1}}, V_{i-1}^{k_{i-1}}/P_{i-1}^{k_{i-1}}),$$

which follows from the consistency requirement that P_i^0 is independent of d_{i-1}, and hence $P_{i-1}^{k_{i-1}} = P_i^0$. Also, from the consistency requirement, and using Lemma 8.9,

$$\begin{aligned} \mathop{\mathbf{M}}_{d_{i-1}} \Phi[T_i^0] &= \mathop{\mathbf{M}}_{d_{i-1}} \Phi[T_{i-1}^{k_{i-1}}]\phi_{C_{d_{i-1}}} \\ &= \Phi[T_{i-1}^{k_{i-1}}] \mathop{\mathbf{M}}_{d_{i-1}} \phi_{C_{d_{i-1}}}. \end{aligned}$$

- Assertion 7 is simply Assertion 1 with i decremented by 1, so follows from Assertion 6 on noting that $e(d_{i-1}) \in T_{i-1}^{k_{i-1}}$. Hence, Assertion 1 is true by induction.
- Assertion 8: This is simply Assertion 5 at the end of the i loop, noting that $P_1^0 = 1$.

Thus, because all assertions hold true during all phases of the algorithm, the correctness the algorithm is proven. □

8.5 Example: OIL WILDCATTER

We present a small example of a problem posed by Raiffa (1968) in the variation given by Shenoy (1992), which we shall refer to as OIL WILDCATTER.

> An oil wildcatter must decide either to drill (yes) or not drill (no). He is uncertain whether the hole is dry (Dry), wet (Wet) or soaking (Soak). ... At a cost of 10,000, the wildcatter could take seismic soundings which help determine the geological structure at the site. The soundings will disclose whether the terrain below has no structure (NoS) — that's bad, or open structure (OpS) — that's so-so, or closed structure (ClS) — that's really hopeful.

8.5.1 *Specification*

The decision network for OIL WILDCATTER representing the joint probability and utility is shown in Figure 8.1. (An equivalent influence diagram representation is shown in Figure 8.2.) It has six nodes in total: two utility nodes U_T (*Utility of testing*) and U_D (*Utility of drilling*); two decision nodes, T (*Test?*) and D (*Drill?*); and two random variable nodes, O (*Amount of oil*) and R (*Results of test*). The decision sequence for this

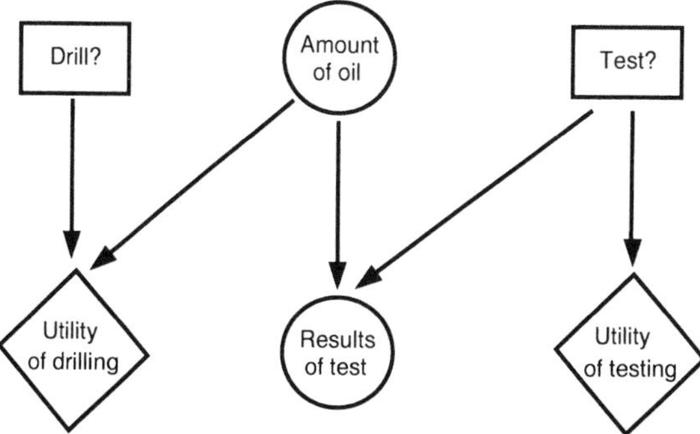

FIGURE 8.1. Oil wildcatter problem represented by a decision network; variables will be denoted by T (*Test?*), D (*Drill?*), O (*Amount of oil*), and R (*Results of test*); and utility nodes by U_T (*Utility of testing*) and U_D (*Utility of drilling*). There is a unique decision sequence given by $(TRDO)$.

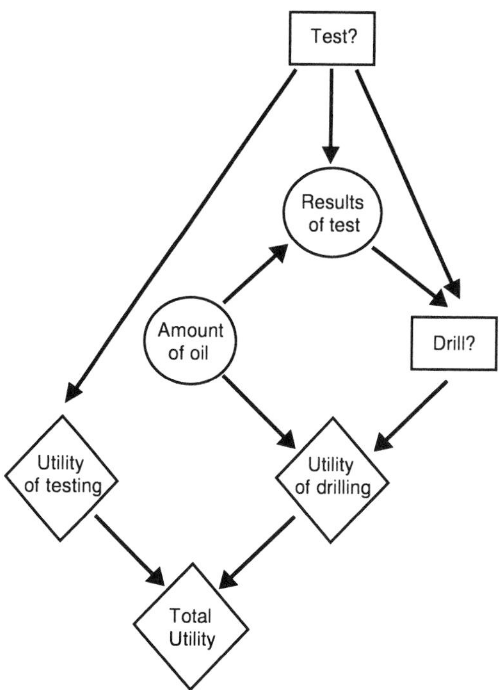

FIGURE 8.2. Oil wildcatter problem represented by an influence diagram. The total utility is represented as dependent upon the two components of the utility.

example is given uniquely by $(TRDO)$. That is, one first decides whether to test the ground or not. If the ground is tested, one knows the ground structure, otherwise one has no test result. Using this information one decides next whether to drill or not. If one drills the amount of oil in the ground is then revealed. The utility associated with T is the (negative) cost of the test, and this depends upon no other variables. The utility associated with the drilling decision D is the difference of the revenue from the amount of oil found and the cost of drilling, and thus is zero if no drilling is done. It depends upon O and D only. There is a given prior associated with the amount of oil in the ground, and a conditional distribution associated with the test result R, which depends upon the amount of oil and whether the ground test is performed.

Logically, if no ground test T is performed, then there can be no test result. The traditional approach adopted in influence diagrams is to add a fourth state, "no result" or NoR, to the state space of R, to reflect this logical requirement. This renders asymmetric problems symmetric, but is also more inefficient than decision trees, because of the larger state spaces which need to be considered. We shall return to this point in Section 8.8. With this extra state for R, the conditional probabilities and utility factors are given in Table 8.1 and Table 8.2 respectively, using the numbers taken from Shenoy (1992), but with utilities scaled down by a factor of 1000.

TABLE 8.1. Conditional probabilities for OIL WILDCATTER.

$P(O)$	O
0.5	Dry
0.3	Wet
0.2	Soak

$P(R \mid T, O)$ R					
NoS	OpS	ClS	NoR	T	O
0.6	0.3	0.1	0	yes	Dry
0.3	0.4	0.3	0		Wet
0.1	0.4	0.5	0		Soak
0	0	0	1	no	Dry
0	0	0	1		Wet
0	0	0	1		Soak

TABLE 8.2. Utility component tables $U_T(T)$ and $U_D(O, D)$ for OIL WILDCATTER.

$U_T(T)$	T
-10	yes
0	no

$U_D(O, D)$ O			
Dry	Wet	Soak	D
-70	50	200	yes
0	0	0	no

8.5.2 Making the elimination tree

Moralization of the graph in Figure 8.1 is simple. One adds an edge between O and T, and another between O and D, and then deletes the two utility nodes. Triangulating the moral graph using the decision sequence $(TRDO)$ as discussed in Section 8.4.3 leads to the rooted elimination tree shown in Figure 8.3, having root $\emptyset$.

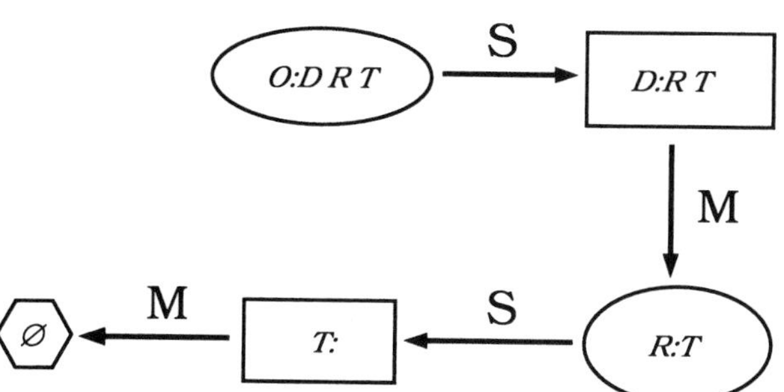

FIGURE 8.3. Elimination tree for the Oil wildcatter decision network. Oval nodes denote elimination sets in which a random variable is eliminated, rectangular nodes denote elimination sets in which a decision node is eliminated. The **S** and **M** symbols indicate respectively sum-marginalization and max-marginalization operations in performing Collect Evidence from the root $\emptyset$. Arrows on the edges indicate the direction of absorption.

TABLE 8.3. Initial non-vacuous elimination set potentials before propagation, zero probability entries removed, for the elimination tree elimination sets $ODRT$ and T. Each pair represents the probability and utility parts. The initialization of the elimination sets DRT and RT leads to vacuous potentials, which are omitted.

		O						
Dry		Wet		Soak		T	R	D
0.3	-70	0.09	50	0.02	200	yes	NoS	yes
0.3	0	0.09	0	0.02	0			no
0.15	-70	0.12	50	0.08	200		OpS	yes
0.15	0	0.12	0	0.08	0			no
0.05	-70	0.09	50	0.1	200		ClS	yes
0.05	0	0.09	0	0.1	0			no
0.5	-70	0.3	50	0.2	200	no	NoR	yes
0.5	0	0.3	0	0.2	0			no

		T
1	-10	yes
1	0	no

8.5.3 Initializing the elimination tree

Table 8.3 shows one possible initialization of the elimination sets of the elimination tree using the values from Table 8.1 and Table 8.2. The only freedom in the initialization process is that of selecting the elimination set into which to incorporate the utility factor $U_T(T)$. This could be put into any elimination set; here we choose that associated with eliminating T.

TABLE 8.4. Decision potentials to two decimal places and with zero probability entries removed on the elimination tree for the root $\emptyset$ and its three nearest neighbours after a collect operation, showing the probability parts γ and utility parts μ. The probability part for T is independent of T, and the probability part of DRT is independent of D given R and T. Note that the elimination set RT is not sum-consistent with T because only a collect operation has been performed, and T was initialized with the utility $U_T(T)$ given in Table 8.3. In contrast the potential on TRD is max-consistent with the potential on TR because before the collect operation the potential on TR was vacuous. Note also that the root $\emptyset$ gives the maximum expected utility, 22.5.

γ	μ
1	22.50

γ	μ	T
1	22.50	yes
1	20.00	no

γ	μ	T	R
0.41	0	yes	NoS
0.35	32.86		OpS
0.24	87.50		ClS
1	20.00	no	NoR

γ	μ	T	R	D
0.41	-30.49	yes	NoS	yes
0.41	0			no
0.35	32.86		OpS	yes
0.35	0			no
0.24	87.50		ClS	yes
0.24	0			no
1	20.00	no	NoR	yes
1	0			no

8.5.4 Collecting evidence

Table 8.4 shows the decision potentials in the elimination sets and separators after a call of COLLECT EVIDENCE from the root $\emptyset$. This corresponds to (i) a summation over O, (ii) maximization over D, (iii) summation over R, and finally (iv) maximization over T. The potential for the leaf elimination set $ODRT$ remains unchanged from its initial value and so is not repeated from Table 8.3.

The optimal decision for T is found from examining the utility parts of the potential in the elimination set T. From Table 8.4 one sees that testing gives the larger expected utility, 22.50, as opposed to 20.00 for not testing. Thus, it is optimal to test the ground, and from the elimination set DRT in which the drilling decision for D is made we find it is optimal to drill if the test result R is OpS or ClS, and not to drill if NoS is the test result.

Note that the probability part of the potential of the elimination set T is independent of T, and that of the elimination set DRT is independent of D given R and T, which is consistent with (8.9).

8.6 Example: DEC-ASIA

We now present a second, larger, example which is the ASIA example from page 20, as extended by Goutis (1995) to include two decisions, which we shall refer to as DEC-ASIA:

> Shortness of breath (dyspnoea) may be due to tuberculosis, lung cancer, bronchitis, none of them or more than one of them but its presence or absence does not discriminate between the diseases. A recent visit to Asia increases the chances of tuberculosis, while smoking is known to be a risk factor for both lung cancer and bronchitis. Suppose a doctor must decide whether a patient arriving at a clinic is to be hospitalized or not. Before taking the decision the doctor can obtain information as to whether the patient has gone to Asia or suffers from dyspnoea, but other relevant factors like smoking history or the presence of any diseases are not known. It has also been suggested that it may be worthwhile to screen the patient by taking chest X-rays. The results of a chest X-ray do not discriminate between lung cancer or tuberculosis. Proponents of the test say that it should be carried out at least for the people that have visited Asia. If a test is carried out, the doctor has access to the results at the time he determines whether to hospitalize or not. If the patient suffers from tuberculosis or lung cancer, he can be treated better in hospital, but hospitalization of healthy individuals should be avoided. Taking X-rays is harmful in itself and the adverse effects are more severe if the patient suffers from tuberculosis.

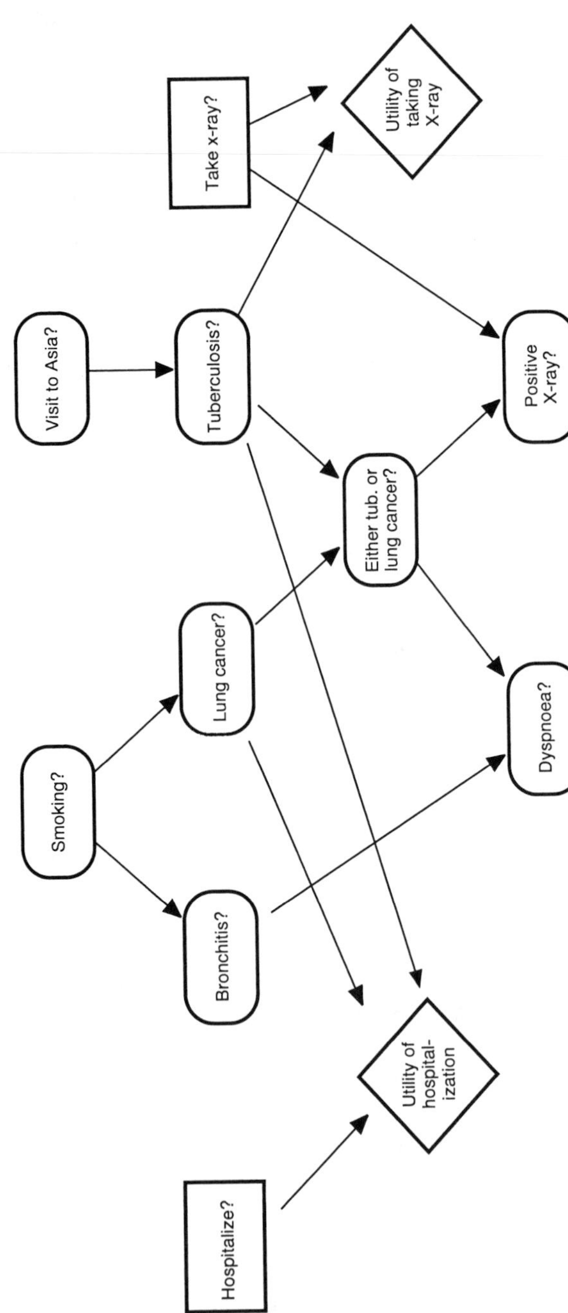

FIGURE 8.4. DEC-ASIA network, having the same form as the ASIA network but with two additional decision nodes and utility nodes.

The decision network for this example is shown in Figure 8.4. It has the same form as the ASIA example of page 20 but with the addition of the two decision nodes, the two utility nodes, and extra edges. All random and decision nodes are binary with the states (yes, no). The random nodes are denoted by upper case italic, and for brevity we use lower case to denote states, as in Section 2.9. Thus, for example, the *Asia?* node will be denoted by A and have the two states $(a, \bar{a})$ with $A = a$ denoting the "yes" state. The two decision nodes are *Hospitalize?*, which we denote by $\mathcal{H}$ with states $(h, \bar{h})$, and *Take X-ray?*, which we denote by $\mathcal{T}$ with states $(\tau, \bar{\tau})$.

The description of the decision problems admits several decision sequences. These depend on what information is available before a decision on taking an X-ray is made, viz. whether the patient has been to Asia, and whether the patient has dyspnoea. The description specifies that information on either of these *can* be obtained.

Let us suppose that it is known before deciding upon an X-ray test whether the patient has been to Asia, but that it it not known until after the test whether the patient has dyspnoea. Then the decision sequence for the problem is given by $(A, \mathcal{T}, \{D, X\}, \mathcal{H}, \{B, E, L, S, T\})$. The elimination tree is unbranched, like that for OIL WILDCATTER, but with 10 elimination sets; we omit a picture.

For the numerical specification we take the numbers from Goutis (1995). The values of probabilities and utilities are given in Table 8.5 and Table 8.6; they are for illustrative purposes only. However, there is one aspect of these tables that requires further elaboration.

TABLE 8.5. Conditional probability distributions for the DEC-ASIA example.

A:	$p(a)$	$=$	0.01	L:	$p(l \mid s)$	$=$	0.1
					$p(l \mid \bar{s})$	$=$	0.01
B:	$p(b \mid s)$	$=$	0.6	S:	$p(s)$	$=$	0.5
	$p(b \mid \bar{s})$	$=$	0.3				
D:	$p(d \mid b, e)$	$=$	0.9	T:	$p(t \mid a)$	$=$	0.05
	$p(d \mid \bar{b}, e)$	$=$	0.7		$p(t \mid \bar{a})$	$=$	0.01
	$p(d \mid b, \bar{e})$	$=$	0.8				
	$p(d \mid \bar{b}, \bar{e})$	$=$	0.1				
E:	$p(e \mid l, t)$	$=$	1	X:	$p(x \mid e, \tau)$	$=$	0.98
	$p(e \mid \bar{l}, t)$	$=$	1		$p(x \mid \bar{e}, \tau)$	$=$	0.05
	$p(e \mid l, \bar{t})$	$=$	1		$p(x \mid e, \bar{\tau})$	$=$	0.5
	$p(e \mid \bar{l}, \bar{t})$	$=$	0		$p(x \mid \bar{e}, \bar{\tau})$	$=$	0.5

TABLE 8.6. Utility tables for the DEC-ASIA example.

	$\mathcal{H}$	T	L
180	h	t	l
120			$\bar{l}$
160		$\bar{t}$	l
15			$\bar{l}$

	$\mathcal{H}$	T	L
2	$\bar{h}$	t	l
4			$\bar{l}$
0		$\bar{t}$	l
40			$\bar{l}$

	$\mathcal{T}$	T
0	τ	t
1		$\bar{t}$
10	$\bar{\tau}$	t
10		$\bar{t}$

As in the OIL WILDCATTER example, there is a test decision, here $\mathcal{T}$, that of taking an X-ray. Clearly if an X-ray is not taken then there can be no X-ray result. Thus, we have again an asymmetric decision problem, which we could model by giving the node X the three states $(x, \bar{x}, \text{NoR})$. As in the OIL WILDCATTER example, this would increase the complexity of the solution process, here by increasing the size of the state spaces of the elimination sets containing X by 50 percent. An alternative method avoiding this increase, due to Goutis (1995), is adopted here. This is to note that if no X-ray is taken, that is, $\mathcal{T} = \bar{\tau}$, then the value of X will never be known, and therefore its value cannot play a part in finding the optimal hospitalization strategy. Thus, following Goutis, we keep X a binary variable with states $(x, \bar{x})$, but make the conditional probability for $p(X \mid E, \mathcal{T} = \bar{\tau})$ independent of the state of E. Then if an X-ray test is not performed the computation of the optimal strategy will be independent of the value of X (because the product $p(A, B, D, E, L, S, T, X : \mathcal{T} = \bar{\tau}, \mathcal{H}) U(L, T, \mathcal{H}, \mathcal{T})$ will then be independent of X for fixed values of the remaining decision and random variables, and hence sum- and max-marginalizations will also be so independent). For illustration we set $p(x \mid e, \mathcal{T} = \bar{\tau}) = p(x \mid \bar{e}, \mathcal{T} = \bar{\tau}) = 0.5$, although any other non-zero value within the range (0,1) is permissible.

With these conditional probabilities and additive components of the utilities the elimination tree may be initialized and a call to COLLECT EVIDENCE initiated at the root. We refrain from listing all of the potentials on the tree. Instead we give in Table 8.7 the potentials of the root and its two nearest neighbours. From the root we see that the maximum expected utility is 47.49. From the set eliminating T we see that it it optimal to take an X-ray ($\mathcal{T} = \tau$) if and only if the patient has been to Asia.

Suppose, however, that the patient has not been to Asia, and (therefore optimally) that an X-ray was not taken. Then from Table 8.8, which lists the potential entries associated with the elimination set ($\mathcal{H}ADX\mathcal{T}$) having non-zero probability part, we can deduce the optimal decision regarding hospitalization. In this case it is optimal not to hospitalize regardless of whether the patient has dyspnoea or not. This deduction is independent of the value of X, consistent with our earlier setting $p(x \mid e, \mathcal{T} = \bar{\tau}) = p(x \mid \bar{e}, \mathcal{T} = \bar{\tau})$.

TABLE 8.7. Root elimination set and two nearest neighbours for the DEC-ASIA example, after initialization and a call to COLLECT EVIDENCE. The maximum expected utility is given by $0.01 \times 49.98 + 0.99 \times 47.46 = 47.49$.

γ	μ
1	47.49

γ	μ	A
0.01	49.98	a
0.99	47.46	$\bar{a}$

γ	μ	A	$\mathcal{T}$
1	49.98	a	τ
1	48.43		$\bar{\tau}$
1	46.99	$\bar{a}$	τ
1	47.46		$\bar{\tau}$

TABLE 8.8. Non zero probability potential entries, to two decimal places, on the elimination set $(\mathcal{H}ADX\mathcal{T})$ for which $A = \bar{a}$ and $\mathcal{T} = \bar{\tau}$. The optimal strategy is not to hospitalize, regardless of the value of D. Note that this strategy is independent of the state of X.

γ	μ	X	D	$\mathcal{H}$
0.22	41.72	x	d	h
0.22	45.27			$\bar{h}$
0.28	28.00		$\bar{d}$	h
0.28	49.15			$\bar{h}$
0.22	41.72	$\bar{x}$	d	h
0.22	45.27			$\bar{h}$
0.28	28.00		$\bar{d}$	h
0.28	49.15			$\bar{h}$

Consider now the alternative scenario in which, before making the X-ray test decision, the doctor has information about whether the patient has been to Asia *and* also whether the patient has dyspnoea. We may use the same decision network, Figure 8.4, but $(\{A, D\}, \mathcal{T}, X, \mathcal{H}, \{B, E, L, S, T\})$ is now the appropriate decision sequence. Under this decision sequence the elimination tree is different from before. In Table 8.9 we show the root elimination set and its three nearest neighbours after initialization and a call to COLLECT EVIDENCE. Note that the maximum expected utility has increased slightly, to 50.95, over the previous scenario value of 47.49. This is reasonable because we would not expect to make worse decisions on the basis of extra information, here knowing the state of D. Moreover, if the patient has been to Asia, then it is optimal to take an X-ray only if the patient has dyspnoea, in contrast to the previous scenario in which it was optimal to take an X-ray if and only if the patient had been to Asia.

TABLE 8.9. Root elimination set and three nearest neighbours after initialization and a call to COLLECT EVIDENCE for the decision network in Figure 8.4 with the decision sequence $(\{A, D\}, \mathcal{T}, X, \mathcal{H}, \{B, E, L, S, T\})$. The maximum expected utility in this case is given by $0.44 \times 53.29 + 0.56 \times 49.14 = 50.95$.

γ	μ
1	50.95

γ	μ	D
0.44	53.29	d
0.56	49.14	$\bar{d}$

γ	μ	D	A
0.00	58.04	d	a
0.43	53.24		$\bar{a}$
0.01	48.61	$\bar{d}$	a
0.56	49.15		$\bar{a}$

γ	μ	D	A	$\mathcal{T}$
0.45	58.04	d	a	τ
0.45	48.22			$\bar{\tau}$
0.44	53.24		$\bar{a}$	τ
0.44	45.27			$\bar{\tau}$
0.55	43.37	$\bar{d}$	a	τ
0.55	48.61			$\bar{\tau}$
0.56	42.16		$\bar{a}$	τ
0.56	49.15			$\bar{\tau}$

8.7 Triangulation issues

The constraints imposed by the decision sequence lead to less choice in triangulating the moral graph, thus making it a simpler problem to find an optimal tree for propagation. However, it is in the nature of multistage decision problems of this sort that the elimination set associated with the last decision variable, d_n, will contain most if not all of the other random and decision variables occurring before d_n in the decision sequence. This is because the final decision is made upon the basis of all previous decisions and revealed random variables and one would expect these to appear in the elimination set in order for d_n to be maximized over all configurations of these variables. Thus, one typically ends up with a large elimination set containing all of the decision variables, which is 'thinned out' in going to the root, as in Figure 8.3 and also the DEC-ASIA example.

However, not all problems yield elimination trees with such an elimination set. Some asymmetric problems, as treated in Section 8.8, provide one counterexample; another is the decision network shown in Figure 8.5 (Jensen et al. 1994). Nodes *d1* to *d4* are the decision nodes, *u1* to *u4* the additive components of the utility, and the remaining nodes the random variables. Using the elimination sequence deployed by Jensen et al. (1994) results in the elimination tree of Figure 8.6, a branched structure. This arises because, once the first decision is made for *d1*, the decision for *d3* is independent of the remaining decisions, as can be seen from the elimination tree. Similarly, once the decision for *d2* is taken, the optimal decision for *d4* then becomes independent of *d1*. An extreme case of such independent decisions occurs in Markov decision problems (Howard 1960; Puterman 1994). The elimination tree is constructed to have one elimination set for each random variable node or decision node in the network. This will generally lead to some elimination sets being subsets of others, and hence some redundancy. Such elimination sets can always be deleted and a junction tree of cliques constructed instead. This is the approach taken by Jensen et al. (1994). However, on such a rooted junction tree the marginalization of a clique to its separator closest to the root becomes more complicated when it involves marginalization over both decision and random variables. One must then perform the marginalization operations in stages within the cliques in a manner that respects the decision sequence. Thus, one saves storage space at the expense of extra bookkeeping. An intermediate approach is to merge sets if this will not lead to marginalization over both decision and random variables when propagating on the junction tree, which will save some storage.

It may also be possible to cope with large elimination sets or cliques by using clique-splitting approximations as discussed in Section 6.6. This would inevitably mean approximations are made, and hence possibly lead to sub-optimal strategies, but they might be close enough to optimal in some applications to be useful.

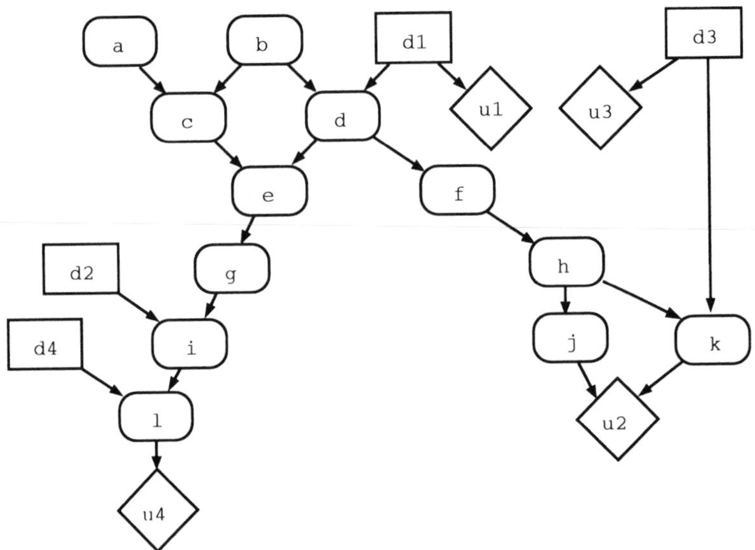

FIGURE 8.5. Decision network of the example due to Jensen et al. (1994).

8.8 Asymmetric problems

Shenoy (1996) has proposed an extension of the valuation approach for asymmetric decision problems which leads to more localized operations. This is achieved by exploiting extra factorization properties of valuations in asymmetric problems. We shall show how similar results can be obtained in the decision network/elimination tree, which we illustrate using the OIL WILDCATTER example.

Consider again the conditional probabilities $P(R \mid O, T)$. To cope with the possibility of not performing a test, T = no, we introduced an extra state for R, viz. "no result" NoR. This symmetrization procedure increases the size of the conditional probability table, which now has essentially two distinct parts depending upon the value for T. If T = no, we know that $P(R \mid O, T = \text{no})$ is zero unless $R = \text{NoR}$, and conversely that $P(R \mid O, T = \text{yes})$ is zero when $R = \text{NoR}$. Alternatively, one can say that knowing not to have a test result, i.e., $R = \text{NoR}$, implies knowing that a test has not been done, T = no, regardless of O, and conversely. We can represent this numerically by a factorization of the conditional probability table for $P(R \mid O, T)$ into two factors:

$$P(R \mid O, T) = f(R, T) g(R, O). \tag{8.19}$$

Table 8.10 shows one such factorization.

This factorization can be turned to advantage in generating the elimination tree. Recall that the first step in forming the elimination tree is

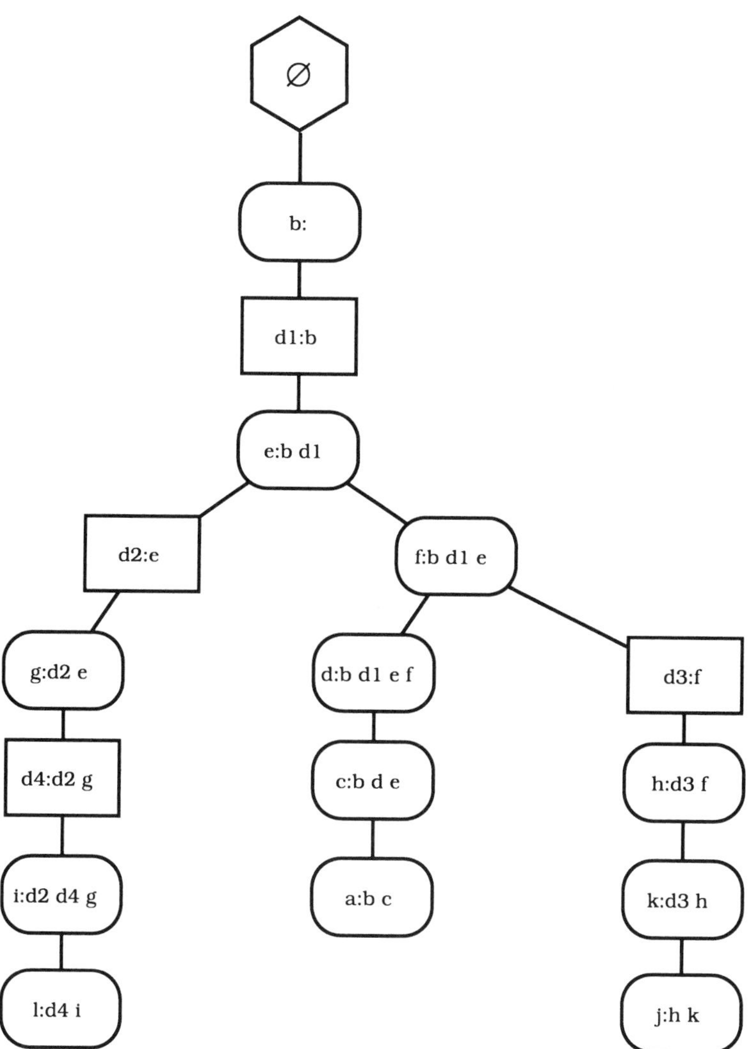

FIGURE 8.6. Elimination tree obtained from Figure 8.5 using the elimination ordering taken from Jensen et al. (1994).

TABLE 8.10. Possible factors $f(R,T)$ and $g(R,O)$ of $P(R \mid O,T)$ for the OIL WILDCATTER problem.

$f(R,T)$	R			
T	NoS	OpS	ClS	NoR
yes	1	1	1	0
no	0	0	0	1

$g(R,O)$	R			
O	NoS	OpS	ClS	NoR
Dry	0.6	0.3	0.1	1
Wet	0.3	0.4	0.3	1
Dry	0.1	0.4	0.5	1

to moralize the decision network. Now the reason for moralizing is to ensure that each conditional probability table or utility table can extend to an elimination set potential, so that it is possible to represent all of the factors of the overall decision potential on the elimination tree. However, because of the extra factorization of (8.19), it is unnecessary for R and both its parents O and T all to be in the same elimination set in order for $P(R \mid O,T)$ to be represented on the elimination tree potentials; it is sufficient for the pairs $\{O,R\}$ and $\{R,T\}$ each to be subsets of possibly distinct elimination sets. That is, in the moralization process for this example we can omit adding the link $O \sim T$ normally induced by R.

Proceeding with the triangulation leads to the elimination tree of Figure 8.7, in which we see T is now absent from two of the elimination sets. This result accords with the statement of the problem in Section 8.5: when the oil wildcatter comes to make a decision whether to drill or not, it is only on the basis of the test result, if one is available; it is irrelevant whether a test has or has not been performed (though this can be deduced logically from the type of test result). Thus, we obtain a more localized computation, and in the process reduce the sizes of the state spaces of the two largest elimination sets of Figure 8.3 by a factor of two. This observation will clearly extend to other problems in which symmetrization of asymmetric problems introduces extra states of variables. When this happens one checks for factorizations of the conditional probabilities of the chain components of random variables and adds only sufficient moral edges to ensure each factor can appear in a clique of the resulting moral graph. This in turn may lead to smaller elimination sets in the elimination tree. Clearly this idea is not restricted to decision problems; one could apply this economy to the simpler probability models of Chapter 6, where the general idea is to exploit any known factorization of prior conditional probabilities to reduce the number of edges required in the moralization stage.

The DEC-ASIA example of Section 8.6 showed yet another way of avoiding the introduction of a "no result" state; however, that approach might not carry over readily to chain graphs that have more complicated factorizations. Which of these methods is optimal can be expected to be problem–dependent.

8.9 Background references and further reading

Decision trees have their origin in game theory as developed by von Neumann and Morgenstern (1944). They are one of the oldest graphical approaches in decision analysis; for a thorough development of the theory see Raiffa and Schlaifer (1961) and Raiffa (1968). They are fine for small problems, but for larger problems they can become unwieldy, because they explicitly display all possible scenarios. In addition they require that the problem specification be first transformed into the decision tree representation, generally a non–trivial process.

Influence diagrams are directed acyclic graphs which represent decision problems more compactly than decision trees (see Howard and Matheson (1984); Shachter (1986); Oliver and Smith (1990); Smith (1989a); Smith (1989b); Tatman and Shachter (1990); and references therein). They model probabilistic dependence, utility dependence, and flow of information in a compact and intuitive graphical form. Originally developed as a structuring tool, the development of algorithms for their evaluation (using barren node elimination and arc-reversal) came later (Olmsted 1983; Shachter 1986). Solution of the influence diagram representation by transformation to a junction tree was described by Jensen et al. (1994), with an essentially identical solution method described by Cowell (1994). Shachter and Kenley (1989) extended influence diagrams from discrete probability problems to multivariate-Gaussian distributions and utilities of quadratic and exponential-quadratic form, and they were later extended by Poland (1994) to the mixed-discrete Gaussian case.

Goutis (1995) also represented a decision problem by a directed acyclic graph whose nodes are either random variables or decision variables. Decision variables can appear more than once in the graphical representation, either to represent a local factor of the overall utility or to represent probabilistic dependencies of other variables on decisions. In contrast to influence diagrams there are no information arcs. Instead, dynamic programming

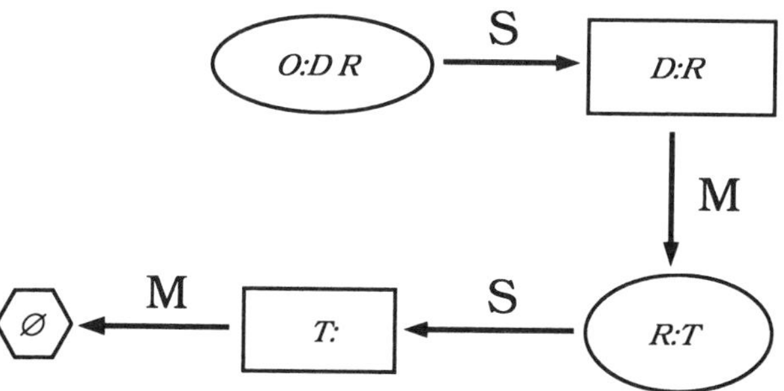

FIGURE 8.7. Elimination tree for the asymmetric oil wildcatter decision network using the modified moralization algorithm.

ideas are applied to solve the problem, using the ancestral subgraphs of decision variables to compute efficiently the contributions to expected utility.

Shenoy (1992) developed an alternative formulation based upon valuation systems. The representation and solution process takes place on a hypergraph tree structure called a *rooted Markov tree*, similar in many respects to the elimination tree, using marginalization and combination of valuations (which represent the probabilities and utilities) according to a *fusion* algorithm described in Shenoy (1992). The algorithm is slightly more efficient computationally than the standard arc-reversal and node-elimination method for solving influence diagrams. Later Shachter and Ndilikilikesha (1993) and Ndilikilikesha (1994) showed how to modify these standard algorithms to make them as computationally efficient as the fusion algorithm.

Which of the various approaches is the most efficient seems to be proble dependent; see Shenoy (1994) and Call and Miller (1990) for discussions of these issues. More recently Charnes and Shenoy (1997) have proposed an efficient forward–sampling method for solving decision problems.

Current research interest is focused on finding more efficient solutions for asymmetric decision problems. The valuation network method of Shenoy (1996) adapted to asymmetric problems is one such proposal; others include *contingent influence diagrams* (Fung and Shachter 1990) and *sequential decision diagrams* (Covaliu and Oliver 1995). A comparison of some methods is made by Bielza and Shenoy (1997).

A good introductory tutorial on decision theory is given by Henrion et al. (1991), while Chapter 3 of Shafer and Pearl (1990) contains some early papers on decision theory, with an historical overview by Shafer.

9
Learning About Probabilities

Previous chapters have been concerned with methods for probabilistic analysis in models for which both the graphical structure and the numerical inputs are completely specified. This and subsequent chapters are concerned with adapting models in the light of observed cases. Learning from data is at the forefront of much current research in graphical models. There are two distinct major tasks: estimating the probabilities in a given model, and searching among a collection of candidate models for one or more graphical structures consistent with the conditional independence relationships suggested by the data. In this chapter we deal with the case of a fixed graphical model, relaxing the requirement that the probabilities be specified precisely. Most of the discussion concerns discrete distributions over directed acyclic graphs.

9.1 Statistical modelling and parameter learning

A limitation of the theory presented so far is that it assumes that the prior probabilities necessary to initialize a system are specified precisely. This is liable to be unrealistic, whether the component conditional and unconditional probabilities are derived from subjective assessments or from specific data, and there are a number of reasons for wishing to be able to retain an explicit representation of the inevitable imprecision. First, this will allow the probabilistic predictions of a system to be tempered by an allowance for possible error. Secondly, it is only by allowing imprecision that the probabilities can be updated in the light of new data. Finally, the procedure for obtaining the probabilities becomes more acceptable if doubt about subjective assessments, or sampling error in data-based estimates, can be recorded and incorporated.

We take a standard statistical modelling approach. The distribution of the data is unknown, but is assumed to belong to some given family of possible distributions. We label the distinct members of this family by the value of a *parameter*, θ say. Thus, θ determines the unknown probabilities, and our task is to use the data to learn about θ. There are a variety of statistical methods for doing this, which can be classified according to the fundamental logical viewpoint out of which they arise. We shall focus here on two methods: *maximum likelihood estimation*, which is a 'classical' (frequentist) estimation method, relying on the data alone; and *Bayesian inference*, which incorporates any additional information there may be, and aims to quantify the remaining uncertainty about θ in probabilistic terms.

After describing the statistical structure of the directed Markov models on which we concentrate, we first address the case of learning from complete databases (where exact solutions are available), describing maximum likelihood in Section 9.3 and Bayesian inference in Section 9.4. Sections 9.6 and 9.7 extend these approaches to the case of learning from incomplete databases, while Section 9.8 describes how the stochastic simulation method of Gibbs sampling may be applied to such problems. Finally, undirected decomposable graphical models are briefly considered in Section 9.9. We do not treat more general (e.g., chain graph) models;these raise problems both of specification and of analysis, and parameter learning in such structures is a field very much in its infancy.

9.2 Parametrizing a directed Markov model

Suppose that we have a DAG $\mathcal{D}$ with node set V, a set of random variables $X = (X_v : v \in V)$, and a joint distribution for X which is directed Markov with respect to $\mathcal{D}$, so that its density factorizes as

$$p(x) = \prod_{v \in V} p(x_v \mid x_{\mathrm{pa}(v)}). \tag{9.1}$$

We are interested in the situation where the conditional probability tables $p(x_v \mid x_{\mathrm{pa}(v)})$ are not specified precisely, but are to be updated in the light of data. We thus consider the joint distribution as belonging to some specified family of possible distributions, labelled by an overall parameter θ. For any node v and configuration ρ of $X_{\mathrm{pa}(v)}$, we shall denote by $\theta^{v,\rho}$ the minimal function of θ determining the conditional distribution of X_v given $X_{\mathrm{pa}(v)} = \rho$; and by θ^v the minimal such function determining the whole collection of these conditional distributions for all possible configurations ρ of $X_{\mathrm{pa}(v)}$. Thus, we can consider $\theta = (\theta^v : v \in V)$, $\theta^v = (\theta^{v,\rho} : \rho \in \mathcal{X}_{\mathrm{pa}(v)})$.

Now (9.1) becomes

$$p(x \mid \theta) = \prod_{v \in V} p(x_v \mid x_{\mathrm{pa}(v)}, \theta^v) \tag{9.2}$$

$$= \prod_{v \in V} p(x_v \mid x_{\mathrm{pa}(v)}, \theta^{v, x_{\mathrm{pa}(v)}}), \tag{9.3}$$

the last expression containing no term involving $\theta^{v,\rho}$ unless the parent configuration is $\rho = x_{\mathrm{pa}(v)}$.

We note that essentially the same factorizations hold for an incompletely observed case, where the value x_v of X_v is observed only for $v \in A$, with A an ancestral set in $\mathcal{D}$. In this case, for $v \notin A$ the terms involving θ^v in (9.2), or $\theta^{v,\rho}$ in (9.3), are simply replaced by the constant 1.

In general, the various component parameters may be related. For example, genetic pedigree networks may require the parameters (θ^v) to be identical for different nodes, to describe identical inheritance patterns in different families. Or, with continuous variables, the distributions of X_v given $X_{\mathrm{pa}(v)}$ might be modelled by a homoscedastic linear regression, so that for every configuration ρ of $X_{\mathrm{pa}(v)}$, $\theta^{v,\rho}$ would consist of the identical set of regression and residual variance parameters. However, analysis is greatly simplified when we can treat the component parameters as distinct. We say that the model exhibits *global meta independence* (Dawid and Lauritzen 1993) if there are no logical constraints linking the $(\theta^v : v \in V)$; and *local meta independence* if, further, this property extends to the collection of all the $(\theta^{v,\rho} : v \in V,\ \rho$ a configuration of $X_{\mathrm{pa}(v)})$.

The following simple example will form the basis of most of our further analysis in this chapter.

Example 9.1 (*General discrete DAG model*). This model, with all variables discrete, consists of *all* possible distributions P having density p factorizing, as in (9.1), according to the DAG $\mathcal{D}$. We can take the parameter θ as P itself, with $\theta^{v,\rho} = P(X_v = \cdot \mid X_{\mathrm{pa}(v)} = \rho)$, etc. For a particular configuration ρ of $X_{\mathrm{pa}}(v)$, we let $\theta_j^{v,\rho}$ denote $P(X_v = j \mid X_{\mathrm{pa}(v)} = \rho, \theta)$, so that, if X_v has states labelled $1, \ldots, k$, then $\theta^{v,\rho} = (\theta_1^{v,\rho}, \ldots, \theta_k^{v,\rho})$. It is readily seen that this model exhibits local (and thus global) meta independence.

Let $x \in \mathcal{X}$ denote a typical configuration of states for all the variables X. Suppose we have a sample of n independent and identically distributed cases, with $x \in \mathcal{X}$ having multiplicity $n(x)$; for $A \subseteq V$ we similarly write $n(x_A)$ for the number of cases observed to have $X_A = x_A$. The numbers $n(x)$ and $n(x_A)$ will be referred to as *counts* and *marginal counts* respectively. Then we may write the likelihood of θ given the data as

$$\begin{aligned} L(\theta) &= \prod_{x \in \mathcal{X}}^{n} p(x)^{n(x)} \\ &= \prod_{x \in \mathcal{X}} \prod_{v \in V} p(x_v \mid x_{\mathrm{pa}(v)})^{n(x)} \\ &= \prod_{v \in V} \prod_{x_{\mathrm{fa}(v)} \in \mathcal{X}_{\mathrm{fa}(v)}} p(x_v \mid x_{\mathrm{pa}(v)})^{n(x_{\mathrm{fa}(v)})} \\ &= \prod_{v \in V} \prod_{x_{\mathrm{fa}(v)} \in \mathcal{X}_{\mathrm{fa}(v)}} (\theta_{x_v}^{v, x_{\mathrm{pa}(v)}})^{n(x_{\mathrm{fa}(v)})}. \end{aligned} \tag{9.4}$$

This factorizes into terms for each of the parameters $\theta^{v,\rho}$.

We note that the identical formula will apply if the cases are possibly incomplete, but each is complete on some ancestral subset. □

9.3 Maximum likelihood with complete data

A very general classical statistical technique for estimating the parameters of a model from data is the *maximum likelihood* procedure, which consists of calculating the likelihood on the basis of the data (i.e., the probability of the observed data as a function of the unknown parameters) according to the model, and finding the values of the parameters that maximize it. For complete data with no constraints relating the component parameters, the calculation for finding the maximum decomposes into a collection of local calculations, as we shall see.

Consider the directed Markov model of Section 9.2. Under global meta independence, for a given complete case x the factors in the likelihood (9.2) vary independently, so that the product will be maximized by maximizing each term separately; a corresponding property holds for local meta independence, using the factorization (9.3). These properties will extend to the situation that the likelihood is based on a database of cases which are complete or observed on an ancestral set.

Under global meta independence the maximum likelihood estimate of θ^v can thus be found by restricting attention to the component likelihood consisting of the product, over the different cases, of terms of the form $p(x_v \mid x_{\mathrm{pa}(v)}, \theta^v)$. Under local meta independence, for estimation of $\theta^{v,\rho}$ the relevant component likelihood is just the product of terms $p(x_v \mid x_{\mathrm{pa}(v)}, \theta^{v,\rho})$ over those cases x having $x_{\mathrm{pa}(v)} = \rho$.

Example 9.2 The model of Example 9.1 exhibits local meta independence. So the maximum likelihood estimate $\hat{\theta}^{v,\rho}$ can be found locally. From (9.4) it is readily seen that the local estimates, based on a database of (complete or ancestrally observed) cases, are given for all $v \in V$ and $x_{\mathrm{fa}(v)} \in \mathcal{X}_{\mathrm{fa}(v)}$ by

$$\hat{\theta}_{x_v}^{v,x_{\mathrm{pa}(v)}} = \frac{n(x_{\mathrm{fa}(v)})}{n(x_{\mathrm{pa}(v)})}, \tag{9.5}$$

i.e., the conditional probabilities are estimated by the ratio of the corresponding counts. □

9.4 Bayesian updating with complete data

The Bayesian approach to learning about the probabilities for a model is to express uncertainty about the probabilities in terms of a *prior distribution* for parameters determining those probabilities, and from the data and the prior calculate the *posterior distribution* using Bayes' theorem. Often a *conjugate prior* is taken as a convenient approximation to facilitate analysis. Informally, a prior distribution for the parameter of a statistical model is said to be conjugate when, for any data, the posterior distribution is of the same functional form. In general, an exponential family of distributions for complete data admits tractable conjugate priors (DeGroot 1970), such that the prior-to-posterior calculation can be performed easily. The family of Dirichlet distributions is conjugate for multinomial sampling, and its specialization, the Beta family, for binomial sampling. Prior-to-posterior analysis using these distributions is illustrated in Appendix A. In general, conjugate priors provide a convenient representation of prior information, readily amenable to updating with complete data. For some problems mixtures of conjugate priors may provide better modelling of prior information, and for complete data these are again reasonably tractable.

9.4.1 Priors for DAG models

Consider the DAG model of Section 9.2. For a Bayesian analysis, we need to specify an appropriate prior for the parameter θ. A major simplifying assumption (Spiegelhalter and Lauritzen 1990) is that of *global independence*: the parameters $(\theta^v : v \in V)$ are taken to be *a priori* probabilistically independent, so that the joint prior density factorizes as

$$p(\theta) = \prod_v p(\theta^v). \tag{9.6}$$

Of course, this assumption only makes sense when the model is itself global meta Markov.

With the global independence assumption the joint density of case-variables X and parameters θ is given by

$$p(x,\theta) = \prod_v p(x_v \mid x_{\mathrm{pa}(v)}, \theta^v) p(\theta^v). \tag{9.7}$$

It readily follows that, in the posterior distribution given a complete case $X = x$, the (θ^v) are still independent, with marginal densities

$$p(\theta^v \mid x) \propto p(\theta^v) p(x_v \mid x_{\mathrm{pa}(v)}, \theta^v). \tag{9.8}$$

The parameter θ^v may be considered, formally, as another parent of v in the network. This is illustrated in Figure 9.1 for the ASIA example (see page 20); for example, θ^B is a random quantity whose realization determines the conditional probability table $p(B \mid S)$, i.e., the incidence of bronchitis in smokers and non-smokers. Using such an extended DAG the posterior independence of all the parameter nodes, given all the case variables X, readily follows from Corollary 5.11, as does the dependence of the posterior distribution of θ^v on the values for $X_{\mathrm{fa}(v)}$ alone, since the Markov blanket $\mathrm{bl}(\theta^v)$ of θ^v is just $\mathrm{fa}(v)$.

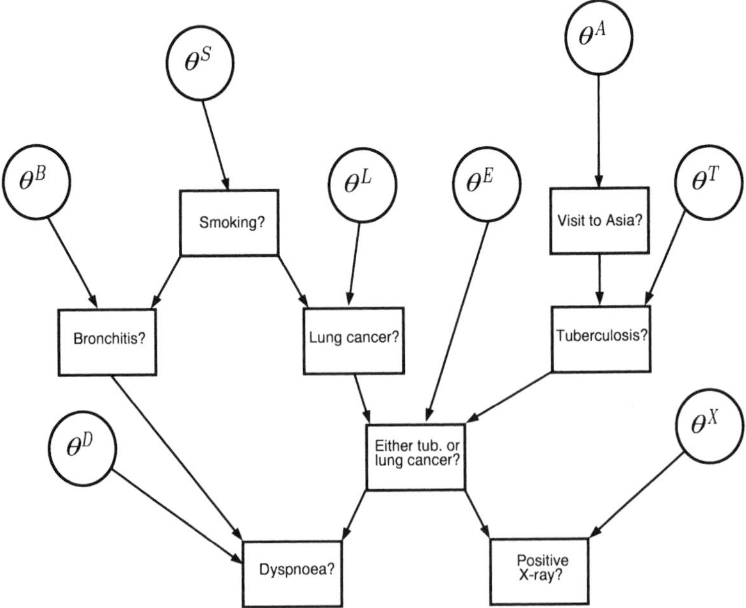

FIGURE 9.1. ASIA network with supplementary 'parameter' nodes representing marginally independent random quantities $\theta^v, v \in V$, whose realizations specify the conditional probability tables for the network.

Now under global independence

$$\begin{aligned} p(x) &= \int p(x, \theta)\, d\theta \\ &= \int \prod_v p(x_v \mid x_{\mathrm{pa}(v)}, \theta^v) p(\theta^v)\, d\theta^v = \prod_v p(x_v \mid x_{\mathrm{pa}(v)}), \end{aligned} \tag{9.9}$$

where

$$\begin{aligned} p(x_v \mid x_{\mathrm{pa}(v)}) &= \int p(x_v \mid x_{\mathrm{pa}(v)}, \theta^v) p(\theta^v \mid x_{\mathrm{pa}(v)})\, d\theta^v \\ &= \int p(x_v \mid x_{\mathrm{pa}(v)}, \theta^v) p(\theta^v)\, d\theta^v \end{aligned} \tag{9.10}$$

is the expectation of the conditional probability table for node v. For the last equality we have used the property $\theta^v \perp\!\!\!\perp \mathrm{pa}(v)$, which again follows on applying Corollary 5.11 to an extended DAG such as that of Figure 9.1. Hence, when processing a new case we can simply use the current 'mean probabilities' to initialize the standard evidence propagation techniques (see Chapter 6). This will apply both before observing any data, in which case the expectation in (9.10) is taken over the prior distribution for θ^v; and after observing data, using the current posterior distribution, so long as this exhibits global independence.

A further simplification is obtained if the model exhibits local meta independence, and in the prior we are willing to assume *local independence* (Spiegelhalter and Lauritzen 1990), meaning that all local parameters $(\theta^{v,\rho})$ in the network are mutually probabilistically independent. In this case, since the likelihood (9.3) exhibits a complete factorization in terms of the $(\theta^{v,\rho})$ while the prior density for the $(\theta^{v,\rho})$ likewise factorizes, it is straightforward to see that the posterior also retains local independence, again with localized updating (and, in fact, no updating at all of $\theta^{v,\rho}$ when processing a case that does not have ρ as the parent configuration of node v). If in addition the local priors are chosen to have conjugate form, then under complete data the local posteriors can be calculated by standard conjugate analysis.

If, rather than being complete, the set of observed nodes for each case forms an ancestral set, then updating can still be done locally, with the distributions for the parameters not involved in the ancestral set unchanged; if either global or local independence holds in the prior, the same property will be retained in the posterior. However, if the data are not complete on an ancestral set, then the posterior will not in general preserve prior local or global independence.

We illustrate these points with a simple example.

Example 9.3 Consider the two-node network shown in Figure 9.2. Each variable X and Y is binary with states $(x, \bar{x})$ and $(y, \bar{y})$, respectively, and the parameters are θ, ϕ, and ψ.

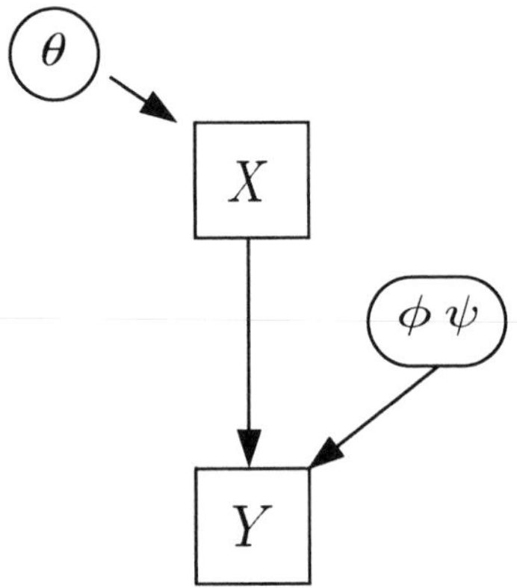

FIGURE 9.2. Simple two-node network showing parametrization.

The parametrization of the probabilities is given by

$$\begin{aligned} P(X = x \mid \theta, \phi, \psi) &= \theta \\ P(Y = y \mid X = x, \theta, \phi, \psi) &= \phi \\ P(Y = y \mid X = \bar{x}, \theta, \phi, \psi) &= \psi. \end{aligned}$$

Thus, in the general notation, $\theta^X = \theta$, $\theta^Y = (\phi, \psi)$, with $\theta^{Y,x} = \phi$ and $\theta^{Y,\bar{x}} = \psi$. A prior for the parameters will exhibit global independence if $\theta \perp\!\!\!\perp (\phi, \psi)$; and local independence if also $\phi \perp\!\!\!\perp \psi$, so that its density factorizes as $p(\theta, \phi, \psi) = p(\theta)p(\phi)p(\psi)$. For illustration we shall assume prior local independence with the following specific (conjugate) Beta marginals (see Appendix A):

$$\begin{aligned} p(\theta) &= b(\theta \mid 2, 3) &\propto& \; \theta(1-\theta)^2, \\ p(\phi) &= b(\phi \mid 4, 2) &\propto& \; \phi^3(1-\phi), \\ p(\psi) &= b(\psi \mid 1, 2) &\propto& \; 1-\psi, \end{aligned}$$

so that the prior density has the form

$$p(\theta, \phi, \psi) = b(\theta \mid 2, 3)\, b(\phi \mid 4, 2)\, b(\psi \mid 1, 2).$$

There are four possibilities for observing complete data $\mathcal{E}$. One of these is $\mathcal{E} = (\bar{x}, y)$. The likelihood for such a case is given by $(1-\theta)\psi$, so that the posterior is given by

$$\begin{aligned} p(\theta, \phi, \psi \mid \mathcal{E}) &\propto (1-\theta)\, \psi\, p(\theta)\, p(\phi)\, p(\psi) \\ &= b(\theta \mid 2, 4)\, b(\phi \mid 4, 2)\, b(\psi \mid 2, 2), \end{aligned}$$

again exhibiting local independence. □

The above analysis is a special case of the following more general one.

Example 9.4 In the discrete model of Example 9.1, we take a locally independent prior, with $\theta^{v,\rho}$ having a Dirichlet distribution (Beta, if $k = 2$), $\mathcal{D}(\alpha_1^{v,\rho}, \ldots, \alpha_k^{v,\rho})$. We can think of the vector $\alpha^{v,\rho} := (\alpha_1^{v,\rho}, \ldots, \alpha_k^{v,\rho})$ as representing counts of past cases which are stored as a summary of our experience. The quantity $\alpha_+^{v,\rho} = \sum_{j=1}^{k} \alpha_j^{v,\rho}$ is the *precision* of this prior distribution.

We denote the density of the overall locally independent prior distribution by $k(\theta \mid \alpha)$. It is specified by the hyperparameter $\alpha = (\alpha^v : v \in V)$, with each α^v a table having an entry $\alpha_j^{v,\rho}$ for each configuration (j, ρ) of the variables $(X_v, X_{\text{pa}(v)})$. We call such a distribution for θ *product Dirichlet.*

We note that the marginal distribution of the next case X, integrating over the prior distribution of θ, is directed Markov and has conditional probability tables

$$p(X_v = j \mid X_{\text{pa}(v)} = \rho, \alpha) = \bar{\alpha}_j^{v,\rho}, \tag{9.11}$$

where $\bar{\alpha}_j^{v,\rho} := \alpha_j^{v,\rho} / \alpha_+^{v,\rho}$.

If we observe a complete case x, the posterior distribution is again product Dirichlet, with hyperparameter $\alpha \circ x$ given by

$$(\alpha \circ x)_j^{v,\rho} = \begin{cases} \alpha_j^{v,\rho} + 1, & \text{if } x_v = j, x_{\text{pa}(v)} = \rho \\ \alpha_j^{v,\rho}, & \text{otherwise.} \end{cases} \tag{9.12}$$

We can perform sequential Bayesian updating by modifying the hyperparameter using (9.12) at the introduction of each new case.

Alternatively we can perform batch Bayesian updating. Combining the likelihood (9.4) with the prior using Bayes' theorem will result in a product Dirichlet posterior, with hyperparameter $\alpha \circ n$, say, where

$$(\alpha \circ n)_{x_v}^{v, x_{\text{pa}(v)}} = \alpha_{x_v}^{v, x_{\text{pa}(v)}} + n(x_{\text{fa}(v)}).$$

Of course, this agrees with the final outcome of sequential Bayesian updating. □

9.4.2 *Specifying priors: An example*

In the analysis of the CHILD example in Chapter 3, Table 3.1 displayed point estimates obtained from the expert. In fact she provided, in addition, *ranges* to reflect her perceived imprecision. These are shown in Table 9.1. The judgments shown in Tables 3.1 and 9.1 can be thought of as representing prior beliefs concerning unknown frequencies, and we need to transform these to parametric prior distributions. A somewhat simplistic procedure is as follows.

The point values given in Table 3.1 are taken as the prior means and shown as the first column of Table 9.2. The range for each response is assumed to represent a two-standard deviation interval. Other values have been tried but good predictive performance has been found with this assumption (Spiegelhalter et al. 1994). So, for example, we assume that the range 0.05 to 0.20 given for $P(LVH? = \text{Yes} \mid \text{PFC}) = 0.10$ in Table 3.1 corresponds to a mean of 0.10 and standard deviation of 0.075. This would be obtained with a Beta distribution with parameters $(1.50, 13.50)$. Similarly, the range 0.70 to 0.99 for $P(LVH? = \text{No} \mid \text{PFC}) = 0.90$ translates to a $\text{Be}(0.33, 2.95)$ distribution. We take the minimum precision over the list of responses to decide the overall precision of the Beta or Dirichlet distribution, and so adopt the parameters shown in Table 9.1 — these are thought of as the implicit sample sizes. This approach is conservative; we want to use the expert's judgments to give our system a 'head start', but also to allow the accumulating data to be able to adapt those judgments reasonably rapidly if necessary. The interpretation of the ranges through standard deviations as described is somewhat questionable in cases where the Beta distributions are particularly skew, such as here, but the procedure is simple and seems to give plausible performance.

TABLE 9.1. Subjective assessments of conditional probability tables, with expressed imprecision, with their translation into implicit samples underlying a $\text{Be}(\alpha_1, \alpha_2)$ distribution (values given to one decimal place).

	LVH?			
Disease?	Yes		No	
	range	α_1	range	α_2
PFC	0.05–0.20	0.3	0.70–0.99	3.0
TGA	0.05–0.20	0.3	0.70–0.99	3.0
Fallot	0.05–0.20	0.3	0.70–0.99	3.0
PAIVS	0.70–0.99	3.0	0.05–0.20	0.3
TAPVD	0.02–0.08	1.2	0.90–0.99	21.3
Lung	0.05–0.20	0.3	0.70–0.99	3.0

	LVH Report?			
LVH?	Yes		No	
	range	α_1	range	α_2
Yes	0.80–0.99	18.3	0.02–0.15	2.0
No	0.01–0.10	1.2	0.90–0.99	21.3

9.4.3 Updating priors with complete data: An example

Suppose we accept the assessments made in Table 9.1, and then observe a case with *Disease?* = PFC who has *LVH?* = yes. By the results in Appendix A, the Be(0.3, 3.0) prior distribution is updated to a Be(1.3, 3.0) posterior distribution, which has mean 0.30 and standard deviation 0.20. If necessary new ranges for the conditional probabilities could be obtained using the same scheme as interpreted above. This would let $p(LVH? = \text{Yes} \mid \text{PFC})$ range in the interval 0.1 to 0.50 which is thus shifted quite a bit to the right of the original interval.

It is clear that a sequence of complete cases could be handled either sequentially or as a batch — the results would be identical.

Table 9.2 shows the results of updating the prior distributions shown in Table 9.1 after observing a hypothetical set of data in which the proportion of $LVH? = \text{Yes}$ is 20 percent for all diseases, so that the final estimates for $p(LVH? = \text{Yes} \mid \mathit{Disease?})$ indicate a range of possible consequences from observing a common proportion. The results for PFC, TGA, and Fallot, all of which have the same prior distribution, show the increasing influence of the data as the sample size increases, while the major shift in the estimated frequency of $p(LVH? = \text{Yes} \mid \text{PAIVS})$, from 0.90 to 0.48, reveals the strong influence of even a small sample when it conflicts with a weak prior.

TABLE 9.2. Results from updating conditional probabilities in the light of observed data. The first column contains the point estimates shown in Table 3.1, and the next two columns show data with common observed proportion (20 percent) of $LVH? = \text{Yes}$, but varying sample sizes. The parameters of the posterior Beta distributions are obtained by combining the prior distributions from Table 9.1 with the data, as described in Appendix A, while the final column shows the posterior means of the conditional probabilities, which are also the predictive probabilities, for each disease, that the next case will exhibit *LVH?* = Yes.

		Hypothetical data		Posterior parameters		
Disease?	Prior mean	yes	no	α_1	α_2	Posterior mean
PFC	0.10	1	4	1.3	7.0	0.16
TGA	0.10	4	16	4.3	19.0	0.18
Fallot	0.10	20	80	20.3	83.0	0.20
PAIVS	0.90	1	4	4.0	4.3	0.48
TAPVD	0.05	1	4	2.2	25.3	0.08
Lung	0.10	1	4	1.3	7.0	0.16

9.5 Incomplete data

We now turn to methods for learning from incomplete databases.

The quality of data in databases varies enormously. A case is called *complete* if every random variable has a state or value assigned to it. A database is complete if all cases in the database are complete. For an incomplete case, one or more of the random variables does not have a value associated with it; the data are *missing* or *unobserved* or *censored.* There may be structure, known or unknown, in the manner in which the data are missing. Some variables may never be observed, their values always missing from the database; such variables are called *hidden* or *latent* variables. In other cases, values of variables may be sometimes observed, sometimes missing; the assumption is that, while all variables do have a value for that case, for some reason some values were not recorded (or were recorded but lost). In such situations the data-gathering process is said to be subject to a *data-censoring* or *missing-data* mechanism. If the probability of observing the particular set of variables observed depends only upon the values of those variables, then the missing data are said to be *missing at random* (MAR). If this probability does not even depend on the observed values, then the data are *missing completely at random* (MCAR). Note that values for latent variables are MCAR because they are always missing, independent of the values of any variables. On the other hand some states of some variables may be more prone to being missing than others — as in the everyday expression "no news is good news". If there is such structure in the manner in which data are missing, then the data-censoring mechanism is called *informative*; otherwise, it is *non-informative.* In any database there may be cases of each type.

Most methods that have been developed for learning in probabilistic expert systems assume the MAR property for all incomplete cases. In this case (at any rate for the methods we consider here) the data can be analysed exactly as if the unobserved variables would never have been observed. However, the inference can be altered if one has information about the nature of the censoring mechanism. For example, suppose one is given the values of N throws of a loaded die, of which n_0 values are specified as missing, and one is interested in estimating the probabilities of throwing each of the various faces of the die. With the MAR assumption one would discard the n_0 missing values and base inferences on the remaining $N - n_0$ cases. However, if one were to be told that only throws showing an odd number were ever subject to censoring, the appropriate inference would be quite different.

The censoring mechanism is 'non-ignorable', or 'informative', when the fact that a case or variable is or is not observed is itself influenced by some unobserved quantity, over and above any dependence of this process on observed variables. For example, in a clinical trial some patients may be lost to follow-up because they are too ill to attend clinic. Treating the observed data as MAR will lead to biased conclusions. In principle it is possible to incorporate additional "selection variables" into one's model, thus modelling explicitly the dependence of the observation process on the values, observed or unobserved, for the variables in the case (Diggle and Kenward 1994; Best et al. 1996). One then needs to take account of the fact that the probability structure of the observed data has been modified by conditioning on the fact of selection. However, any conclusions drawn can be very sensitive to the specific modelling assumptions made concerning the selection process, and external evidence about this may be very weak. Cowell et al. (1993a) consider the implementation of directed graphical models incorporating selection in the context of biased reporting of adverse drug reactions. Spirtes et al. (1995) treat some general aspects of causal inference from observational data in the presence of latent variables and selection bias. This is currently an active area of research.

9.5.1 Sequential and batch methods

There are two main strategies for approaching the analysis of an incomplete database, according as the data are to be treated in a *sequential* or in a *batch* manner. Each approach has its merits and deficiencies.

Generally, updating with incomplete data using sequential methods will necessitate the introduction of approximations after updating with each new incomplete case, in order to make calculations tractable. The errors introduced by such approximations will tend to accumulate, and can render the final results dependent upon the order in which the cases in a database are processed, even though with a fully correct sequential updating the final result can not have such dependence.

In contrast, iterative batch methods tend to give more robust results, but they typically require many iterations and can take a long time to converge, and it may not be possible to demonstrate that convergence has occurred. Also, in general it is not possible to incorporate new cases efficiently into current estimates — if new cases come in then the whole iterative procedure may have to be repeated from the beginning. By their nature sequential updating methods do not have this problem.

Although there is no necessary connection between the strategy chosen and the method of inference, for our incomplete data problems it turns out that maximization of the likelihood or the posterior density is best approached by batch methods, whereas a fully Bayesian inference lends itself naturally to sequential methods.

9.6 Maximum likelihood with incomplete data

9.6.1 *The EM algorithm*

In this section we address the problem of maximum likelihood estimation from incomplete databases with data missing at random. In this case the maximum likelihood estimate typically cannot be written in closed form.

A general numerical iterative batch procedure for finding maximum likelihood estimates with incomplete data is the *EM algorithm*. Suppose that an underlying random variable X has density $f(x \mid \theta)$, but observation on X is incomplete, consisting only of the value $Y = y$ for a known function $Y \equiv g(X)$. Given an initial estimate θ, the 'E-step' entails forming the current expected value of the log-likelihood function

$$Q(\theta' \mid \theta) = \mathrm{E}_\theta\{\log f(X \mid \theta') \mid g(X) = y\}. \tag{9.13}$$

The 'M-step' then maximizes Q over θ', yielding the next estimate. The algorithm alternates between these two steps, until convergence is attained.

Lauritzen (1995) showed how evidence propagation on the junction tree can be exploited to perform the E-step of the EM algorithm for the discrete DAG models of Example 9.1, when not all variables are observed. This method generalizes the Baum–Welch algorithm (Baum 1972), which was introduced for the special case of *hidden Markov models*.

Let our initial estimated probability distribution be P. Then the E-step is equivalent to calculating the current expected marginal counts

$$n^*(x_a) := \mathrm{E}_P\{n(x_a) \mid \text{observed data}\}, \tag{9.14}$$

where the index a ranges over the set $\mathcal{A} = \{\mathrm{fa}(v) : v \in V\}$ of the node families of the DAG, and i ranges over all complete configurations, so that i_a ranges over all configurations of X_a. This is because the distributions form an exponential family, with log-likelihood linear in the set of marginal counts $\{n(x_a)\}$.

To perform this calculation, assume that, underlying the observed but incomplete data, there are N independent full cases $X^1, X^2, \ldots, X^N$. Let y^i denote the possibly incomplete observation on X_i. Define

$$\chi^\nu(x_a) = \begin{cases} 1, & \text{if } X^\nu_a = x_a \\ 0, & \text{otherwise.} \end{cases}$$

Then we have

$$\begin{aligned} n^*(x_a) &= \mathrm{E}_P\left\{\sum_{\nu=1}^{N} \chi^\nu(x_a) \,\middle|\, y^1,\dots,y^N\right\} \\ &= \sum_{\nu=1}^{N} \mathrm{E}_P\{\chi^\nu(x_a) \mid y^\nu\} \\ &= \sum_{\nu=1}^{N} P(X_a^\nu = x_a \mid y^\nu). \end{aligned} \tag{9.15}$$

Let $\mathcal{T}$ be a junction tree constructed from the DAG $\mathcal{D}$. Initialize its potentials with the conditional probability tables corresponding to the distribution P. For each case ν, one can efficiently calculate the relevant summand in (9.15) simultaneously for all a and x_a using the propagation algorithms of Chapter 6 to condition on the relevant evidence $g(X^\nu) = y^\nu$ for that case. Then the $(n^*(x_a))$ are obtained on accumulating the resulting clique probability tables over all cases, thus giving the estimated counts. For computational efficiency one would collect cases with identical observations y^ν, and only calculate the terms of (9.15) once for each of these identical cases.

The M-step then uses these estimated counts to construct a new distribution P', with density p' satisfying

$$Np'(x_a) = n^*(x_a),\ a \in \mathcal{A},\ x_a \in \mathcal{X}_a. \tag{9.16}$$

As in (9.5), we obtain the explicit solution

$$p'(x_v \mid x_{\mathrm{pa}(v)}) = n^*(x_{\mathrm{fa}(v)})/n^*(x_{\mathrm{pa}(v)}). \tag{9.17}$$

(In this formula $n^*(x_\emptyset) = N$ will appear in the denominator whenever the variable v has no parents.) These new conditional probabilities are then used to initialize the junction tree $\mathcal{T}$ for another estimation of the complete-data counts.

A procedure analogous to the above can also be used for conditional-Gaussian distributions, in which the propagation scheme of Chapter 7 can be used to calculate the conditional moments required in the E-step.

The EM algorithm is known to converge relatively slowly. The process can be speeded up by using gradient-descent search near the maximum, a process of the same order of complexity as calculating the E-step of the EM algorithm (Thiesson 1997).

The EM algorithm has many generalizations, including GEM algorithms (Thiesson 1991), which operate by not necessarily maximizing Q, but only finding a value of θ' that makes $Q(\theta' \mid \theta)$ strictly increase over $Q(\theta \mid \theta)$. Also, one can consider the 'complete data' for each case just to include the values for the ancestral set of the observed variables, an idea related to that of Geng et al. (1996) and Didelez and Pigeot (1998).

9.6.2 Penalized EM algorithm

When there are hidden variables, such as in the CHILD example, the likelihood function typically has a number of local maxima (Settimi and Smith 1998). Also, maximum likelihood can give results with unsuitably extreme probabilities at the hidden nodes in the network. In such cases it seems better to exploit prior information, and penalize the likelihood for deviating from the prior assessments. The EM algorithm can be adapted for maximizing such penalized likelihoods (Green 1990).

In the model analysed above, take a locally independent product Dirichlet prior distribution, so that the $p(\cdot \mid x_{\mathrm{pa}(v)})$ are independent and Dirichlet-distributed with parameters $\alpha(x_{\mathrm{fa}(v)})$. On multiplying the likelihood function by the prior density, the posterior mode can be found iteratively by replacing (9.17) with

$$p'(x_v \mid x_{\mathrm{pa}(v)}) = \frac{n^*(x_{\mathrm{fa}(v)}) + \alpha(x_{\mathrm{fa}(v)}) - 1}{n^*(x_{\mathrm{pa}(v)}) + \alpha(x_{\mathrm{pa}(v)}) - k},$$

(where k denotes the number of states of node v). If any $\alpha(x_{\mathrm{fa}(v)})$ is less than 1, the posterior distribution may not have a mode in the interior and the above expression can turn negative. It might thus seem more suitable to penalize the likelihood directly by interpreting the $\alpha(x_{\mathrm{fa}(v)})$ values as counts in a likelihood, leading to

$$p'(x_v \mid x_{\mathrm{pa}(v)}) = \frac{n^*(x_{\mathrm{fa}(v)}) + \alpha(x_{\mathrm{fa}(v)})}{n^*(x_{\mathrm{pa}(v)}) + \alpha(x_{\mathrm{pa}(v)})}.$$

The resulting estimates can also be interpreted as posterior modes if densities are calculated with respect to an underlying measure that is uniform on a logarithmic probability scale; alternatively, they are approximately equal to posterior means.

Table 9.3 shows some results using both batch and sequential learning procedure for direct posterior modes and approximate posterior means for the CHILD example reported in Spiegelhalter et al. (1993). Sequential learning is treated in detail in Section 9.7 below. We note the tendency for the posterior modes to be at extreme values.

9.7 Bayesian updating with incomplete data

When the data are incomplete, exact Bayesian learning of the parameters is typically computationally infeasible. In this case some approximations are required. In this section we develop the theory of sequential Bayesian updating with incomplete data in discrete directed graphical models, using Dirichlet priors. After presenting the general theory, and approximations

TABLE 9.3. Estimates, taken from Spiegelhalter et al. (1993) for the CHILD example, of conditional probabilities obtained from batch learning (using direct posterior mode and approximate posterior mean) and sequential learning (using a moment-matching approximation).

	LVH? = yes			
Disease?	Dir. post. mode	Appr. post. mean	Sequential	Seq. precision
PFC	0.056	0.088	0.062	16.4
TGA	0.000	0.016	0.011	40.1
Fallot	0.003	0.026	0.020	30.8
PAIVS	1.000	0.959	0.965	13.1
TAPVD	0.004	0.035	0.035	32.1
Lung	0.079	0.139	0.016	11.4

	LVH Report? = yes			
LVH?	Dir. post. mode	Appr. post. mean	Sequential	Seq. precision
Yes	0.937	0.916	0.913	29.6
No	0.020	0.018	0.021	112.9

for making the analysis tractable, we examine some results from numerical simulations.

With complete data, and assuming local independence in the prior, updating is straightforward whether done in batch or sequentially. However, with incomplete data, an exact Bayesian analysis would require consideration of all possible completions of the incomplete cases; even with a Dirichlet prior for a discrete model, a small network and a moderate amount of data, this would result in a mixture of Dirichlets with a computationally unmanageable number of terms.

Example 9.5 In Example 9.3 there are two ways in which a case can be incomplete, but complete on an ancestral set; this happens when only X is observed, either as x or as $\bar{x}$. Thus, suppose that $\mathcal{E} = x$ is observed; the likelihood is given by θ (under the MAR assumption), and the posterior density is then

$$\begin{aligned} p(\theta, \phi, \psi \mid \mathcal{E}) &\propto \theta\, p(\theta)\, p(\phi)\, p(\psi) \\ &= b(\theta \mid 3, 3)\, b(\phi \mid 4, 2)\, b(\psi \mid 1, 2). \end{aligned}$$

In either case the likelihood factorizes and local independence is retained.

However, this is not the case when we observe Y only, which can happen in two possible ways. Suppose we observe $\mathcal{E} = y$. The likelihood (again under the MAR assumption) is now given by $\theta\phi + (1-\theta)\psi$ and does not factorize into functions of the individual parameters. The posterior density is given by

$$\begin{aligned}p(\theta,\phi,\psi\,|\,\mathcal{E}) &\propto \{\theta\phi+(1-\theta)\psi\}\,p(\theta)\,p(\phi)\,p(\psi)\\ &\propto \frac{2}{5}\cdot\frac{2}{3}\,b(\theta\,|\,3,3)\,b(\phi\,|\,5,2)\,b(\psi\,|\,1,2)\\ &\quad+\frac{3}{5}\cdot\frac{1}{3}\,b(\theta\,|\,2,4)\,b(\phi\,|\,4,2)\,b(\psi\,|\,2,2)\\ &\neq p(\theta\,|\,\mathcal{E})\,p(\phi\,|\,\mathcal{E})\,p(\psi\,|\,\mathcal{E}).\end{aligned}$$

The posterior distribution of any parameter is now a proper mixture of Beta distributions; in addition, local independence has been lost. A similar result occurs for the incomplete observation $Y=\bar{y}$, which has likelihood $\theta(1-\phi)+(1-\theta)(1-\psi)$. Although the number of terms in the mixture is small in this example, in a general DAG there will be as many terms as there are completions of the minimal ancestral set of the case variables in the observed data, and approximations are necessary to make the analysis tractable. Some of these we discuss below in Sections 9.7.2, 9.7.3 and 9.7.4. □

9.7.1 *Exact theory*

Consider the discrete directed Markov model with a conjugate product Dirichlet prior density $k(\theta\,|\,\alpha)$, as in Example 9.4. Suppose we observe incomplete evidence $\mathcal{E}$ on a case. Under the MAR assumption the posterior density is given by

$$p(\theta\,|\,\mathcal{E},\alpha)=\sum_{x\,\Rightarrow\,\mathcal{E}} k(\theta\,|\,\alpha\circ x)\,p(x\,|\,\alpha)/p(\mathcal{E}\,|\,\alpha),$$

where the sum is over all complete configurations x consistent with $\mathcal{E}$.

We see that if we start with a single conjugate prior, after observing a case having incomplete evidence we will be left with a finite mixture of conjugate densities. Hence, we need to consider the more general case of updating such a mixture.

A *mixture prior* for θ has density of the form

$$p(\theta)=\sum_{\alpha\in\mathcal{A}} w(\alpha)k(\theta\,|\,\alpha). \tag{9.18}$$

It requires a finite collection $\mathcal{A}$ of hyperparameter values, with a weight $w(\alpha)$ for each $\alpha\in\mathcal{A}$. In general, such a prior does not satisfy global or local independence.

We can interpret a mixture prior as follows. We suppose that the hyperparameter α is itself chosen at random from $\mathcal{A}$, with probability distribution given by w; that, given α, θ has density $p(\theta\,|\,\alpha)=k(\theta\,|\,\alpha)$; and that, given both α and θ, the distribution of X is given by (9.3) so that, in particular,

$X \perp\!\!\!\perp \alpha \mid \theta$. All desired probability calculations can thus be based on the joint density

$$\begin{aligned} p(\alpha, \theta, x) &= p(\alpha)\, p(\theta \mid \alpha)\, p(x \mid \alpha, \theta) & (9.19)\\ &= w(\alpha)\, k(\theta \mid \alpha)\, p(x \mid \theta). & (9.20) \end{aligned}$$

Given a mixture prior (9.18), and possibly incomplete evidence $\mathcal{E}$, the posterior density of θ (under the MAR assumption) is given by

$$p(\theta \mid \mathcal{E}) = \sum_{\alpha \in \mathcal{A}} \sum_{x \,\Rightarrow\, \mathcal{E}} w(\alpha)\, p(x \mid \alpha)\, k(\theta \mid \alpha \circ x)/p(\mathcal{E}). \qquad (9.21)$$

We see that there is in general an increase in the number of terms in the mixture. As we process more and more incomplete cases, this number can increase exponentially. For computational feasibility, some form of approximation is required. Spiegelhalter and Lauritzen (1990) proposed approximating the mixture generated on processing each successive incomplete case by a single conjugate density. Here we present a generalization of their theory in which non-trivial mixtures (possibly with fewer terms) are retained, and in addition we impose on both prior and posterior the requirements of (1) global independence, or (2) local independence. We first discuss the way in which these independence requirements can be effected, returning later (Sections 9.7.4 and 9.7.6) to the issue of limiting the growth in the number terms in the mixture.

9.7.2 Retaining global independence

Suppose that the prior satisfies global independence, each component parameter θ^v having a marginal distribution of mixed conjugate form. After observing an incomplete case, the posterior will, typically, no longer be globally independent. However, we propose to reintroduce global independence by approximating the true joint posterior density by the product of the posterior marginal densities for the (θ^v). Thus, in Example 9.3, for the incomplete data $\mathcal{E} = \bar{y}$ this would mean approximating the true joint posterior density, given by the mixture

$$\begin{aligned} p(\theta, \phi, \psi \mid \mathcal{E}) \;\propto\;& b(\theta \mid 3,3)\, b(\phi \mid 4,3)\, b(\psi \mid 1,2)\\ &+ 3\, b(\theta \mid 2,4)\, b(\phi \mid 4,2)\, b(\psi \mid 1,3), \end{aligned}$$

by the product of its margins for θ and (ϕ, ψ):

$$\begin{aligned} p(\theta \mid \mathcal{E})\, p(\phi, \psi \mid \mathcal{E}) \;\propto\;& \{b(\theta \mid 3,3) + 3\, b(\theta \mid 2,4)\}\\ &\times \{b(\phi \mid 4,3)\, b(\psi \mid 1,2) + 3\, b(\phi \mid 4,2)\, b(\psi \mid 1,3)\}\,. \end{aligned}$$

In general, we consider a globally independent prior, each factor being a mixture of product Dirichlet distributions (although before processing a database of cases there might be only one term in each mixture). That is, we take a prior of the form

$$p(\theta) = \prod_v p(\theta^v),$$

with

$$p(\theta^v) = \sum_{\alpha^v} w(\alpha^v)\, k(\theta^v \mid \alpha^v). \tag{9.22}$$

Each $k(\theta^v \mid \alpha^v)$ consists of a product of Dirichlet densities, one for each $\theta^{v,\rho}$, where ρ is a configuration of $X_{\mathrm{pa}(v)}$. Although the weight function w may depend on the node v, we omit that from the notation.

After observing possibly incomplete evidence $\mathcal{E}$ on a case, the true posterior density, given by (9.21), will be approximated by the globally independent one given by the product of its marginal densities:

$$p(\theta \mid \mathcal{E}) \simeq \prod_v p(\theta^v \mid \mathcal{E}). \tag{9.23}$$

We thus need to compute $p(\theta^v \mid \mathcal{E})$ for each v. We restrict attention to a specific node v^*, and for notational simplicity introduce $\phi = \theta^{v^*}, \beta = \alpha^{v^*}$.

We can write the joint density of $(\beta, \phi, X_{\mathrm{fa}(v^*)})$ given the evidence $\mathcal{E}$, evaluated at $X_{\mathrm{fa}(v^*)} = f$, as

$$p(\beta, \phi, f \mid \mathcal{E}) = p(f \mid \mathcal{E})\, p(\beta|f)\, p(\phi \mid \beta, f), \tag{9.24}$$

where we have used the property $(\beta, \phi) \perp\!\!\!\perp \mathcal{E} \mid \mathrm{fa}(v^*)$, which is easily derived by applying Corollary 5.11 to the DAG now extended with nodes for both the (θ_v) and the (α_v). The term $p(\theta \mid \beta, f)$ is just the conjugate posterior $k(\phi \mid \beta \circ f)$, as given by (9.12); the other terms provide a posterior weight to be attached to this term in the overall mixture posterior for ϕ.

We now address calculation of these other terms. Clearly $p(f \mid \mathcal{E}) = 0$ for any configuration f of $X_{\mathrm{pa}(v^*)}$ which is inconsistent with the evidence $\mathcal{E}$, and such terms may be omitted. Otherwise, we can calculate this term as follows. Because the prior incorporates global independence, the marginal distribution of X, on collapsing out over both θ and α, is itself directed Markov, with conditional probability tables given by

$$P(X_v = j \mid X_{\mathrm{pa}(v)} = \rho) = \sum_{\alpha^v} w(\alpha^v) P(X_v = j \mid X_{\mathrm{pa}(v)} = \rho, \alpha^v), \tag{9.25}$$

where $P(X_v = j \mid X_{\mathrm{pa}(v)} = \rho, \alpha)$ is given by (9.11). In (9.25) we have used the property $\alpha^v \perp\!\!\!\perp \mathrm{pa}(v)$, which is easily seen (for example, by again applying Corollary 5.11 to the extended DAG). Also, $\mathrm{fa}(v)$ is contained in some clique of the junction tree. Thus, if we use the tables (9.25) to initialize the junction tree, and propagate the evidence $\mathcal{E}$, we can simultaneously calculate the term $p(f \mid \mathcal{E})$ for every configuration f of $X_{\mathrm{pa}(v^*)}$ for every node v^*.

Now consider the term $p(\beta \mid f)$ in (9.24). Express f as (j, ρ), with j the value of X_{v^*} and ρ the configuration of $X_{\mathrm{pa}(v^*)}$. Then, by Bayes' theorem and (9.11), and using $\beta \perp\!\!\!\perp \mathrm{pa}(v^*)$,

$$p(\beta \mid f) = p(\beta \mid j, \rho) \propto p(\beta|\rho)\, p(j \mid \rho, \beta) = w(\beta)/\bar{\beta}_j^\rho, \tag{9.26}$$

where $\bar{\beta}_j^\rho := \beta_j^\rho / \beta_+^\rho$. Now $p(\beta \mid f)$ can be found by normalizing (9.26) to sum to one over the collection of values for β.

Finally, we obtain the marginal posterior for $\phi = \theta^{v^*}$ as the mixture

$$p(\phi \mid \mathcal{E}) = \sum_f \sum_\beta p(f \mid \mathcal{E})\, p(\beta|f)\, k(\phi \mid \beta \circ f), \tag{9.27}$$

where the sum over f is restricted to configurations for $X_{\mathrm{pa}(v^*)}$ consistent with $\mathcal{E}$, and $p(f \mid \mathcal{E})$ and $p(\beta|f)$ are calculated as described above. Inserting this into (9.23) yields our globally independent approximate joint posterior for θ, and, using this as the updated prior, we can now process the next case.

9.7.3 Retaining local independence

We can alternatively approximate the true density of θ by the product of its marginals for each of the parameters of each parent configuration on each node, so imposing local independence. Thus, in Example 9.3 for incomplete data $\mathcal{E} = \bar{y}$ this would mean approximating the posterior density,

$$\begin{aligned} p(\theta, \phi, \psi \mid \mathcal{E}) \quad \propto \quad & b(\theta \mid 3, 3)\, b(\phi \mid 4, 3)\, b(\psi \mid 1, 2) \\ & + 3\, b(\theta \mid 2, 4)\, b(\phi \mid 4, 2)\, b(\psi \mid 1, 3), \end{aligned}$$

by the product

$$\begin{aligned} p(\theta \mid \mathcal{E})\, p(\phi \mid \mathcal{E})\, p(\psi \mid \mathcal{E}) \quad \propto \quad & \{b(\theta \mid 3, 3) + 3\, b(\theta \mid 2, 4)\} \\ & \times \{b(\phi \mid 4, 3) + 3\, b(\phi \mid 4, 2)\} \\ & \times \{b(\psi \mid 1, 2) + 3\, b(\psi \mid 1, 3)\}. \end{aligned}$$

In general, at any stage, the joint distribution of the parameters is approximated by one satisfying local independence and having marginals that are mixtures of Dirichlet distributions.

Now consider again the general discrete DAG model of Example 9.1, with a locally independent prior

$$p(\theta) = \prod_{v \in V} \prod_{\rho \in \mathcal{X}_{\mathrm{pa}(v)}} p(\theta^{v,\rho}), \tag{9.28}$$

each factor now being of mixed conjugate form

$$p(\theta^{v,\rho}) = \sum_{\alpha^{v,\rho}} w(\alpha^{v,\rho})\, k(\theta^{v,\rho} \mid \alpha^{v,\rho}). \tag{9.29}$$

We propose a similar locally independent approximation (9.28) to the posterior given incomplete evidence $\mathcal{E}$, with (9.29) replaced by the correct marginal posterior for $\theta^{v,\rho}$. We fix attention on a specific node v^* and a specific parent configuration $\rho^* \in \mathcal{X}_{\mathrm{pa}(v^*)}$, and, changing notation, now denote θ^{v^*,ρ^*} by ϕ, and α^{v^*,ρ^*} by β. Similar to (9.24), we have the joint density for $(\beta, \phi, X_{\mathrm{fa}(v^*)})$ given $\mathcal{E}$

$$p(\beta, \phi, f \mid \mathcal{E}) = p(f \mid \mathcal{E})\, p(\beta | f)\, p(\phi \mid \beta, f), \tag{9.30}$$

with $f = (j, \rho)$ a configuration for $X_{\mathrm{fa}(v^*)}$.

By local independence, $p(\phi \mid \beta, f)$ is just the component conjugate prior density $k(\phi \mid \beta)$ if $\rho \neq \rho^*$; while for $\rho = \rho^*$ this term becomes $k(\phi \mid \beta \circ f)$, where

$$(\beta \circ j)_{j'} = \begin{cases} \beta_{j'} + 1, & \text{if } j' = j \\ \beta_{j'}, & \text{otherwise.} \end{cases} \tag{9.31}$$

The term $p(f \mid \mathcal{E})$ can again be found simultaneously for all nodes v^* and all configurations f of $X_{\mathrm{fa}(v^*)}$ by propagation in the DAG, initialized as in (9.25). Also, $\beta \perp\!\!\!\perp \mathrm{pa}(v^*)$ as before, and in addition, by local independence, $\beta \perp\!\!\!\perp v^* \mid (X_{\mathrm{pa}(v^*)} = \rho)$, for $\rho \neq \rho^*$. It follows that $p(\beta \mid f) = p(\beta \mid j, \rho) = w(\beta)$ for $\rho \neq \rho^*$, while $p(\beta \mid j, \rho^*) \propto w(\beta)\, \bar{\beta}_j$, where $\bar{\beta}_j := \beta_j / \beta_+$.

Hence, (9.30) becomes

(i) $p(j, \rho \mid \mathcal{E})\, w(\beta)\, k(\phi \mid \beta)$ for $f = (j, \rho), \rho \neq \rho^*$;

(ii) $p(j, \rho^* \mid \mathcal{E})\, p(\beta \mid j, \rho^*)\, k(\phi \mid \beta \circ j)$ for $f = (j, \rho^*)$.

For fixed β, the sum of the weights on $k(\phi \mid \beta)$ over all terms of type (i) is $w(\beta)\,\{1 - P(X_{\mathrm{pa}(v^*)} = \rho^* \mid \mathcal{E})\}$, where $P(X_{\mathrm{pa}(v^*)} = \rho^* \mid \mathcal{E})$ is readily obtained from the same propagation used to calculate the $\{p(f \mid \mathcal{E})\}$ (or as $\sum_j p(j, \rho^* \mid \mathcal{E})$).

So finally the posterior $p(\phi \mid \mathcal{E})$ for $\phi = \theta^{v^*,\rho^*}$ is the mixture

$$\sum_j \sum_\beta p(j, \rho^* \mid \mathcal{E})\, p(\beta \mid j, \rho^*)\, k(\phi \mid \beta \circ j) + \{1 - p(\rho^* \mid \mathcal{E})\}\, p(\phi), \tag{9.32}$$

$p(\phi)$ being the original mixture prior for ϕ.

Note that if $\mathcal{E}$ implies a known configuration $\rho \neq \rho^*$ for $X_{\mathrm{pa}(v^*)}$, then the distribution of ϕ is not updated; nor is it updated if in $\mathcal{E}$ we observe $X_{\mathrm{pa}(v^*)} = \rho^*$, but we do not observe X_v for $v = v^*$ or any descendant of v^*.

We now obtain our locally independent approximation to the posterior density of θ by multiplying the marginals (9.32) over all parent configurations of all nodes, and we are then ready to process the next case.

For the conjugate case (i.e., a single term in the prior) (9.32) reduces to equation (11) of Spiegelhalter and Lauritzen (1990). For a node v with k states, and initially m terms in the mixture for each parent configuration, after processing an incomplete case this could grow to up to $(k+1) \times m$ terms. Taking all the parent configurations into account, the overall complexity is thus much greater than when retaining global independence only.

9.7.4 Reducing the mixtures

With general patterns of missing data (9.27) or (9.32) will need further approximation to prevent an explosion of terms. We concentrate here on some detailed suggestions for reducing the mixture (9.32); similar ideas apply to (9.27).

A related problem arises in the area of 'unsupervised learning': see, for example, Titterington et al. (1985) and Bernardo and Girón (1988), in which a Dirichlet mixture is approximated by a single Dirichlet distribution $\mathcal{D}(\alpha_1^*, \ldots, \alpha_k^*)$ with suitably chosen parameters. Techniques include the 'probabilistic teacher', in which the missing variables are randomly fixed according to their current posterior distribution, and 'quasi-Bayes' (Smith and Makov 1978) or 'fractional updating' (Titterington 1976). However, one can criticize these procedures on the grounds that, if the parent configuration of a node is observed, but there is no evidence observed on that node or any of its descendants, then the associated precision is nevertheless increased by one, even though in the exact analysis there should be no change (cf. Bernardo and Girón (1988)).

An alternative approach is that of the 'probabilistic editor' (Titterington et al. 1985), in which the approximating distribution attempts to match the moments of the correct mixture. There is a degree of arbitrariness as to exactly how this is done, since, while there are no constraints on the mean structure of a Dirichlet distribution, once this is specified the covariance structure is then determined by a single quantity, the overall precision.

Consider a density for $\phi = \theta^{v^*,\rho^*}$ of the general mixed conjugate form

$$p(\phi) = \sum_{\alpha} w(\alpha)\, k(\phi \mid \alpha). \tag{9.33}$$

This could be the prior, or the exact posterior (9.32). Let $m_j(\alpha)$, $\nu_j(\alpha)$ be the mean and variance of $\phi_j = \theta_j^{v^*,\rho^*}$ under $k(\phi \mid \alpha)$. Then the overall mean and variance of ϕ_j are given by

$$m_j = \sum_{\alpha} w(\alpha)\, m_j(\alpha)$$

and

$$\nu_j = \sum_{\alpha} w(\alpha) \left[\nu_j(\alpha) + \{m_j(\alpha) - m_j\}^2 \right].$$

Spiegelhalter and Lauritzen (1990) approximated the mixture (9.33) by a single conjugate term, chosen to have the same mean vector and the same value of the "mean-weighted average variance" $\overline{\nu} = \sum_j m_j \nu_j$. This reduction is applied after introducing each new case. One attractive feature of the method is that if no relevant evidence about ϕ is obtained in a case, so that the true posterior is the same as the prior, then this will also be the approximating posterior. Thus, the technique obeys the desiderata formulated by Bernardo and Girón (1988). It is even possible for the precision to *decrease* on obtaining evidence. This phenomenon occurs when an event that has a high estimated prior probability is contradicted by the available evidence, and this behaviour seems not unreasonable.

While the need to contain the combinatorial explosion of the mixtures is urgent, one might nevertheless feel that collapsing a mixture down to a single term might be too extreme; in particular, multi-modality and skewness information can be lost. Hence, we might instead attempt to reduce the mixture to a smaller but still non-trivial mixture. A related problem is encountered in sequential analyses of multi-process models for time series (West and Harrison 1989), for which good approximations have been achieved by collapsing a mixture with a large number of terms down to a mixture with far fewer terms by merging 'similar' components (West 1992). This approach was examined in the current context by Cowell et al. (1995) using a simple test model consisting of a single node having three possible states, for which incomplete information on a case consists of a finding that one of the three states has *not* occurred. For this very simple model an exact analysis could be performed for a database of 200 randomly incomplete cases, and compared with various mixture approximation methods. A general sequential reduction procedure based on the clustering of West (1992) was employed as follows.

Suppose a fixed limit of N is set for the number of terms allowed in a mixture, and that an exact update with an incomplete case results in a mixture with $n > N$ terms. To reduce the number of terms, repeatedly: (i) identify the term with smallest weight, (ii) pair it with the 'closest' of the remaining terms, then (iii) combine this pair of terms into single term (adding the weights). This is repeated until only N terms remain in the mixture.

Step (ii) requires some notion of a distance between two Dirichlet distributions. There is some arbitrariness in this; Cowell et al. (1995) used the simple squared Euclidean distance between the predictive means, $\sum_j (\bar{\alpha}_{1j} - \bar{\alpha}_{2j})^2$. For step (iii) there is a variety of combination methods that may be employed. Ideally the method should be *associative*, so that the final result does not depend on the order in which terms were collapsed. Since the moment-matching method described above does not have this property, it was replaced by a variant in which the precision is determined by matching the *unweighted* average second moment about zero. This is then associative. If the conjugate priors $k(\phi \mid \alpha)$ are Dirichlet distributions this leads to the choice $\mathcal{D}(\alpha_1, \dots, \alpha_k)$ where

$$\alpha_j = m_j \, \alpha_+, \quad \alpha_+ = \frac{\sum m_j(1 - m_j)}{\sum \nu_j} - 1. \tag{9.34}$$

Cowell (1998) has proposed another method for reducing mixtures, less ad hoc than moment-matching methods, but computationally more expensive. It is based on a minimizing procedure involving the Kullback–Leibler divergence between the prior and posterior predictive distributions of the two mixtures; for combining Beta distributions to a single Beta distribution the method reduces to the above variant of the moment-matching method. The theory is a special case of that presented by Cowell (1996) for matching priors between graphical models, as discussed in Section 11.6.3 below.

9.7.5 *Simulation results: full mixture reduction*

Spiegelhalter and Cowell (1992) examined the consequences, both theoretically and through simulation experiments, of the approximation method of Spiegelhalter and Lauritzen (1990), which collapses a mixture down to a single conjugate term using (mean-weighted) moment-matching. The simulations suggest the following: the procedure works well in the standard unsupervised learning problem in which the parent nodes are observed; may be inaccurate when trying to learn conditional probabilities with unobserved parent nodes; and could be potentially misleading for trying to learn about a link in a network on which no direct evidence is being observed. In the last case of systematically missing data the estimation procedure can be inconsistent, and the prior knowledge can have a strong influence. It then seems more appropriate to collapse out the nodes which are not observed, learn on the collapsed graph, and then transform back if necessary; see the discussion by Smith of Spiegelhalter and Cowell (1992) for details. However, for randomly missing data the approximation seemed at the time to work quite well. Similar conclusions were drawn by Olesen et al. (1992), who add an extra feature which they call *fading*: each time a new case is taken into account to update the prior, the equivalent sample size is discounted by a factor q, a positive real number less than but typically close to one. This gives a maximal asymptotic sample size of $1/(1-q)$. This extra factor essentially corresponds to flattening the density by raising it to the power q, known as power-steady dynamic modelling (Smith 1979, 1981). It makes the system forget the past at an exponential rate, thus helping it to adapt in changing environments.

Figure 9.3 shows results given by Spiegelhalter and Cowell (1992) for learning on the ASIA network for the *Bronchitis–Smoking* link. One thousand cases were generated at random, using the probabilities of Lauritzen and Spiegelhalter (1988), with the adaptation that the probability of bronchitis in a non-smoker was changed from 0.3 to 0.05. All the conditional probabilities are assumed correctly known, except those specified by θ^B, i.e., the incidence of bronchitis in smokers and in non-smokers. Initial prior means of 0.5 and precisions equal to 1 were used, and for each node each case was randomly and independently set missing with probabilities 0.2, 0.4, 0.6, and 0.8 respectively.

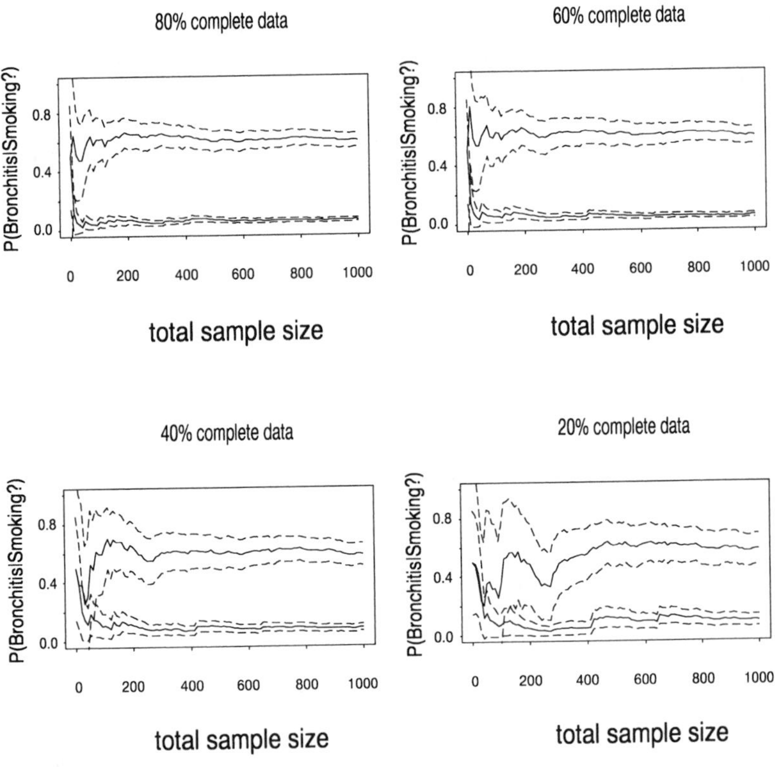

FIGURE 9.3. Simulation results based on different proportion of completeness of data for sequential learning reported by Spiegelhalter and Cowell (1992). Plots show the estimated proportions, with $\pm$ 2 posterior standard deviations, of bronchitis in smokers (upper curves) and non smokers (lower curves). The prior mean for each proportion was 0.5, with precision equal to 1.

9.7.6 Simulation results: partial mixture reduction

As described in Section 9.7.4, the simulation experiments of Cowell et al. (1995) were based on the simple test case of incomplete observations in the trinomial distribution, using sequential reduction of a mixture until it had a manageable number of terms. Various combination methods were investigated, including the deterministic methods of fractional updating and (unweighted raw) moment-matching, together with a number of stochastic methods. The stochastic methods were introduced as a possible way of avoiding the bias of the deterministic methods, but generally their performance proved disappointing. The simulation results indicated that direct moment-matching using (9.34) generally gives a very good approximation to the exact mixture distribution.

It is also of interest to examine the effects of the approximations of retaining global, or local, independence between cases. Intuitively, retaining only global independence should be better, but the complexity of the required calculations then becomes much greater in general. This may be surprising in view of the remark at the end of Section 9.7.3, that more terms are generated overall when local independence is enforced. But one must now consider the complexity of the reduction process. The sequential reduction of Section 9.7.4 requires an ordering of terms, which has complexity $O(n \log n)$, followed by a set of n pairwise comparisons for each clustering of two terms, hence $O(n^2)$ pairwise comparisons in achieving the desired clustering. Hence, the updating process may be dominated by this mixture reduction phase which will have complexity $O(n^3 \log n)$. However, the complexity will usually be much smaller if one retains local independence, because although the numbers of terms over all the parent configurations can be larger, each mixture that requires reduction will have a number of terms of the order of the number of states of the variable, instead of the number of configurations of the family. That is, instead of reducing one large mixture, one reduces a number of much smaller mixtures with, for each node v, a reduction in complexity given by a factor approximately equal to the square of the number of parent configurations of v.

We now describe a simulation study of approximate mixture learning to compare the effects of retaining global independence only with those of insisting on local independence. We use the Asia network (see page 20), in which all parent-child conditional probability tables are regarded as unknown parameters. From a complete database of 100 cases, simulated using the probability values of Lauritzen and Spiegelhalter (1988), incomplete databases corresponding to 75 percent complete, 50 percent complete and 25 percent complete were generated by randomly altering the known state on each variable to unknown with probabilities of 0.25, 0.5 and 0.75 respectively. Locally independent $\mathrm{Be}(1,1)$ priors were placed on all of the parameters of the network. If the size of the mixtures exceeded a certain pre-assigned threshold, then the (unweighted raw) matched-moment sequential reduction method in (9.34) was applied to bring it back to the threshold.

The Asia network, despite consisting of only eight binary nodes, is too large for an exact analysis with all but the smallest datasets. Thus, the quality of the predictions was judged against estimated posterior expectations and their standard deviations obtained by Gibbs sampling (discussed below in Section 9.8) using the BUGS software program (see Appendix C).

Figure 9.4 shows plots of kernel density estimates for the conditional probabilities $\theta_{b \mid s}$, $\theta_{b \mid \bar{s}}$, $\theta_{l \mid s}$, and $\theta_{l \mid \bar{s}}$, together with the posterior distributions obtained for retaining one term or up to twenty terms in the mixtures using sequential updating. For each dataset, the surprising thing to note is that there is very little difference between the sequential updating results: keeping more terms does not seem to affect the learning behaviour. In contrast, when local independence is dropped, but only global independence is retained, keeping more terms in the mixture does make a difference and generally does improve the predictive performance, as illustrated in Figure 9.5 for the same datasets.

The following heuristic is helpful to understand why, when retaining local independence, increasing the number of terms in the mixtures has little effect. Consider again Equation (9.32). The last, unupdated, term arises from those assumed completions of the case involving some other parent configuration $\rho \neq \rho^*$. If there are many parent completions consistent with the case, $p(\rho^* \mid \mathcal{E})$ (and thus each $p(j, \rho^* \mid \mathcal{E})$) will typically be close to zero. Then the posterior will tend to differ little from the prior on updating with such a case. Hence, the incomplete cases in the data will have little effect on the learning in comparison to the complete cases (except possibly for learning on a node all of whose parents are observed).

In conclusion, for relatively small amounts of missing data (75 percent complete), it appears that retaining local independence and collapsing mixtures to a single term using simple moment-matching gives reasonably good performance, and this can be done efficiently. However, as the percentage of missing data increases, loss of local independence becomes more severe and retaining locally independent mixtures makes little difference over a full reduction to a single term. In this case a better approximation is to retain mixtures having global independence only, again reduced using simple moment matching, but this implies a significant increase in computational complexity which may be impractical for some applications.

9.8 Using Gibbs sampling for learning

As we have seen in Section 9.7, when learning with incomplete data on discrete models, in an exact Bayesian analysis one encounters mixtures with many terms, each requiring drastic pruning approximations. For models with continuous variables the situation becomes worse — if data are complete then conjugate prior-to-posterior analysis is tractable, but for incomplete data, consideration of all possible completions leads to intractable integrations.

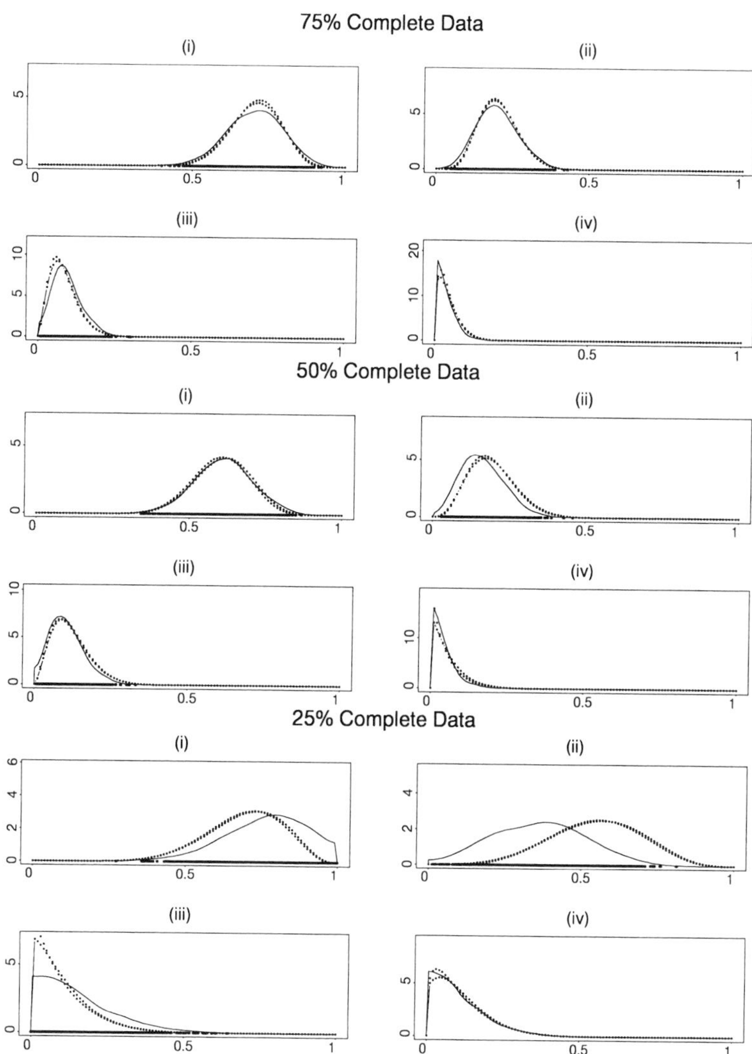

FIGURE 9.4. Kernel density estimates of the marginal posterior densities obtained from BUGS (solid line) for each dataset, together with posterior densities obtained from sequential updating retaining one term and twenty terms (dotted lines) in mixtures retaining local independence. Plots (i) = $\theta_{b\,|\,s}$; plots (ii) = $\theta_{b\,|\,\bar{s}}$; plots (iii) = $\theta_{l\,|\,s}$; and plots (iv) = $\theta_{l\,|\,\bar{s}}$.

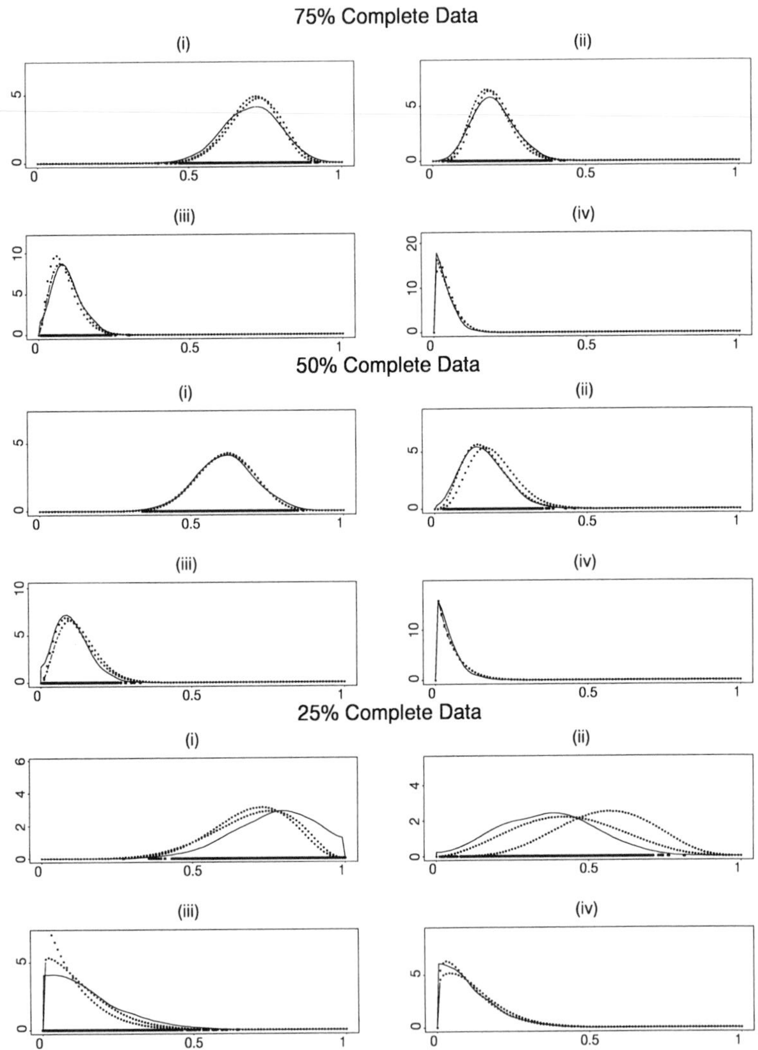

FIGURE 9.5. Kernel density estimates of the marginal posterior densities obtained from BUGS (solid line) for each dataset, together with posterior densities obtained from sequential updating retaining one term and twenty terms (dotted lines) in mixtures retaining only global independence. Plots (i) = $\theta_{b\,|\,s}$; plots (ii) = $\theta_{b\,|\,\bar{s}}$; plots (iii) = $\theta_{l\,|\,s}$; and plots (iv) = $\theta_{l\,|\,\bar{s}}$.

Gibbs sampling (Appendix B) can be used to perform learning on probabilistic networks in this situation. The crucial idea is that the graph of the observables is repeated for each case, each of these graphs 'hanging off' a set of common conditional probability parameters. Buntine (1994) gives a comprehensive review of the theory; here we present an outline.

The learning problem is illustrated by the graphical network of Figure 9.6 in which the observables are replicated, once for each case in the database. In this example there is an unknown parameter ϕ to be estimated from the

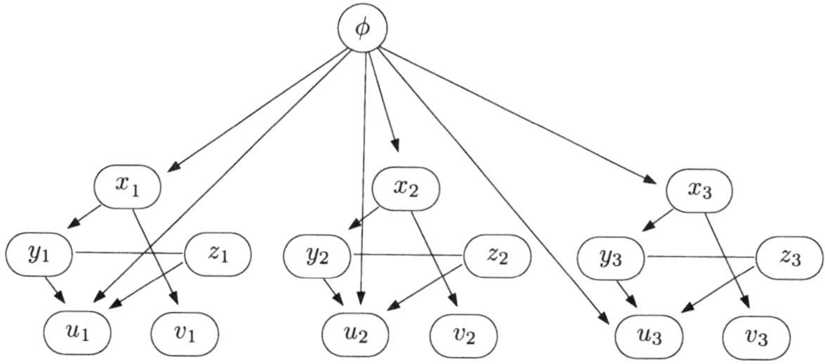

FIGURE 9.6. Replicated graph picture with a single parameter ϕ.

data (in general there will be many), and the observables are the indexed variables u, v, x, y, and z. Clearly for large databases this type of representation is impractical. The model and data can be compactly represented by using the notion of a *plate* to represent the observables; the plate is used to represent a set of replicate cases, as illustrated in Figure 9.7.

One can now run the Gibbs sampler on the plated graph Figure 9.7, generating samples by cycling through unobserved nodes, using the data to fix or clamp the observed nodes permanently. From the stored samples generated one estimates the quantities of interest, typically the posterior distributions of the unknown parameters or predictive distributions for new cases.

We illustrate this with Figure 9.8, the ASIA network in which θ^B represents the unknown conditional probabilities of *Bronchitis?* given *Smoking?*, and the observed part of the network is represented by the replicated plates. We use a dataset of five cases (omitting details here), in which the true value for *Smoking?* is not observed for case 2, who has bronchitis and dyspnoea, and case 3, whose only positive feature is an X-ray. With independent uniform priors on each of the unknown conditional distributions $P(\textit{Bronchitis?} = \text{yes} \mid \textit{Smoking?} = \text{no})$ and $P(\textit{Bronchitis?} = \text{yes} \mid \textit{Smoking?} = \text{yes})$, using BUGS we obtain after 10,000 iterations with a 1,000 iteration burn-in posterior mean estimates (standard deviations) of $P(\textit{Bronchitis?} = \text{yes} \mid \textit{Smoking?} = \text{no}) = 0.48\ (0.21)$ and $P(\textit{Bronchitis?} =$

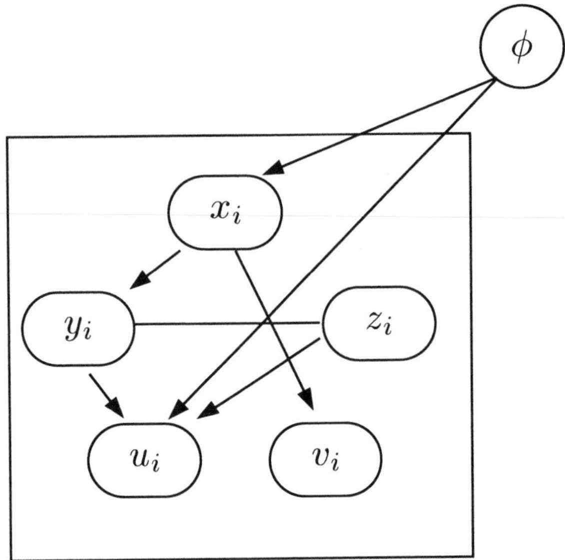

FIGURE 9.7. Plated version of Figure 9.6.

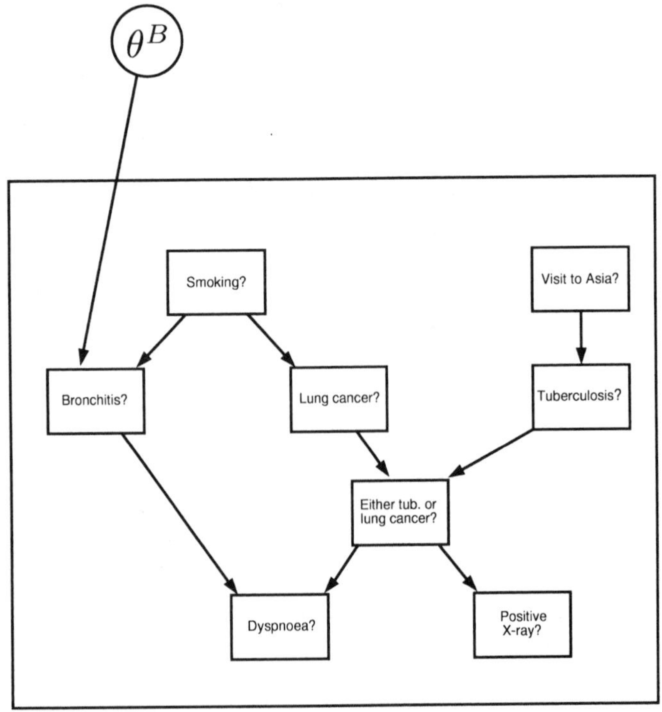

FIGURE 9.8. Plated illustration of ASIA network.

yes | *Smoking?* = yes) = 0.66 (0.23). In addition, we estimate that for cases 2 and 3 respectively, there is a probability 0.56 and 0.37 that the persons observed are smokers.

We emphasize again the generality of this procedure, in that models of almost arbitrary complexity can be considered when densities are positive. However, for learning in networks that allow exact propagation it is clearly more efficient to alternate between updating *all* the unobserved data for fixed parameters θ by a single sweep for each case of SAMPLE RANDOM-CONFIGURATION described in Section 6.4.3, and updating the parameters for fixed θ. If conjugate prior distributions are used and global or local independence prevails, the last step can be done very efficiently. This scheme is a variant of *blocking Gibbs sampling*described in Appendix B. In genetic pedigrees this scheme is absolutely necessary as Gibbs sampling with sitewise updating leads to reducible Markov chains.

9.9 Hyper Markov laws for undirected models

The learning methods discussed above are mostly tailored to directed graphical structures. The theory for Bayesian learning on undirected graphs or general chain graphs is less well developed, whereas the maximum likelihood theory is described in detail in Lauritzen (1996) and fully implemented in the software program CoCo (see Appendix C). However, for Bayesian learning with complete case data on undirected decomposable graphs Dawid and Lauritzen (1993) have presented a general theory. This theory is based on the junction tree representation $\mathcal{T}$ of a decomposable graph $\mathcal{G}$.

If C is any clique of $\mathcal{G}$, there is an unknown joint distribution θ_C governing the observables $X_C = (X_v : v \in C)$. If one assumes the full distribution θ to be Markov over $\mathcal{G}$, then θ is determined by $(\theta_C : C \in \mathcal{C})$, and thus the prior distribution is expressible as a joint distribution for these quantities. Now consider any separator S, and let A and B denote the node-sets contained in the two parts of $\mathcal{T}$ remaining when the link through S is removed. The Markov property for X implies $X_A \perp\!\!\!\perp X_B \mid X_S$ (and given θ). There is an analogous condition one might impose on the prior, viz. $\theta_A \perp\!\!\!\perp \theta_B \mid \theta_S$. If this holds for all $S \in \mathcal{S}$, the prior distribution is termed *hyper Markov*. This often reasonable condition turns out to streamline the Bayesian analysis greatly. In particular, if we specify the marginal prior distributions for each θ_C, then (subject to certain obvious compatibility conditions) there will be exactly one hyper Markov prior for θ having these marginals — clearly this greatly simplifies prior specification. The posterior distribution (based on complete data) will again be hyper Markov, so that we only need to find the marginal posterior for each θ_C.

In general, the posterior for θ_C will depend on data about variables outside, as well as inside, C. However, we can strengthen the hyper Markov property to require that $\theta_A \perp\!\!\!\perp \theta_{B\,|\,S}$. Under this *strong hyper Markov* assumption structure, the posterior for θ_C will only involve data on variables in C, and thus Bayesian analysis is entirely localized, proceeding quite independently within each clique.

Bayesian inference in the general (weak) hyper Markov case can in principle be performed using a propagation algorithm similar to the algorithms discussed in Chapters 6 and 7, which allows information about θ_C from data outside C to filter via the passage of suitable flows. However, this propagation generally involves integrals of complicated functions, and is thus not easy to implement. The problem is simplified if the prior can be expressed as a mixture of strong hyper Markov priors. This property is preserved on observing possibly incomplete data, thus allowing Bayesian inference from such data. However, the number of terms in the mixture can grow exponentially with the sample size, and approximation methods are needed.

Dawid and Lauritzen (1993) give examples of the construction of a range of specific hyper Markov laws, including the hyper normal, hyper Multinomial, hyper Dirichlet, the hyper Wishart, and hyper inverse Wishart laws. These laws occur naturally in connection with the analysis of decomposable discrete log-linear and Gaussian covariance selection models. We refer the reader to the paper for more details.

9.10 Current research directions and further reading

Learning from data is at the centre of current research in Bayesian networks. Buntine (1994, 1996) provides extensive reviews with many references. Jordan (1998) presents a recent collection, which contains both tutorial introductions to Bayesian networks and also recent developments about learning. The proceedings of the yearly UAI conferences are a good source to keep up with recent developments.

Buntine (1994) emphasized the importance of exponential families for learning about probabilities in Bayesian networks. Thiesson (1997) defined a class of *recursive exponential models* for learning in directed Markov models. Geiger and Meek (1998) showed formally that curved exponential families arise for some probabilistic networks.

The treatment of sequential learning presented in Section 9.7 is the natural generalization of the theory of Spiegelhalter and Lauritzen (1990), who also developed other representations of priors for networks. Consonni and Giudici (1993) presented learning using hierarchical priors. Heckerman et al. (1995b) looked at parameter and structural learning for discrete directed networks using Dirichlet priors; their work was extended to directed Gaussian networks by Geiger and Heckerman (1994). Heckerman et al. (1995b) referred to global and local independence as *global parameter independence* and *local parameter independence* respectively.

The paper of Dawid and Lauritzen (1993) is the standard reference to hyper Markov laws for undirected graphs. Giudici (1996) applied the hyper Markov ideas of Dawid and Lauritzen (1993) to structural learning for undirected Gaussian graphical models using inverse Wishart distributions to represent the prior (see also Chapter 11).

Rubin (1976) introduced the term 'missing at random', and this paper is a standard reference on the treatment of missing data. A recent accessible treatment of the analysis of missing data from a Bayesian perspective can be found in Gelman et al. (1995).

Maximum likelihood estimation is a standard technique treated in most statistics textbooks. The EM algorithm was promoted by Dempster et al. (1977) for finding maximum likelihood estimates involving missing data or incomplete information in some sense. In models with latent structure (Lazarsfeld and Henry 1968) data are missing systematically. The algorithm was used in a special case of this by Dawid and Skene (1979). In connection with log-linear models for contingency tables, the algorithm was studied by Fuchs (1982) extending work of Chen and Fienberg (1974, 1976) and Hocking and Oxspring (1974). Generalized EM algorithms are discussed by Thiesson (1991). The EM algorithm can be slow to converge, but can be speeded up near the maximum using gradient descent methods, which can be performed locally (Russell et al. 1995; Thiesson 1995, 1997). Zhang (1996) showed that increases in efficiency for sequential learning, the EM algorithm, and gradient descent can be achieved by exploiting independence relationships. Bauer et al. (1997) also discussed ways to speed up the convergence of the EM algorithm.

Ramoni and Sebastiani (1997b, 1997c) have proposed a batch learning method for analysing incomplete data for discrete models, which they called *bound and collapse*, in which upper and lower bounds are calculated and combined using prior information regarding the missing data mechanism to yield point estimates and credible intervals. Ramoni and Sebastiani (1997a) extended the method to structural learning (see Chapter 11).

Although Gibbs sampling can be applied to arbitrary networks and distributions, it can be slow. There is much activity on developing efficient analytic techniques for learning in classes of models that can be represented by specialized graphical structures: neural networks, Kalman filters, and hidden Markov models are examples. Reviews of these techniques — such as mean field and mixture approximation — for these and other models may be found in Jordan (1998) and the references therein.

10

Checking Models Against Data

Testing the compatibility between data and an assumed model is a central activity in Statistical Science. For models that are specified by means of prior information elicited from experts, it is particularly important that the initial assumptions be critically examined in the light of data obtained (Box 1980, 1983). In probabilistic expert systems, we need to check both the qualitative structural assumptions and the quantitative prior assessments.

One can take either a Bayesian or classical (frequentist) approach, and process the data either sequentially or as a single batch. Here we concentrate on some sequential techniques for model checking, using both a classically motivated 'absolute' approach and a Bayesianly motivated 'relative' approach.

Four types of diagnostic monitor for probabilistic networks will be described. A *parent-child monitor* provides a direct check on the adequacy of the beliefs about the conditional probability distribution of a chain component given its parents. An *unconditional node monitor* checks the suitability of the probability distribution at a node given the evidence on previous cases, while a *conditional node monitor* checks how well the node is predicted, given, in addition, all the other available evidence on the current case. A *global monitor* measures the overall degree of support supplied by the data for the model. Many other types of monitor could be envisaged, all aimed at checking on how well the system is predicting the incoming data.

Our constructions follow the *prequential* approach of Dawid (1984) in basing criticism of a model on the quality of the predictions it makes sequentially. We illustrate our monitors using the CHILD example. Cowell et al. (1993b) undertake an empirical investigation of the behaviour of these monitors using simulated data; a summary is given in Section 10.5 below.

10.1 Scoring rules

A *scoring rule* is a procedure for evaluating the quality of a probability statement about an event or variable in the light of its outcome (Dawid 1984). Scoring rules have been widely used in areas such as probabilistic weather forecasting (Murphy and Winkler 1977, 1984). The basic idea is that a high penalty is incurred if a low probability was given to the outcome that occurs. But if a forecaster is to be motivated by the scoring rule to provide an honest assessment of his or her uncertainty, then the scoring rule must have certain additional specific mathematical properties. Examples of such so-called *proper* scoring rules include the quadratic *Brier score* (Brier 1950), and the *logarithmic score* (Good 1952).

Here we shall adopt the logarithmic score, although other scoring rules could be used. There are numerous arguments in favour of the logarithmic score. It is (for multi-attribute variables) the only proper scoring rule whose value is determined solely by the probability attached to the outcome that actually occurs (Bernardo 1979), and for a multivariate outcome it is just the sum of the component logarithmic scores based on sequential observation of the individual variables. The logarithmic score also plays a fundamental role in likelihood and Bayesian inference, and in information and communication theory.

Let (Y_m) denote discrete random variables which are to be observed for a sequence of cases $m = 1, \ldots, M$, and let $p_m(\cdot)$ denote the predictive probability (density) for Y_m, the next Y, after the first $m-1$ cases have been observed. Suppose that $Y_m = y_m$ occurs. Then we associate with this observation a *logarithmic score* S_m given by

$$S_m = -\log p_m(y_m). \tag{10.1}$$

By cumulating over the series of M cases we obtain a total penalty $S = \sum_{m=1}^{M} S_m$. Note that, if we are coherently updating our probability judgments on the basis of a joint probability distribution for the Y's, then $p_m(y_m) = p(y_m \mid y_1, \ldots, y_{m-1})$, and hence

$$\begin{aligned} S &= -\log \prod_{m=1}^{M} p_m(y_m) = -\log \prod_{m=1}^{M} p(y_m \mid y_1, \ldots, y_{m-1}) \\ &= -\log p(y_1, \ldots, y_M), \end{aligned} \tag{10.2}$$

which is just the negative logarithm of the overall probability of all the data observed, i.e., the logarithmic score based on the full observation

$(Y_1, \ldots, Y_M) = (y_1, \ldots, y_M)$; in particular, S is invariant to the order in which the data are observed. (This property would not hold, in general, if we had used a scoring rule other than the logarithmic).

Example 10.1 Suppose we observe an independent and identically distributed sequence $(Y_1, \ldots, Y_M)$, each term of which is a discrete observation in one of K categories, with true response probabilities $p_1, \ldots, p_K$ $(\sum_k p_k = 1)$, which are assigned the Dirichlet prior $\mathcal{D}(\alpha_1, \ldots, \alpha_K)$, $(\sum_k \alpha_k = \alpha_+)$. If n_k of the Y's fall in category k, $(\sum_k n_k = n)$, then the conditional probability of observing this particular sequence is $\prod_k p_k^{n_k}$. The total logarithmic penalty is (see Appendix A, (A.14)):

$$S = -\log\left[\frac{\Gamma(\alpha_+)}{\prod_{k=1}^K \Gamma(\alpha_k)} \frac{\prod_{k=1}^K \Gamma(\alpha_k + n_k)}{\Gamma(\alpha_+ + n)}\right]. \tag{10.3}$$

□

10.1.1 *Standardization*

We need to decide whether the total observed penalty is a cause for alarm or not, and to do this we require some method of calibrating its value. Two approaches are possible.

The *relative* approach explicitly sets up some alternative 'reference system', giving rise to a total penalty S^{ref}. The sign of $S^{\text{ref}} - S$ determines the better predictor. Also, we have

$$\exp(S^{\text{ref}} - S) = \frac{p(\text{all data} \mid \text{expert's prior})}{p(\text{all data} \mid \text{reference prior})}, \tag{10.4}$$

and the right-hand side is just the *Bayes factor* (see, e.g., Kass and Raftery (1995)) in favour of the hypothesis that the expert's prior is the more appropriate.

By contrast, *absolute* standardization does not involve any alternative predictive system. We construct a suitable statistic to test the null hypothesis that the observed events are occurring with the probabilities stated by the system. Under this hypothesis, just before Y_m is observed the penalty to be incurred is a random quantity with expectation E_m and variance V_m given by

$$E_m = -\sum_{k=1}^K p_m(d_k) \log p_m(d_k), \tag{10.5}$$

$$V_m = \sum_{k=1}^K p_m(d_k) \log^2 p_m(d_k) - E_m^2, \tag{10.6}$$

where $(d_1, \ldots, d_K)$ are the possible values of Y_m.

We may now construct a 'standardized' test statistic:

$$Z_M := \frac{\sum_{m=1}^{M} S_m - \sum_{m=1}^{M} E_m}{\sqrt{\sum_{m=1}^{M} V_m}}. \tag{10.7}$$

Note particularly that we are not requiring the (Y_m) to be modelled as independent, so that E_m and V_m may depend on earlier observations. Because of this possible data-dependence of E_m and V_m, the standardization can be regarded as performed in the *prequential frame of reference* (Dawid 1984). Thus, Z_m may be termed a *prequential monitor*. It can be shown that, under broad conditions not requiring independence, as $M \to \infty$ the distribution of Z_M will be asymptotically standard normal under the null hypothesis that the data arise from the model (Seillier-Moiseiwitsch and Dawid 1993).

Example 10.2 Suppose that each Y_m can take values 0 or 1, and that, for a sequence of eight cases, the system's predictive probabilities (π_m), where $\pi_m = p_m(1)$, and the corresponding outcomes (y_m), are as follows:

m:	1	2	3	4	5	6	7	8
π:	0.4	0.5	0.3	0.6	0.2	0.1	0.2	0.7
y:	0	1	1	0	1	0	0	1

Note that $\pi_4 = 0.6$ means that, after any processing it may perform of the outcomes of the first three cases, the system outputs a probability 0.6 that case 4 will have $Y_4 = 1$. This is calculated before the actual outcome $y_4 = 0$ is observed. The system has thus assigned probability $p_m(0) = 0.4$ to the actual observation, and so the associated logarithmic penalty score for this case is $S_4 = -\log 0.4 = 0.916$.

Before observing Y_4, the system assigned a probability of 0.6 to getting $Y_4 = 1$, and thus a predictive score of $S_4 = -\log 0.6 = 0.511$; and a probability of 0.4 to getting the (actual) value $S_4 = 0.916$. Its expected score on this case was thus $E_4 = 0.6 \times 0.511 + 0.4 \times 0.916 = 0.673$; similarly, the variance of this distribution was $V_4 = 0.0395$.

In this way we can calculate the following values for S_m, E_m, and V_m.

m:	1	2	3	4	5	6	7	8	Total
S:	0.511	0.693	1.204	0.916	1.609	0.105	0.223	0.357	5.618
E:	0.673	0.693	0.611	0.673	0.500	0.325	0.500	0.611	4.586
V:	0.0395	0	0.151	0.0395	0.307	0.435	0.307	0.151	1.430

The total logarithmic penalty is $S = \sum_{m=1}^{8} S_m = 5.618$. If we had used an alternative reference system, which assigned, say, $\pi_m = 0.5$ in all cases, that would have yielded $S^{\text{ref}} = 8 \times -\log 0.5 = 5.545$. The difference $S^{\text{ref}} - S = -0.073$, being negative, indicates a very slightly better overall performance for the reference system. The Bayes factor in favour of the reference system is $e^{0.073} = 1.08$.

The absolute goodness of fit between the system's probability forecasts and the actual outcomes may be assessed using the test statistic $Z = (5.618 - 4.586)/\sqrt{1.43} = 0.863$. Were the asymptotic theory of Seillier-Moiseiwitsch and Dawid (1993) to be applicable to this short sequence, we could regard this as an observation on a random variable which has an approximate standard normal distribution, under the null hypothesis that the data are consistent with the assigned probabilities. Since $|0.863| < 2$, we cannot regard this as a surprising value from that distribution and this test thus does not indicate any misfit between system and data. □

10.2 Parent-child monitors

We now extend the above methods to cases where each observation is itself multivariate. We consider a general chain graph model, with possibly incomplete observation on each case.

With each conditional probability specification $p(X_k \mid X_{\mathrm{pa}(k)})$ of each chain component k under the model we associate a set of *parent-child monitors*. These are intended to detect discrepancies between prior beliefs in X_k and the observed value x_k of X_k, when any particular parent configuration $X_{\mathrm{pa}(k)} = \rho$ obtains. Hence, these monitors apply only in cases in which the chain component and its set of parents (for a DAG, a node, and its parents) are all observed. For such a case let $p_m(\cdot \mid X_{\mathrm{pa}(k)})$ denote the conditional probability distribution for X_k given its parents after $m-1$ cases have already been processed. This formulation allows for probability updating, as described in Chapter 9; if probability updating is not being undertaken, then the conditional table will remain unchanged as the data are processed. Suppose that, in the mth case, configuration $X_k = x_k$ is observed for chain component k and that $X_{\mathrm{pa}(k)} = \rho$ is the observed configuration of its parents. Then the relevant conditional distribution for X_k is $p_m(\cdot \mid X_{\mathrm{pa}(k)} = \rho)$, and so the parent-child monitor for parent configuration ρ of chain component k is incremented by the associated logarithmic score

$$-\log p_m(x_k \mid X_{\mathrm{pa}(k)} = \rho). \tag{10.8}$$

We may also cumulate with this monitor its expected score and its variance, allowing calculation of the standardized statistic on substituting the conditional probability distribution $p_m(\cdot \mid X_{\mathrm{pa}(k)} = \rho)$ into (10.5), (10.6) and (10.7) respectively. Note that the parent-child monitor, and its standardization, can be calculated directly from the conditional probability tables: evidence propagation is not required.

Spiegelhalter et al. (1994) describe standardization of parent-child monitors. We illustrate this for the CHILD example introduced in Chapter 3,

TABLE 10.1. Prior assessments for $p(Disease? \mid Birth\ asphyxia? = \text{yes})$, both raw (columns 1 and 2) and transformed (final column).

Disease?	Prior estimate	Prior range	Dirichlet parameter α_k
PFC	0.20	0.05–0.30	0.85
TGA	0.30	0.10–0.50	1.28
Fallot	0.25	0.15–0.35	1.06
PAIVS	0.15	0.10–0.30	0.64
TAPVD	0.05	0.02–0.10	0.21
Lung	0.05	0.02–0.10	0.21
	1.00		4.25

using data on 168 cases which became available subsequent to the construction of the network (see Spiegelhalter et al. (1993) for additional discussion of this dataset).

Example 10.3 Consider the node *Disease?* for the parent configuration *Birth asphyxia?* = yes (see Figure 3.1). Table 10.1 shows the initial prior estimates and ranges and their transformation to a Dirichlet distribution as described in Section 9.4.2 (Spiegelhalter et al. 1994). We note that were *Birth asphyxia?* to be observed on all cases, only those cases for which *Birth asphyxia?* = yes, and *Disease?* is observed, would be relevant for the parent-child monitor of interest. However, since for some cases *Birth asphyxia?* was not observed, there is some indirect learning of $p(Disease? \mid Birth\ asphyxia? = \text{yes})$ and so the whole series of 168 cases has been used in the following analysis. The approximate learning method of Spiegelhalter and Lauritzen (1990) was used — see Section 9.7.4.

The relative approach contrasts the total penalty S with the penalty S^{ref} that would have been obtained had the expressed prior opinion been irrelevant and, instead, a *reference* prior assumed. Taking as reference prior the Dirichlet distribution $\mathcal{D}(\frac{1}{K}, \dots, \frac{1}{K})$, Figure 10.1 shows that the reference predictions initially incur smaller penalty than those based on the expert prior. However, allowing the probabilities to be updated by the data gives $S = 42.8$ and $S^{\text{ref}} = 43.8$, after accumulating over all 31 cases who had birth asphyxia, providing a Bayes factor of 2.7 in favour of the prior provided by the expert as against the reference prior. Figure 10.1 also shows the substantially higher penalty incurred by using the initial prior estimates but without learning from the data, illustrating the danger of uncritically holding onto precise prior assumptions.

Absolute standardization gives the following results. By the time the first case with birth asphyxia arrives, indirect evidence from cases for which *Birth asphyxia?* was missing has slightly revised the initial distribution

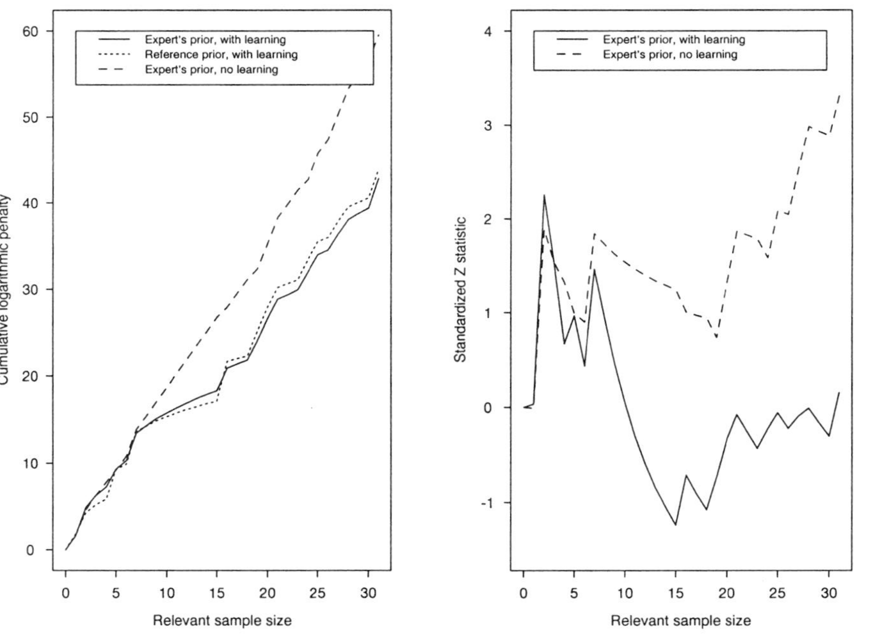

FIGURE 10.1. Parent-child monitor for the distribution of *Disease?* for parent configuration *Birth asphyxia?* = yes. On the left are the cumulative logarithmic penalties for three sets of predictions: based on the unchanged point prior estimates shown in Table 10.1 (expert's prior, no learning); based on the Dirichlet distribution in Table 10.1 (expert's prior, with learning); and based on a reference Dirichlet distribution $\mathcal{D}(0.16, \ldots, 0.16)$. On the right are standardized monitors for the expert's prior, with and without learning, showing early indications of poor predictions which are eventually overcome by learning from the accumulating data.

TABLE 10.2. Observed frequency distribution for *Disease?* when *Birth asphyxia?* = yes, with counts expected under the initial prior estimates.

Disease?	Expected count	Observed count	Observed proportion
PFC	6.20	19	0.61
TGA	9.30	3	0.10
Fallot	7.75	1	0.03
PAIVS	4.65	2	0.06
TAPVD	1.55	0	0.00
Lung	1.55	6	0.20
	31	31	1.00

to $(0.196, 0.295, 0.263, 0.147, 0.049, 0.049)$, from which we calculate that the penalty has expectation $E_1 = 1.609$ and variance $V_1 = 0.272$. In fact the observed disease for this first case was PFC, yielding penalty $S_1 = -\log 0.196 = 1.628$. Thus, $Z_1 = (1.628 - 1.609)/\sqrt{0.272} = 0.035$. The second case with birth asphyxia had disease TAPVD, which received a penalty of 3.281 and gave $Z_2 = 2.25$. Figure 10.1 shows the standardized parent-child penalty with and without learning. If no learning is allowed, the prior assessments remain constant in time and the standardized penalty increases; with learning the predictions adapt to the data after a while, and the standardized penalty eventually stabilizes in the neighbourhood of zero. In fact, as may be seen from Table 10.2, the observed data indicated that PFC and lung disease were both considerably more common in cases with birth asphyxia than initially expected.

Only a restricted set of nodes was observed on the 168 cases available for analysis: overall there were 21 observed combinations of parent configuration and child node. In general, the relative approach gave results similar to the absolute significance tests: for 3 out of 21 parent-child monitors the reference prior would have given better predictions, and these had Z statistics of 2.44, 1.84 and 2.30. The highest Z was for the assessment of *Sick?* given *Disease?* = Lung: for this case a prior assessment of 0.70 (range 0.50 – 0.90) was made for the probability of being sick, translating to a $\text{Be}(3.0, 1.25)$ distribution, while in fact only 6/16 cases were reported as sick. This gave $Z = 2.44$, while $S^{\text{ref}} - S = -0.02$, showing a marginal preference for the reference $\text{Be}(0.5, 0.5)$ prior. □

10.2.1 Batch monitors

We have so far concentrated on sequential monitoring. It is also possible to consider *batch* monitors, calculated on the basis of the total observed counts as shown in Table 10.2 (and thus ignoring any indirect evidence from cases where *Birth asphyxia?* was not observed). Since the statistic $S^{\text{ref}} - S$ is invariant to the order of the data, the sequential relative monitor can equally serve as a batch relative monitor. From a frequentist perspective, however, in order to obtain an absolute standardization we need to calculate the chance of getting such an extreme value of S, or one still more extreme, under the null hypothesis that the initial prior assessment is appropriate. Spiegelhalter et al. (1994) relate this to the proposal of Box (1980) for comparing prior with data. The resulting test can be approximated by an adjusted Pearson's X^2 statistic. Specifically, we calculate the expected counts under the point prior estimates, as shown in Table 10.1. We then calculate a X^2 statistic based on the observed and expected counts, which in our example gives $X^2 = 54.14$ on five degrees of freedom. To allow for the prior imprecision, this must be discounted by a factor $(\alpha_+ + 1)/(\alpha_+ + n)$, which is $(4.25+1)/(4.25+31) = 0.149$ in our case. The final test statistic is thus $54.14 \times 0.149 = 8.07$, which on five degrees of freedom gives $P = 0.17$, slight evidence against the expert's prior. The apparent discrepancy between this finding and the support for the hypothesis that the expert is correct under the relative approach is an illustration of *Lindley's paradox* (Lindley 1957), whereby a null hypothesis, even though not itself a particularly good explanation for the data in an absolute sense, can still be better than a profession of 'ignorance' over a wide range of possible parameter values.

10.2.2 Missing data

If we observe a batch of complete data, the comparison of the prior distribution and the observed counts by means of significance testing, as described in Section 10.2.1, is essentially straightforward. With incomplete data, however, even if missing completely at random, there can be difficulties in deriving the appropriate null sampling distribution for the batch test statistic. The sequential monitors are largely immune to this problem; a suitable version of (10.7) may be referred, asymptotically, to a standard normal null distribution, under very broad conditions.

The reference approach based on the logarithmic score is in principle invariant to the order of the data, and hence there should be no difference between analysing the data as a batch and using sequential techniques. In practice, however, missing data may require approximations in the learning process, as discussed in Chapter 9, and the inferences may then depend on the order in which the cases are processed.

10.3 Node monitors

We now define two types of node monitor, *unconditional* and *conditional*, which are designed to identify problems of poor structure and/or poor probabilities associated with a node. Both types of node monitor for a node v are updated if and only if the state of v is observed.

Let now P_m, with density $p_m(\cdot)$, denote the marginal distribution of X_v after $m-1$ cases with respective evidences $(\mathcal{E}_1, \ldots, \mathcal{E}_{m-1})$ have been processed. Suppose that in the mth case $X_v = x_v$ is observed. Then the *unconditional node monitor* for v is given a score

$$-\log p_m(x_v). \tag{10.9}$$

Also calculated from the probability distribution P_m are the expected score and its variance, together with the standardized statistic, as in (10.5) to (10.7). These calculations ignore any evidence on the mth case. The probabilities needed to evaluate the unconditional node monitors are readily calculated from the junction tree by local propagation and marginalization.

The *conditional node monitors* are defined in a similar manner, but using probabilities further conditioned on the other evidence in the mth case. Suppose that in the mth evidence set $\mathcal{E}_m$ the state $X_v = x_v$ is observed. Now suppose that in the junction tree all the evidence in $\mathcal{E}_m$ is propagated *except* the piece of evidence $X_v = x_v$; denote this evidence by $\mathcal{E}_m \backslash X_v$. Then we can calculate the probability distribution for X_v given all this other evidence. It is this distribution that is used to calculate the contribution

$$-\log p_m(x_v \mid \mathcal{E}_m \setminus X_v) \tag{10.10}$$

to the score, together with its expectation and variance, and hence the standardized test statistic.

At first sight it may seem that, if a dataset has evidence on K nodes, then K propagations will be necessary to evaluate all the conditional node monitors, a process which could be very time consuming. However, by using the fast retraction algorithm described in Section 6.4.6, each of the K conditional node monitors can be calculated in one propagation in the junction tree. (Exceptions can arise if the joint probability is zero for some configurations.)

Example 10.4 Table 10.3 shows unconditional and conditional standardized monitors for all observed nodes in the CHILD example based on the 168 observed cases. Examining the unconditional monitors it is clear that the overall incidence of *Birth asphyxia?* and *Disease?* are poorly modelled even after learning. Conditional monitors for *Lower body* O_2*?* and *RUQ* O_2*?* suggest the assessments in this part of the graph should be carefully examined. The remaining conditional monitors appear reasonable: although

TABLE 10.3. Final conditional and unconditional standardized monitors for observed nodes, both with and without parameter learning.

Node	N	With learning Cond.	With learning Uncond.	Without learning Cond.	Without learning Uncond.
n1: Birth asphyxia?	120	0.38	1.96	1.55	5.78
n2: Disease?	168	0.64	2.61	1.39	5.93
n3: Age at presentation?	165	1.41	-0.59	0.47	-3.39
n9: Sick?	168	1.47	0.02	-0.18	0.48
n15: LVH report?	141	-1.17	0.06	-3.01	-4.11
n16: Lower body O_2?	45	-2.13	-0.98	-2.26	-1.39
n17: RUQ O_2?	120	-1.91	0.36	-1.22	1.96
n18: CO_2?	146	-1.25	-1.55	-5.10	-5.57
n19: X-ray report?	168	-0.98	0.99	-2.74	-1.89
n20: Grunting report?	165	-0.52	-0.62	-3.99	-3.79

the performance of the monitors without learning suggests the initial prior estimates were rather poor, with learning the considerable prior imprecision ensures rapid adaptation of the conditional probabilities to more appropriate values. The large number of negative Z-values suggests that the probability assignments have been conservative, in that the observed penalty has been much less than that expected, with the events assigned high probabilities occurring still more often than predicted. □

Diagnostics such as those shown in Table 10.3 can only be a first step toward improving the system. Since errors in specification at the top of the graph may filter through to affect all the unconditional monitors, it is appropriate to adjust the priors in sequence in order to identify additional poor assessments. When reasonable unconditional monitors are obtained then aberrant conditional monitors should be a better reflection of poor structure.

10.4 Global monitors

We define the contribution of the mth case to the *global monitor* for a model to be the logarithmic score based on the total evidence observed, i.e., $-\log P_m(\mathcal{E}_m)$, where $P_m(\mathcal{E}_m)$ denotes the probability of the evidence $\mathcal{E}_m$ for the mth case after the previous $m-1$ cases have been processed. As shown in Chapter 6, this probability is obtainable as the normalization of any clique or separator potential table in the junction tree after the evidence $\mathcal{E}_m$ has been propagated in a representation of P_m.

Calculation of E_m, V_m, and the standardized statistic Z_m for the global monitor is in general quite laborious, except for the case of complete data, for which a local propagation scheme exists (see Section 6.4.7), or for models having small total state space. Thus, we confine attention to the relative approach, comparing the overall predictive quality of competing models.

If we define the overall global monitor to be $G := \sum_m -\log P_m(\mathcal{E}_m)$, we have, as for (10.2):

$$\begin{aligned} G &= -\log \prod_{m=1}^{M} P_m(\mathcal{E}_m) = -\log \prod_{m=1}^{M} P(\mathcal{E}_m \,|\, \mathcal{E}_1, \ldots, \mathcal{E}_{m-1}) \\ &= -\log P(\mathcal{E}_1, \ldots, \mathcal{E}_M) = -\log P(\mathcal{E}), \end{aligned} \tag{10.11}$$

the logarithmic score based on all the evidence $\mathcal{E}$ observed. If no approximations have been made, this will be independent of the ordering of the cases, and can be regarded as the negative log-likelihood for the system based on the evidence. In particular, if two different systems yield respective global monitors G_M^1 and G_M^2 on the same evidence, the *log-likelihood ratio*, or *log-Bayes factor*, in favour of system 1 as against system 2 is simply $\Delta_M = G_M^2 - G_M^1$, which thus supplies an immediately interpretable comparison of the two models. If initially the prior log-odds in favour of system 1 are λ, then the posterior log-odds, taking all the evidence into account, will be $\lambda + \Delta_M$.

10.4.1 Example: CHILD

We take as our baseline model the CHILD network, with the expert's priors and parameter learning using all available data. We first contrast it with a naïve Bayes network (see Section 2.8), using priors derived by the process of *expansion and contraction* (see Section 11.6.4). Additional comparisons are made with the naïve Bayes network and the CHILD network without learning (i.e., keeping the expert's point estimates).

Figure 10.2 shows the log-Bayes factors (differences in global monitors over the baseline model) for the three alternative models, based on sequential observation of the 168 cases. Allowing learning from the data about the parameters starts to show benefit after about 25 cases, and thereafter the without learning model is clearly inferior. The naïve model with learning is initially poor, but starts to outperform CHILD without learning after about 52 cases (although it becomes increasingly inferior to the more structured baseline model). The naïve model learning with reference priors almost catches up with the model given an expert 'head-start' after about 120 cases. Overall, the data strongly support the more structured model. Jeffreys' rule-of-thumb that a log-Bayes factor of more than 4.6 constitutes 'decisive' evidence is applicable after only 11 cases. Further comparison with more local structural adjustments should now be investigated using these techniques.

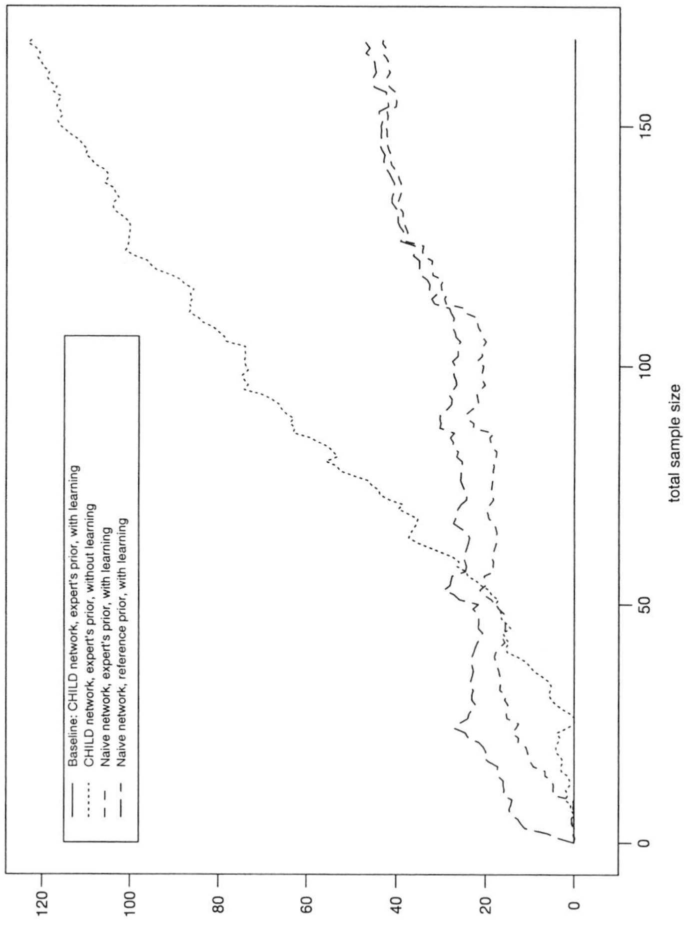

FIGURE 10.2. Global monitors for three alternative models compared to a baseline assumption. The ordinate also expresses the log-Bayes factor in favour of the baseline model.

TABLE 10.4. Measures of the diagnostic accuracy of four models.

Model	Conditional penalty S	Mean posterior prob. ($e^{-S/N}$)	Accuracy in 168 cases
Baseline: CHILD, expert's priors, learning	156.3	0.40	110
CHILD, expert's priors, no learning	197.4	0.31	99
Naïve network, expert's priors, learning	172.4	0.36	113
Naïve network, reference priors, learning	176.4	0.35	111

The above model comparison has focused on the ability of each model to predict the totality of data observed. It may be appropriate to place more weight on a model's ability to predict the disease node. Table 10.4 summarizes the diagnostic performance of the four models being examined. The conditional monitor S is the cumulated logarithmic penalty given to the node *Disease?* conditional on all other evidence; hence, $\exp(-S/N)$ ($N = 168$) is the geometric mean of the probabilities of the true diseases. The accuracy is simply the number of cases for which the disease receiving the highest probability, given the evidence, was the true disease.

Based solely on the scoring rule, the baseline model is superior. However, this benefit of additional quantitative and structural input is not translated into increased diagnostic accuracy. Further analysis reveals that this is primarily due to the incoming evidence being diluted by the conservative probability assignments in CHILD, making it harder to push the rarer diseases into first place. It is also worth noting that all the models in Table 10.4 do considerably worse, in terms of simple accuracy, than a simple algorithmic approach; the algorithm of Franklin et al. (1991) described in Section 2.2 attained an accuracy of 132. The clinicians in GOS had an accuracy of 114, and the referring paediatricians 76. Thus, a probabilistic expert system can currently match the quality of middle-level paediatric cardiologists at GOS. Clearly, further development is required to exploit better the interactions that the algorithmic approach incorporates.

10.5 Simulation experiments

In this section we briefly summarize the results of extensive numerical simulations performed by Cowell et al. (1993b) in an investigation of the ability of the monitors to detect model specification errors. The reader is referred to that paper for further details.

The experiments involved two structurally distinct models: Model 0, the *true model*, and Model 1, the *candidate model*. Graphical representations of these models are shown in Figure 10.3. Model 1 is the Asia network (see page 20). As can be seen from the figure, in going from Model 1 to Model 0 the structure of the model is changed by (i) the addition of the link $S \to D$ and (ii) the removal of the link $S \to B$. For the without learning experiments, the probability tables for Model 1 were as given in Table 2.3, with the exception of $p(b \mid \bar{s}) = 0.05$. For Model 0, each node which has the same parents in both models (i.e., A, S, T, L, E, and X) was assigned the same conditional distribution as in Model 1. This leaves the nodes *Bronchitis?* and *Dyspnoea?*. At B, the value $p(b) = 0.5$ was assigned, while values were chosen for $p(D \mid E, B, S)$ such that the implied distribution of D given only E and B agreed with the relevant conditional probability table in Model 1. Then the marginal distributions at each node agree across the models for nodes A, S, T, L, E, and X, but differ for B and D, while the conditional distributions at each node, given the values at all the others, agree for A, T, L, E, and X, but differ for B, S, and D.

Model 0, with the above probabilities, was used to generate the Monte-Carlo data for the numerical experiments. One thousand cases were generated, and information on nodes was randomly withheld so as to form one complete and two incomplete datasets. For each model two classes of monitoring experiments were performed. In the first class of experiments (without learning) the probabilities were fixed throughout the analysis. In the second class, the matched-moment approximate learning scheme of Spiegelhalter and Lauritzen (1990) was employed to update probabilities sequentially. In the class of learning experiments, all probabilities were taken as initially uniformly distributed on the interval $[0, 1]$, with the exception of the tables for the E-node in both models; this is a logical node, and this feature was retained in the probability distributions. The same simulated data were used for the with learning and the without learning experiments.

The prequential monitors are intended to draw attention to various indications of incorrect topology or probabilities in models; also they should not complain unnecessarily. The simulation study confirmed the value and usefulness of the monitors. When the model used (Model 0) was a good description of the data-generating process, the Z values tended to be below 2 in modulus; exceptions to this could be satisfactorily explained as due to small sample size or non-applicability of approximate standard normality. On the other hand, when assessing the performance of an incorrect model (Model 1) the Z statistics clearly highlighted its inadequacies. The node-monitor results for the complete data experiments without learning indicated problems in the link $S \to B$, and at D, and the same conclusions were drawn from the incomplete datasets (without learning). The global monitors indicated a clear preference for the true Model 0 over Model 1 without learning.

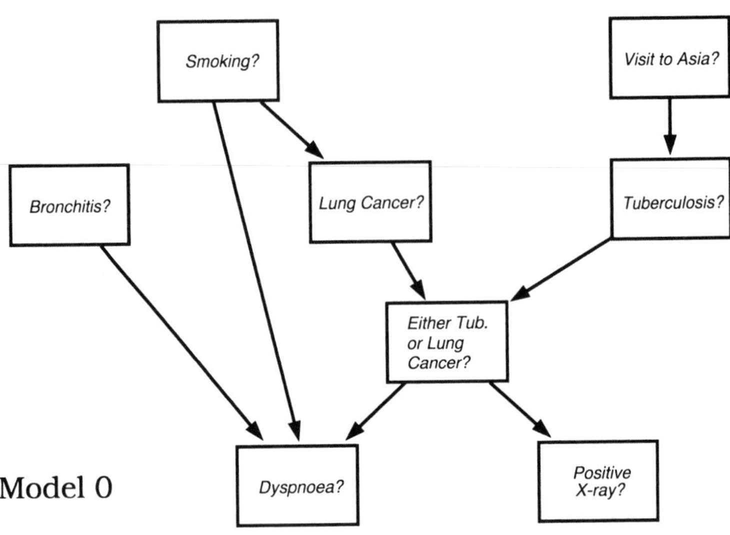

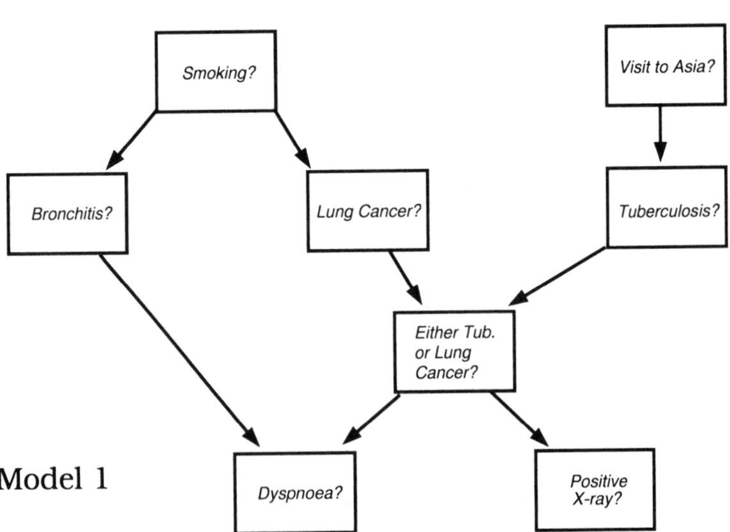

FIGURE 10.3. True model (Model 0) and candidate model (Model 1). Nodes are also denoted by labels: A (*Visit to Asia?*), B (*Bronchitis?*), D (*Dyspnoea?*), E (*Either Tuberculosis or Lung cancer?*), L (*Lung cancer?*), S (*Smoking?*), T (*Tuberculosis?*), X (*X-ray?*).

When learning was allowed, however, it became much more difficult to detect poor structure. The Z statistics essentially failed to discriminate Model 1 from Model 0, and indeed the global monitors indicated that — at any rate with incomplete data — the incorrect but more parsimonious Model 1 was yielding more successful predictions. This behaviour is perfectly acceptable, even desirable. In spite of its incorrect model structure, Model 1 will learn those probability values that provide the best fit possible to the actual data. The correctly structured Model 0 will eventually learn the true probability values, but, since it has more values to learn, will do so more slowly. From a purely 'technological' or predictive viewpoint (see Section 11.1), Model 1, even though incorrect, can provide a better explanation, at least for small data-sets.

10.6 Further reading

There appear to have been few systematic studies of diagnostics for detecting errors in the specification of probabilistic expert systems; most of the literature on model comparison is based on learning graphical structure from the data, as discussed in the next chapter. Exceptions include Riva and Bellazzi (1996), who use *scoring metrics* to compare pre-specified models for medical monitoring applications using one-step-ahead predictions, and Lam (1998), who uses a *Minimum Description Length* approach to refine a network on the basis of new data.

11
Structural Learning

So far in this book it has been assumed that the appropriate graphical structure has been provided by a domain expert, and that, subject to the diagnostic checks outlined in Chapter 10, this will be retained for future use. However, it is more realistic to admit that any such structure suggested can only be tentative, and ought to be subject to revision in the light of further data. There are also many situations in which the appropriate domain expertise is not initially available, so that right from the beginning we are obliged to looked to data to suggest what qualitative modelling assumptions — for example, conditional independence assumptions — might be appropriate. The prevalence and importance of such problems is reflected in the rapidly expanding interest and activity in the general area of structural learning of probabilistic networks from data, an area that merges naturally with a number of more general fields of research, including statistical modelling, machine learning, and neural networks. Because of the rapid pace of development, we shall not, except to some extent in Section 11.6, go into any detail of specific methods. Instead, our objective is to provide an overall framework within which the variety of work in the area can be organized, and to provide pointers to some important publications. General reviews of structural learning in graphical models have been provided by Buntine (1994, 1996), Sanguesa and Cortes (1997) and various papers in the collection edited by Jordan (1998), particularly that by Heckerman (1998).

In this chapter we first give an account of some very general aspects of statistical model assessment. Then we consider problems specific to graphical modelling, reviewing some of the relevant work that has been done for various different graphical structures.

11.1 Purposes of modelling

There are two fundamentally different attitudes one can take to the purpose of modelling.

Scientific modelling is concerned with attempting to understand some assumed 'true' objective process underlying the generation of the data. This process, if it were known, could be used for purposes of explanation, or causal understanding, as well as prediction. By definition, there can only be one true process, and the purpose of inference is to say something about what it might be.

Technological modelling has less grandiose aims. The purpose of such a model is to provide a good explanation of past data, and good forecasts for future data, irrespective of whether it corresponds to any underlying 'reality'. Thus, the object of inference is now the usefulness, rather than the truth, of a model. From such a viewpoint, one can allow the co-existence of several different models for the same phenomenon.

The interpretation and use we may make of a model will depend very much on whether we regard it as a scientific or a technological one. In principle, our methods of statistical inference should also be sensitive to this distinction. Nevertheless, it turns out that most of the methods that are useful are equally applicable, although with some differences of interpretation, in either setting.

11.2 Inference about models

Traditionally, there have been three main ways of conducting inference about model structure on the basis of data.

Model testing: Assessment of the adequacy of a single candidate model as an explanation of empirical data.

Model choice: Selection of one out of a family of possible candidate models as the best explanation of the data.

Model comparison: Forming a view as to the relative quality of the various candidate models as explanations of the data.

Model testing was the subject of Chapter 10, and we shall have little more to say about it here. For either of the other two approaches, we need to define (explicitly or implicitly) the class of models that we are concerned with. For example these might be all the directed graphical models, or all the undirected or decomposable or chain graph models, over a specified set of (observable or latent) variables, possibly with some restrictions on the directionality of arrows.

Classically, model choice was performed by methods related to those of model testing, using a sequence of significance tests, each designed to see whether it might be possible to replace a more complex model by a simpler one. This approach is still the one typically used in constraint-based graphical modelling (see Section 11.4), where the purpose is to select, on the basis of various tests of conditional independence applied to the data, an appropriate graphical representation of the inferred collection of conditional independence properties. For example, standard statistical hypothesis testing procedures based on likelihood ratio tests can be used for selecting among graphical models (Whittaker 1990; Spirtes et al. 1993; Edwards 1995). Such an approach needs to be guided by an appropriate *search strategy* (see Section 11.6.1) for moving through the typically very large space of all possible graphical models, of the desired structure and semantics, which are entertained as possible representations of the problem.

More recent work approaches model choice by way of model comparison: if we can compare models (in a transitive way), then we can also select the most preferred. Hence, a unified approach to these two problems can be based on the construction of some sort of 'score' for each model in the light of the available data, to provide a basis for the comparison. Such a score can have many interpretations, but a particularly useful one is as a measure of *model uncertainty*; for example, if we are conducting 'scientific modelling' (and hence believe that one of the models in our family of candidates is the 'true model'), then an appropriate measure of uncertainty might be the posterior probability, in the light of the data, that a particular model is the true one.

11.3 Criteria for comparing models

Scoring methods aim to create a single number for each model, expressing the extent to which the data support the truth or usefulness of that model.

Let $\mathcal{M}$ denote the set of all models under consideration. Although we assume that $\mathcal{M}$ is a finite set, it can be of very high cardinality. Each model $m \in \mathcal{M}$ defines an overall structure. The specific numerical values for probabilities within this structure will typically not be given in advance, but will depend on some parameter θ_m ranging over a parameter-space Θ_m. The various parameter-spaces, for distinct models, need not be related. We denote by d_m the effective dimension of Θ_m, i.e., the minimal number of mathematically independent quantities needed to identify a particular member of m; this dimension is not always straightforward to calculate for graphical models involving latent variables (Geiger et al. 1998; Settimi and Smith 1998).

Conditional on both m and θ_m, we have a fully specified probabilistic model, with density $p(\cdot \mid m, \theta_m)$, say. We suppose that we have gathered some data D, and wish to use this to score each model in $\mathcal{M}$ or in a suitable subset of $\mathcal{M}$ to which the data suggest we can confine attention.

The most important scoring methods are: maximized likelihood; predictive assessment; and marginal likelihood. We consider these in turn.

11.3.1 Maximized likelihood

Given data D, one might calculate, for any model m, the maximum likelihood estimate $\hat{\theta}_m$ of its parameter, and the associated maximized log-likelihood $\hat{l}(m) := \log p(D \mid m, \hat{\theta}_m)$. This can be used as a crude measure of the success of model m in describing the observed data D, although it does not directly address the value of m as an explanation. Hence, this approach may be regarded as aimed more at technological than at scientific modelling.

In rare cases it may be that we have such large sample sizes relative to the complexity of the models that sampling error in $\hat{\theta}_m$ can be ignored. Then it would be appropriate simply to score each model by means of its maximized likelihood, and thus to seek (using an appropriate search strategy) the model that maximizes $\hat{l}(m)$. More typically, however, even when sample sizes appear large, sampling error and the sheer size of the parameter space will imply that the maximum likelihood estimate is not really a believable value for the parameter, particularly when the model is complex. A common response to this difficulty is to incorporate some form of penalty for model complexity into the maximized likelihood. There is a wide range of penalty functions that have been suggested, of which the most popular are Akaike's Information Criterion (AIC), which uses $\hat{l}(m) - d_m$, and the Jeffreys–Schwarz criterion, sometimes called the Bayesian Information Criterion (BIC), which, for an independent random sample size n, uses $\hat{l}(m) - \frac{1}{2} d_m \log n$. The importance of avoiding over-fitting by means of some such trade-off between model fit and model complexity has long been recognized in classification problems (Breiman et al. 1984).

We note that none of the above criteria makes any use of any available prior information about either structure or parameters (although qualitative knowledge, for example concerning plausible causal orderings of quantities in the graph, may be useful to restrict the class of models and/or guide the associated search strategy).

11.3.2 Predictive assessment

From a purely technological viewpoint, a good way of scoring a model is in terms of its success in predicting actual outcomes. If we could specify not only the appropriate model m but also the correct parameter value θ_m, one reasonable measure of this success might be the overall *logarithmic score* $\log p(D \mid m, \theta_m)$. This differs by an unimportant sign change from the definition in Section 10.1, since we are now treating the score as a measure of goodness, rather than a penalty. When D consists of a sequence $(x_1, \ldots, x_n)$ of observations on n cases, the total logarithmic score can be decomposed as

$$\log p(D \mid m, \theta_m) = \sum_{i=1}^{n} \log p(x_i \mid x_1, \ldots, x_{i-1}; m, \theta_m), \tag{11.1}$$

with each individual term $\log p(x_i \mid x_1, \ldots, x_{i-1}; m, \theta_m)$ interpretable as a partial score, measuring the quality of the fully specified model's predictive distribution, after observing the first $i-1$ cases, for the next observable X_i, evaluated in the light of its actual outcome x_i. In the common situation that the constituent cases in D arise as a random sample, this term becomes simply $\log p(x_i \mid m, \theta_m)$.

If we do not know θ_m, but must estimate it from the data (using, for example, the maximum likelihood estimate $\hat{\theta}_m$), a crude approximation to the above score might be obtained by simply substituting this estimate. Once again, however, this is likely to be misleading: the effects of sampling error, coupled with the bias induced by the fact that our estimate is itself chosen so as to optimize the performance of the model, will lead to an over-optimistic estimate of performance. In order to counter this, one should ensure that in each individual term in (11.1) the outcome x_i that is being predicted does not in any way contribute to the estimate, $\hat{\theta}^i_m$ say, of θ_m used in that term. A popular approach for the random sampling case is *cross-validation* (Stone 1974), which, in its simplest form, takes as $\hat{\theta}^i_m$ the maximum likelihood estimate based on all observations excluding the ith. However, although this avoids obvious bias in each individual term, it still induces potentially misleading correlations between the different terms. An alternative approach avoiding this problem, which is equally applicable when the cases are not modelled as independent and identically distributed, is *prequential validation* (Dawid 1984), where the estimate $\hat{\theta}^i_m$ is based on the data for just the first $i-1$ observations. Under suitable smoothness conditions, the cross-validatory log-likelihood will be asymptotically equivalent to AIC, and the prequential log-likelihood to BIC. Note however (Geiger and Meek 1998) that the required smoothness conditions may not hold in some graphical models with latent variables, and in such cases these asymptotic equivalences are no longer assured.

The *Minimum Description Length* (MDL) principle or *Stochastic Complexity* criterion (Rissanen 1987) has also been suggested as a model scoring device (Lam and Bacchus 1994). This aims to evaluate a model in terms of its ability to describe the observed data parsimoniously. It can be implemented in a number of asymptotically equivalent variant ways, one of which turns out to be essentially identical to assessment by prequential log-likelihood.

Although we have presented predictive assessment of a model in terms of a criterion function based on the logarithm of the density, there is no difficulty in principle in extending the approach to other measures of goodness of prediction fit (see, for example, Dawid (1992b)). Ideally, such a measure should be directly related to the purpose of the analysis. Sebastiani and Ramoni (1998) have suggested using a formal loss function, while Heckerman (1998) discusses 'local scoring', which confines attention to the marginal density for a particular variable of interest.

11.3.3 Marginal likelihood

If we wish to approach the problem from a Bayesian viewpoint, we are obliged to incorporate additional ingredients into our analysis. In particular, in addition to specifying the possible models and their associated families of distributions, we need, for each model m, to specify a prior distribution over Θ_m for its parameter θ_m, with density $p_m(\theta_m)$, say. We refer back to Chapter 9 for some specific suggestions of convenient forms for this prior density in particular problems. Then, conditional only on the model structure m, the marginal density function is given by

$$p_m(\cdot) \equiv \int_{\Theta_m} p(\cdot \mid m, \theta_m)\, p_m(\theta_m)\, d\theta_m. \tag{11.2}$$

The *marginal likelihood* function over models $m \in \mathcal{M}$, in the light of the data D, is thus given by

$$\tilde{L}(m) := p(D \mid m) = \int p(D \mid m, \theta_m)\, p_m(\theta_m)\, d\theta_m, \tag{11.3}$$

and this, or its logarithm $\tilde{l}(m)$, can be used directly as a score for comparing different models.

If we are taking a scientific view of modelling, and thus believe in the truth of some model m, then $p_m(\theta_m)$ should express our conditional beliefs about the value of the relevant parameter θ_m, given that model m is the true one. Then (11.2) expresses predictive belief about observables, conditional on the truth of model m, and (11.3) supplies an appropriate measure of the support for the truth of model m in the light of the data D. Alternatively, from a technological standpoint, we can treat both the family of distributions specified by m and the prior density for its parameter as essentially arbitrary ingredients whose combination is to be judged in terms of how well it explains the data, as measured by the marginal likelihood (11.3).

Under the standard smoothness conditions (and with the same caveats as before for graphical latent variable models), it can be shown that, for large random samples, marginal likelihood is asymptotically equivalent to prequential likelihood, and hence log marginal likelihood is asymptotically the same as BIC or MDL (Heckerman 1998).

Some difficulties. For directed graphical models with discrete variables, complete data and Dirichlet prior distributions, the marginal likelihood has a simple form that decomposes into factors (of the form given in Appendix A, equation (A.14)) for each parent-child combination (Cooper and Herskovits 1992; Heckerman et al. 1995b). Similar simple formulae can be found in the case of complete data on continuous variables arising from a suitable Gaussian model with a conjugate prior. However, in most other problems an important practical difficulty in implementing marginal likelihood is that the integral in (11.3) is not obtainable in closed form. Various approximations to this integral have been suggested (Kass and Raftery 1995; Heckerman 1998). As we have already seen, one possible approximation to the log marginal likelihood is BIC, but this does not explicitly incorporate the prior density $p_m(\theta_m)$ at all, and is too crude to use except for extremely large samples. Another difficulty which some consider serious is the dependence of the marginal likelihood (11.3) on the specification of the prior density $p_m(\theta_m)$. In large samples, this can be regarded, approximately, as contributing an additive contribution $\log p_m(\hat{\theta}_m)$ to the log marginal likelihood, and this can be highly sensitive to the prior specification, especially to its variance or other measure of spread. Asymptotically, this constant contribution of the prior will be completely swamped by that of the data, which will be approximately proportional to sample size, so that this problem eventually disappears; however, even for large data-sets the contribution of the prior can be significant.

11.3.4 Model probabilities

When we are conducting scientific modelling, it is appropriate to complete the Bayesian probabilistic specification by further assessing a prior probability, $p(m)$, that model m is the true one. Then, by Bayes' theorem, we can calculate the posterior probability for each model by

$$p(m \mid D) = \frac{\tilde{L}(m)\, p(m)}{\sum_{m' \in \mathcal{M}} \tilde{L}(m')\, p(m')}. \tag{11.4}$$

One could then select the model having the largest posterior probability; or, as we consider below, use the posterior probabilities to average the predictions of the individual models.

While straightforward in principle, direct practical implementation of the above fully Bayesian approach is beset by a number of difficulties. In the first place, we have to assign a prior probability $p(m)$ to each possible model in the problem. General scientific principles suggest that simple models should be given higher probability. Various means of implementing such a preference for simplicity have been suggested (Heckerman 1998). A second problem is that, even if we are only interested in the posterior probability of a single model, we still need to calculate the marginal likelihood for every possible model in order to obtain the denominator term in (11.4), and the space of models can be very large.

A simple way of avoiding both these problems is to concentrate on the relative acceptability of two models, m_1 and m_2 say, in $\mathcal{M}$. This can be measured by the *posterior odds* $p(m_1 \mid D)/p(m_2 \mid D)$ in favour of m_1 as against m_2; in virtue of (11.4), this is given by

$$\frac{p(m_1 \mid D)}{p(m_2 \mid D)} = \frac{p(m_1)}{p(m_2)} \times \frac{\tilde{L}(m_1)}{\tilde{L}(m_2)}.$$

The contribution of the data to this comparison is thus entirely localized in the 'ratio of marginal likelihoods' term $\tilde{L}(m_1)/\tilde{L}(m_2)$, the so-called *Bayes factor* in favour of m_1 as against m_2 (see, e.g., Kass and Raftery (1995)). The logarithm of the Bayes factor is simply the difference of the global monitors introduced in Section 10.4; model comparison using Bayes factors was illustrated using the CHILD example in Section 10.4.1.

11.3.5 Model selection and model averaging

Suppose that we wish to predict, or make inferences about, some quantity Y (a future observable, or more generally any quantity which has a clear meaning independent of which model is being considered). If we knew which model m to use, we could apply the usual probability calculations to deduce the relevant inferential distribution $p(y \mid m, D)$ after observing the data D. However, when there is uncertainty about the true model, selecting and using a single 'best' model to make predictions will result in underestimation of the appropriate uncertainty (Draper 1995). The standard Bayesian way of dealing with this is to average over models. Thus, if one wishes to make inferences about Y, taking into account model uncertainty, the correct overall posterior or predictive density is given by the model average, using the posterior model probabilities as weights:

$$p(y \mid D) = \sum_{m \in \mathcal{M}} p(y \mid m, D)\, p(m \mid D), \tag{11.5}$$

with $p(m \mid D)$ given by (11.4). It can be shown (see, e.g., Madigan and Raftery (1994)) that this predictive distribution will give better predictions (as evaluated by means of the logarithmic scoring rule) than are expected from the use of a single model:

$$\log p(y \mid D) \geq \mathrm{E}\left\{\log p(y \mid m, D)\right\},$$

where the expectation on the right hand side is with respect to the posterior distribution of m over $\mathcal{M}$. This property justifies the use of model averaging as a technique to improve predictive performance, even within a purely technological setting.

To implement (11.5), one in principle has to deal with all possible models, of which there may be a very large number. However, many of these will make a negligible contribution. Even averaging over a small number, 10 or 20 say, of the best models can yield a significantly improved predictive performance (Madigan and Raftery 1994).

11.4 Graphical models and conditional independence

In this book, we have considered several kinds of graphical representation of qualitative structure. In all cases, these have coded assumptions of conditional independence between various sets of variables. However, the relationship between a graphical model and a collection of conditional independence relationships is not completely straightforward: it may not be possible to find any graphical representation of all and only the conditional independence relationships desired; or the collection might be graphically representable by any of a whole class of equivalent graphical structures. There is, moreover, an initial choice to make as to just which variables should be included in the representation. Clearly we should include any that are of direct interest, or that are observable and related to quantities of direct interest. But it is often also helpful to include latent variables, which are unobservable but serve to relate the other variables. Typically models including such latent variables will be both more meaningful and simpler than those dealing only with variables of direct interest. Even when some of the latent variables are purely fictitious (such as nodes in internal layers of neural network models), such extended models can be valuable in allowing more compact and powerful representation of the conditional independence relationships holding between the observable variables. However, latent variable models may also impose probabilistic constraints over and above those expressible as conditional independence properties (Spirtes et al. 1993; Settimi and Smith 1998).

When choosing or assessing a graphical model, its intended semantics must be carefully delineated. In this connection it is important to distinguish between *belief networks* and *causal networks*. Belief networks are only intended to represent relationships between variables that can be regarded as arising from some overall joint distribution, this distribution being supposed stable over the various individual cases to which the network might be applied. This is the point of view which we have taken so far in this book. If there are equivalent belief networks, representing the same conditional independence relationships, it does not really matter which one we use (although if we are searching through a collection of possible graphical representations, it is important to be clear which of these are equivalent to each other, so as not to waste effort in trying to select between these).

Networks which are regarded as causal, on the other hand, can express distinctions between graphical representations that are distributionally equivalent. This is because they are intended to express more powerful properties than simple belief networks. Specifically, they assert the stability of some of the probabilistic ingredients when others are disturbed in various ways. For example, in a simple DAG model, the property $v \perp\!\!\!\perp \mathrm{nd}(v) \mid \mathrm{pa}(v)$ might be causally interpreted as asserting that the table of conditional probabilities for X_v given $X_{\mathrm{nd}(v)}$ not only depends solely on the values of $X_{\mathrm{pa}(v)}$, but also that it would still be the appropriate table, even if the values of some or all of the variables in $X_{\mathrm{nd}(v)}$ were set by outside intervention, rather than merely left to arise internally from their own specified joint distribution. In this way a causal network provides a model for, not just one, but a whole constellation of related situations, differing according to their specifications as to which variables are set externally, but related by the assumption of common conditional probability tables for the unset variables. With such extended semantics, two different networks which would be equivalent as belief networks are typically inequivalent as causal networks, since they say different things about the effects of interventions.

We see from this discussion that, even if we have succeeded in divining (e.g., on the basis of extensive data) just which properties of conditional independence hold between the observable variables, there is still further work to do to extract the most appropriate graphical representation. Methods which address this problem have become known as *constraint-based* methods. The principal uncertainty involved relates to the choice of the 'correct' graphical structure to represent the conditional independence or other qualitative inputs, which are taken as given. This approach underlies much of the work in discovering causal structure developed by, for example, Glymour et al. (1987) and Pearl (1995).

11.5 Classes of models

11.5.1 Models containing only observed quantities

In this section we try to put some of the literature on structural learning of graphical models into the context which we have established above. There has been much work concerned with graphical modelling from a standard statistical perspective (see Lauritzen (1996) for a full treatment).

Undirected graphs. Undirected graphical models of various kinds have long been of interest to statisticians. Such statistical models include log-linear models for discrete data (Edwards and Havránek 1985, 1987; Andersen et al. 1991), multivariate normal models for continuous data (Wermuth 1976; Whittaker 1990), and mixed continuous and discrete models (Lauritzen and Wermuth 1989; Edwards 1990, 1995; Lauritzen 1996). The statistical literature has largely used selection strategies based on maximized likelihood to construct on directed graphical models from data. For a marginal likelihood approach, we further need to specify suitable priors for the parameters of our undirected models. In the case of decomposable graphical models, priors having the hyper Markov property of Dawid and Lauritzen (1993) (see Section 9.9) are particularly convenient. Giudici (1996) has applied and extended these ideas to structural learning of undirected Gaussian models, using priors related to the inverse Wishart distribution.

Directed graphs. For discrete directed networks, Cooper and Herskovits (1992) applied the marginal likelihood criterion (11.3) to evaluate and compare different directed belief networks. This was extended by Heckerman et al. (1995b). Because they consider pure belief networks, with no attempt to interpret them as causal explanations, they require that any set of data should assign an identical score to two structurally different but Markov equivalent belief networks, having different directions on some of the arrows but implying identical conditional independence properties. This imposes relations between the prior distributions used for the parameters of Markov equivalent models; more details are given in Section 11.6.3. An alternative approach would be to represent each Markov equivalence class by a suitable representative or characterization (Section 5.5.1) and assign a score directly to this.

In contrast, when a network is intended to be understood causally, the directions of the arrows are crucial to its interpretation. We might wish to try to extract some information on this directionality from data. Ideally for such causal inference we would have data collected under a variety of conditions, in which some of the variables were set and others were not. Then we might be able to distinguish between two distinct causal networks which, as belief networks, were Markov equivalent and so assign them different scores. Commonly, however, the only data available relate to cases in which variables arise naturally from their joint distribution, with no perturbations or interventions. Additional assumptions are then required if causal structure is to be learnable from such data. One approach is to assume that all learned conditional independence properties of the joint distribution are reflected in its underlying graph, and that no additional conditional independence relationships occur due to the particular values of the parameters of the model. This requirement of 'faithfulness' is related to that of 'strong completeness' (Section 5.5). Then any arrow that appears with identical orientation in all graphical representations of the conditional independence relationships discovered is assumed to have a causal interpretation. This controversial area is explored from a maximized likelihood perspective by, for example, Spirtes et al. (1993).

As the number of parents of a node increases, the node's conditional probability table grows exponentially. One can impose some structure on this table in order to reduce the overall complexity of the model (Buntine 1994; Chickering et al. 1997). Suggestions have included structuring the tables as classification trees (Friedman and Goldszmidt 1998) or as neural networks (Monti and Cooper 1997).

Gaussian directed graphical models are considered by Geiger and Heckerman (1994), while Lauritzen and Wermuth (1989) treat directed models containing a mixture of discrete and continuous quantities as a special case of chain graph models.

From one point of view, the whole of statistical regression analysis can be regarded as a special case of the construction of a directed graph to relate observable quantities. There is a huge literature on model selection in linear models, primarily based on maximized likelihood.

Chain graphs. Lauritzen and Wermuth (1989), Wermuth and Lauritzen (1990) and Lauritzen (1996) consider the problem of selecting a chain graph structure based on complete data, based on maximum likelihood approach and substantive knowledge. The programs DiGram (Kreiner 1989) and Bifrost (Højsgaard and Thiesson 1995) select chain graph models based upon maximized likelihood.

There is also a large literature, predominantly in the social sciences, on constructing structural equation models, which are closely related to the alternative interpretations of chain graphs mentioned in Section 5.5.2.

11.5.2 Models with latent or hidden variables

So far we have considered simple graphical representations of the relationships between observable quantities. A very broad and flexible extension is to allow our models to include, in addition, unobserved latent variables. The most utilized latent variable model is the naïve Bayes model, in which observable features are modelled as conditionally independent, given an underlying latent 'cause.' In Section 2.8 we introduced this model as a potential diagnostic tool, assuming that it was possible to observe the underlying disease variable on the training set of cases. However, even when this is not the case, by including a latent variable as a common parent of all features we can obtain a simple explanation, expressed purely in terms of conditional independence, for the observed dependence between the feature variables. This model can be used as a basis for a clustering or unsupervised learning procedure, as embodied in the AUTOCLASS program (Cheeseman and Stutz 1997). In a causal modelling setting, the TETRAD program (Glymour et al. 1987) infers the existence of hidden variables that can explain observed relationships.

When we allow latent variables, we extend enormously the number and the complexity of the models that we can entertain. Indeed, even for a given model structure, we have a good deal of freedom in specifying the number of states, or more generally the state space, of a latent variable. Scoring alternative structures is made difficult by the fact that, even with simple conjugate prior distributions, the marginal likelihood no longer factorizes and exact calculation is generally impossible (see Section 9.5). Again, approximations are necessary: the Cheeseman-Stutz method (Cheeseman and Stutz 1997) has been found useful (Geiger et al. 1998).

A further attractive extension is to allow the conditional independence structure relating the observables to vary, according to the values of the latent variables. Experience with such 'mixture of DAGs' models is described by Thiesson et al. (1998).

For applications involving processes that evolve over time, a natural representation is in terms of hidden Markov models, in which the relationship between adjacent observations is explained through their common dependence on an unobserved Markov process (Smyth et al. 1997). Further generalization leads into the extensive areas of probability models for pattern recognition and neural networks (Ripley 1996).

11.5.3 Missing data

By definition, values of latent variables are never observed. In addition, in a particular set of data, there may be other nodes whose values are missing in some or all cases. As in Section 9.5, we can distinguish between situations where the data are missing at random, and those where the data-censoring mechanism is non-ignorable. In the former case, provided that all factors possibly influencing the pattern of selection are included as variables in our model, any likelihood-based assessment will not be biased by the censoring mechanism. The technical issues then centre on constructing suitable approximations for scores based on incompletely observed cases, exactly as in the context of latent variables (Chickering and Heckerman 1997). Approximation methods such as those of Section 9.5 can again be applied to this problem.

11.6 Handling multiple models

The graphical models which may be under consideration as possible candidate structures can be exceedingly numerous, and it will generally be impossible even to list them, let alone specify any detailed additional structure (such as probability tables or prior distributions) for them all. This raises a number of issues, of which the most important are the need for a suitable data-driven strategy to search through a very large model space, and the need for some fairly automatic way of imposing additional structure.

11.6.1 Search strategies

The cardinality of the space of possible structures grows extremely rapidly — super-exponentially — with the number of nodes in the graph. Robinson (1977) gives the following recursive formula (no closed form formula is known) for calculating the number $f(n)$ of unlabelled DAGs on n nodes:

$$f(n) = \sum_{i=1}^{n} (-1)^{i+1} \frac{n!}{(n-i)!i!} 2^{i(n-i)} f(n-i). \tag{11.6}$$

For example, there are approximately 4.2×10^{18} directed acyclic graphs having 10 nodes. For decomposable graphs the figure is smaller but still immense; for example, there are about 2×10^{11} decomposable graphs with 10 nodes.

Chickering (1996) shows that the general problem of determining the optimal structure is NP-hard, and so since an exhaustive analysis of all possible structures is generally infeasible a selective search algorithm needs to be established. It seems appropriate for this to be local, in some sense, to the original graph. In particular, in examples such as Figure 3.1, which have a clear block structure, we might be convinced by the partial order expressed by the blocks, and hence not entertain thoughts of links going from lower to higher blocks, but might question the existence of links between blocks or from higher to lower blocks.

Heckerman (1998) describes a number of search strategies that have been adopted for comparisons based on scoring, noting that with complete data the factorization of the score associated with any model makes it straightforward to carry out a search from any current model by evaluating the effect of adding or removing all feasible arcs. Thus we perform a walk through the model space by adding or dropping single links at a time. The walk can be performed either deterministically or randomly. Gabriel (1969) proposed two *coherence* rules for pruning this search space. One rule is that if a model m is rejected by some scoring criterion, then all models that are subgraphs of m are also rejected. The reason for this rule is that if m scores badly, then the set of the conditional independence relationships implied by m is rejected; but a sub-model will have even more conditional independence relationships and so it too should be rejected. The second rule is that m is not rejected, then models which have m as a subgraph are also not rejected. The search strategy of Edwards and Havránek (1985) uses both these rules; Madigan and Raftery (1994), undertaking a model-averaging search, use only the first.

Using either or both rules, one can generate from a model m, passing the current test, new models typically differing by one edge, while avoiding generating models that have already failed, or any of their sub-models. With complete data, changes to the posterior probability can be expressed by functions local to the families (in the case of directed graphs) or cliques (in the case of undirected decomposable graphs) where the link-changes have taken place.

There will often be qualitative prior information that specifies constraints on the ordering of the variables, which can be used to reduce the search space. This has been used, for example, by Cooper and Herskovits (1992) in their reconstruction of the ALARM network, and in the BIFROST program (Lauritzen et al. 1994).

Within the constraint-based framework, Spirtes et al. (1993) review a number of search algorithms for recovering an assumed causal structure underlying a joint distribution; for example, their PC algorithm first constructs a preliminary undirected graph, and then systematically attempts to remove edges based on conditional independence tests of increasing order, exploiting the global Markov property or, equivalently, d-separation.

If a search strategy allows the addition of an edge, for directed graphs one must be careful that one does not create a directed cycle. This can be avoided by defining a prior ordering of the nodes, so that any edges added are always directed from, say, the first node to the second node with respect to the ordering. This is used in the K2 search heuristic of Cooper and Herskovits (1992), who survey a number of other approaches.

For decomposable graphs one must be careful that adding or removing an edge does not destroy the decomposability of the graph. Lauritzen (1996) shows that if an edge belongs to only one clique of a decomposable graph $\mathcal{G}$, then on removing that edge the resulting graph is also decomposable (see also Kjærulff (1994)). Also, given two decomposable graphs on n vertices, $\mathcal{G}_1$ and $\mathcal{G}_2$, it is possible to go from one graph to the other by a sequence of single-edge additions and deletions keeping at all intermediate stages within the class of decomposable graphs (Lauritzen 1996).

It would be most efficient if it were possible to conduct both structural learning and parameter learning in a single unified procedure. Since, even for a purely local search, it may be very expensive to evaluate the score for all new candidate structures, approximations using expected scores calculated under the current model may be used to guide the search. One implementation of this idea is the *structural EM algorithm* (Friedman 1998). There have also been developed Markov chain Monte Carlo simulation schemes to explore multiple model structures and their parameter spaces (Madigan and York 1994). Since the structures involved will have different complexity, particular care has to be taken in setting the probabilities of jumps between different models (Green 1995).

11.6.2 Probability specification

Serious practical implementation difficulties arise when we have a very large collection of models and wish to associate a probabilistic structure with each one. Although this problem affects any kind of graphical modelling, we shall here largely restrict attention to discrete DAG models.

The principal interest is in models with unknown parameters, where we wish to specify a prior distribution for the parameter of each model. However, we start by considering the simpler case where we wish to specify exact numerical probability tables for every model considered. It is not realistic to expect that suitable values can be individually assessed for every member of such a large class of models. This problem is especially acute for models generated by an automatic model search procedure.

One approach is to perform a careful assessment for just one baseline model (preferably one which is in some sense as unconstrained as possible), and then attempt to use this to induce the values for other models, so as to be somehow compatible with the baseline assessment. One appealing way is to choose the values in the new structure so as to minimize some measure of the distance of the associated overall joint probability distribution Q for all the variables from P, the joint distribution implied by the baseline assessments. Since the fixed baseline distribution P and the variable distribution Q on the new structure do not play symmetrical rôles in this formulation, we need not insist on symmetry of the distance function. Here we shall confine attention to the Kullback–Leibler divergence:

$$K(P,Q) = \mathrm{E}_P\{\log p(X)/q(X)\} = \sum_x p(x)\log\{p(x)/q(x)\},$$

where p and q are the multivariate probability (density) functions for P and Q. This vanishes when $Q = P$, and is otherwise positive.

Given P, we thus search for a distribution Q, factorizing according to the graphical structure under consideration, so as to minimize $K(P,Q)$; or, equivalently, so as to maximize $\mathrm{E}_P \log q(X)$. Note that, when Q is completely unconstrained, this expression is maximized by taking $Q = P$.

In the case that Q is constrained to be directed Markov over a DAG $\mathcal{D}$, there is another appealing way of approximating P: for any node $v \in V$ we can calculate, from P, the conditional distribution for v given its parents in $\mathcal{D}$, and use the collection of conditional probability tables so formed to define the probabilistic structure of Q. We shall denote this directed Markov distribution on $\mathcal{D}$ by $P^{\mathcal{D}}$. When P is itself directed Markov on a baseline structure, we can use probability propagation in that structure to determine the desired conditional probabilities.

The following theorem shows the equivalence of the above two approaches.

Theorem 11.1 *Among all distributions Q that are directed Markov with respect to $\mathcal{D}$, $K(P,Q)$ is minimized by the choice $Q = P^{\mathcal{D}}$.*

Proof. The proof is by induction on the size of $\mathcal{D}$. The result is trivially true if $\mathcal{D}$ has only one node. Otherwise, let v be a terminal node of $\mathcal{D}$, and $\mathcal{D}'$ the subgraph of $\mathcal{D}$ induced by the set $A = V \setminus \{v\}$. Define $S = \mathrm{pa}(v)$. It is easy to see that

$$\mathrm{E}_P \log q(X) = \mathrm{E}_{P_A} \log q_A(X_A) + \mathrm{E}_{P_S} \mathrm{E}_{P_{v \mid S}} \log q_{v \mid S}(X_v \mid X_S).$$

By the inductive hypothesis applied to $\mathcal{D}'$, the first term is maximized by the choice

$$Q_A = P^{\mathcal{D}'} = P_A^{\mathcal{D}},$$

while the choice $Q_{v \mid S} = P_{v \mid S}^{\mathcal{D}}$ maximizes the second term. Hence, the induction is established. □

This result for the special case that each node in $\mathcal{D}$ has at most one parent is given by Pearl (1988), Theorem 1, §8.2.1. A similar result can be established when we wish to specify a Markov distribution Q on a decomposable undirected graph $\mathcal{G}$, so as to be compatible with a baseline distribution P: the choice of Q that minimizes the Kullback–Leibler divergence $K(P,Q)$ is given by the unique $\mathcal{G}$-Markov distribution $P^{\mathcal{G}}$ whose clique marginals agree with the relevant marginal distributions constructed from P.

In general, there are no similar results that directly pertain to finding the best unknown structure. However, finding the best representation of a probability distribution in terms of an undirected tree has a simple solution, as shown by Chow and Liu (1968). First we define the *cross-entropy* between X and Y as

$$H(X,Y) := \mathrm{E}_{P_{XY}} \log \frac{p(X,Y)}{p(X)\,p(Y)} = K(P_{XY}, P_X \otimes P_Y).$$

We then have:

Theorem 11.2 *Among all undirected trees $\mathcal{T}$ and distributions Q which are Markov with respect to $\mathcal{T}$, $K(P,Q)$ is minimized by the choices $\mathcal{T} = \mathcal{T}^*$, the maximum weight spanning tree with weight $H(X_u, X_v)$ on edge $\{u,v\}$, and $Q = P^{\mathcal{D}^*}$, where $\mathcal{D}^*$ is any rooted version of $\mathcal{T}^*$ with directions on edges pointing away from the root.*

See Pearl (1988) §8.2.1 for a proof of this and other results along these lines.

Causal ordering. It is often appropriate to arrange the variables in the problem in a *causal order*, with arrows only allowed to point from lower ordered to higher ordered variables. Suppose that we have two DAGs, $\mathcal{D}_0$ corresponding to a baseline model, and an alternative structure $\mathcal{D}$, both respecting the causal ordering. Our baseline probability distribution P is supposed to be directed Markov with respect to $\mathcal{D}_0$.

Fix a particular node v. In moving from $\mathcal{D}_0$ to $\mathcal{D}$, v will retain some of the parents it had in $\mathcal{D}_0$, lose others, and gain some entirely new parents. Denote the set of these Intersecting, Old, and New parents of v by I, N, and O, respectively. Because of the common causal ordering, no member of N can be a descendant of v in $\mathcal{D}_0$, whence it follows that, under P, $v \perp\!\!\!\perp N \mid (O, I)$. Hence, denoting a typical configuration of X_N by n, etc., and a typical value of X_v by j, we have

$$\begin{aligned} p^{\mathcal{D}}(j \mid n, i) &= p(j \mid n, i) \\ &= \sum_o p(j \mid o, i)\, p(o \mid n, i), \end{aligned} \tag{11.7}$$

where $p(o \mid n, i)$ is evaluated (for example, by probability propagation on $\mathcal{D}_0$) from the distribution P. The term $p(j \mid o, i)\, p(o \mid n, i)$ can be considered as an 'expansion' of the baseline probability table $p(j \mid o, i)$ in $\mathcal{D}_0$ to incorporate the new parents in $\mathcal{D}$, while the summation 'contracts out' the contribution of the old parents. If the set of parents of v for $\mathcal{D}$ contains that for $\mathcal{D}_0$ (so that O is empty), the original probability table is simply duplicated for each configuration of the new parents: the property $v \perp\!\!\!\perp N \mid I$ is retained, but now encoded into the numerical probabilities rather than into the graphical structure.

11.6.3 Prior information on parameters

Still in the context of discrete DAG models, suppose now that we wish to allow for parameter uncertainty, by associating a prior distribution with each graphical structure. For some purposes it is important to distinguish between the cases for which we do or do not have a pre-specified causal ordering. In the latter case, we can in principle consider all possible DAGs relating the variables; in the former, we restrict attention to those respecting the causal ordering. Of course, intermediate positions, where we only have a partial ordering on the variables, are also possible.

For simplicity, we choose to express prior uncertainty, for any of the models under consideration, by means of a conjugate *product Dirichlet* distribution, as introduced in Example 9.4. For a given graphical structure $\mathcal{D}$, this will have a hyperparameter $\alpha = (\alpha^v : v \in V)$. The marginal distribution of X will again be directed Markov on $\mathcal{D}$, having the mean conditional probability tables given by (9.11). In addition, for each parent configuration ρ of each variable v we have an associated precision $\alpha_+^{v,\rho}$. Thus, such a prior can be specified by its underlying graph, its marginal probability distribution, and its collection of precisions. Given such a specification on a baseline DAG $\mathcal{D}_0$, we require some general procedure for constructing a compatible specification for some other structure $\mathcal{D}$.

For the case of no prior causal ordering, Heckerman et al. (1995b) introduced some simple and appealing qualitative requirements as to how prior distributions for different models should be related. They showed that these can in fact only be satisfied when those priors are all product Dirichlet. In addition, we must require $\alpha_+^{v,\rho} = K\,P(X_{\mathrm{pa}(v)} = \rho)$, where P is the associated marginal distribution, and K is a constant, the *overall precision*, common to all the structures. It then follows that $\alpha_j^{v,\rho} = K\,P(X_v = j, X_{\mathrm{pa}(v)} = \rho)$. Hence, once we have specified the marginal distribution, we have only one degree of freedom left for the precisions. We term such a distribution *directed hyper Dirichlet*.

Under their conditions, if we start with a baseline directed hyper Dirichlet prior over a complete graph, specified by a marginal distribution P and an overall precision parameter K, the compatible specification for any other structure $\mathcal{D}$ will be the directed hyper Dirichlet distribution over $\mathcal{D}$, with the same overall precision K, whose marginal distribution is just the 'minimum Kullback–Leibler divergence' approximation $P^{\mathcal{D}}$ to P, matching the parent-child conditional probability tables, as described in Section 11.6.2. This construction is analogous to that of Dawid and Lauritzen (1993) for the discrete decomposable undirected case, in which the undirected hyper Dirichlet prior is represented by its mean probability values and a single number to represent precision, and indeed the two approaches are essentially equivalent if we confine attention to perfect DAGs, which are Markov equivalent to decomposable undirected graphical models. If we regard hyperparameter values as derived from prior data counts, then this approach can be regarded as using a fixed prior sample to define the hyperparameter for any of the models considered.

11.6.4 *Variable precision*

Because it has so few free hyperparameters, the directed hyper Dirichlet prior might be regarded as too inflexible to represent realistic prior opinion: even in the above situation of no causal ordering, we might wish to use variable precisions, expressing differing degrees of confidence about different parts of the network structure. We then require a method which, starting with a general baseline product Dirichlet distribution on a baseline DAG, allows us to construct a compatible general product Dirichlet distribution for a new DAG structure. As a first step, we might carry out the Kullback–Leibler divergence construction described above to relate the expected probability values across models. However, we also need to relate parameter precisions across different models, and this is less straightforward.

Expansion and contraction. Spiegelhalter et al. (1993) proposed a method of *expansion and contraction*, somewhat analogous to (11.7), to relate precisions. If α is the hyperparameter of the baseline model over $\mathcal{D}_0$, and β that of the new model over $\mathcal{D}$, we take (for a node v, omitted from the notation):

$$\beta_+^{(n,i)} = \sum_o \alpha_+^{(o,i)} p(n \mid o, i), \tag{11.8}$$

where p is the baseline marginal distribution. We can then construct $\beta_j^{(n,i)} = \beta_+^{(n,i)} p(j \mid n, i)$, where $p(j \mid n, i) = p^{\mathcal{D}}(j \mid n, i)$ is given by (11.7) in the case of a common causal ordering.

If the baseline model is directed hyper Dirichlet with overall precision K, we shall have $\alpha_+^{(o,i)} = K\, p(o, i)$, so that (11.8) yields $\beta_+^{(n,i)} = K\, p(n, i)$, and we obtain $\beta_j^{(n,i)} = K\, p(j, n, i)$. However, typically this will not be directed hyper Dirichlet on $\mathcal{D}$, since we need not have $p^{\mathcal{D}}(j, n, i) = p(j, n, i)$.

In general, this procedure gives $\sum_n \beta_+^{(n,i)} = \sum_o \alpha_+^{(o,i)}$. In particular, the total precision, summing over all parent configurations of node v, is the same for both $\mathcal{D}_0$ and $\mathcal{D}$. If, with a common causal ordering, the parent set for v in $\mathcal{D}_0$ is entirely contained in that for $\mathcal{D}$, so that O is empty, we obtain $\beta_j^{(n,i)} = \alpha_j^i\, p(n \mid i)$, and thus the corresponding expected probabilities will be duplicated, as described at the end of Section 11.6.2; while the precision associated with any original set of parents is distributed over the various possible configurations of new parents according to the conditional baseline probability distribution over such configurations. In particular, if the parent sets coincide in the two models, the associated hyperparameter tables will be identical: thus the method has the property of *parameter modularity* (Heckerman et al. 1995b).

Predictive matching. Cowell (1996) suggested that two priors in possibly different models can be considered close if their predictive distributions for a sequence of future observables are close.

This suggests a multi-stage matching procedure. The first stage matches the predictive distribution of the next observation, using minimum Kullback–Leibler divergence. Subject to this, the second stage minimizes the Kullback–Leibler divergence between the predictive distributions for the next two observations. If this does not determine a unique prior, one can go on to consider still more observations. For the problem of selecting a directed hyper Markov prior, it turns out that it is sufficient to look at just two observations.

The second stage can be expressed as searching for a prior, with mean Q given by the first stage, to minimize

$$\mathrm{E}_P\{K(P_X, Q_X)\} = \sum_x p(x) \sum_y p_x(y) \log \{p_x(y)/q_x(y)\}, \tag{11.9}$$

where P_x denotes the baseline posterior predictive distribution for the second future case Y after observing that the next case X takes value x; and similarly for Q_x, based on a candidate prior.

In general, minimizing (11.9) is difficult, but some specific results are available (Cowell 1996). In particular, suppose that our baseline prior is directed hyper Dirichlet over $\mathcal{D}_0$, and our candidate priors are directed hyper Dirichlet over $\mathcal{D}$, where both DAGs share a common causal ordering. When the parent set I for v in $\mathcal{D}_0$ is a subset of its parent set $N \cup I$ in $\mathcal{D}$, it can be shown that the precision in $\mathcal{D}$ to be attached to any parent configuration (n, i) of v is identical to the precision in $\mathcal{D}_0$ associated with the parent configuration i. Taking this in conjunction with the duplication properties of the tables of mean probabilities, we see that each baseline vector $(\alpha_j^i : j = 1, \ldots, k)$ of hyperparameters is duplicated for every configuration n of the originally irrelevant parents. This is in contrast to the behaviour of the expand-and-contract method, where the initial precision is divided up, rather than duplicated. However, the method again satisfies the principle of parameter modularity.

The above result is readily applied to automatic model searches where new candidate models are generated by incrementally adding directed edges to existing DAGs.

Epilogue

It should be clear, particularly from the last chapter of this book, that the field of probabilistic networks and expert systems is developing rapidly, and no overview could hope to stay up-to-date for long. For this reason we have deliberately chosen to focus on a solid exposition of a limited class of issues, with some confidence that the fundamental methodology will remain of interest and continue to provide a basis for insights into the new challenges that will come about.

Such challenges are arising as the ideas associated with probabilistic expert systems spread far beyond their original focus on fixed networks of discrete variables with precisely specified probabilities. Real applications increasingly feature uncertainty about both quantitative and structural aspects of the model, possibly incomplete data, and a mixture of discrete and continuous nodes whose parent-child relationships may need to be expressed as statistical models. Dynamic networks, in which the structure changes in order to model some underlying evolving process, are a particularly important development whose application extends into the broader arena of signal processing and on-line monitoring.

The problems associated with such models are being tackled by a variety of professional disciplines, which makes the research area very exciting but also difficult to predict. Current developments suggest that the future will see an increasingly unified approach to complex stochastic systems that exploit conditional independence both for graphical representation and as a foundation for computation, and this perspective will incorporate artificial intelligence, signal processing, Bayesian statistics, machine learning, and a host of other disparate topics. By laying out a firm foundation for some of the core ideas in probabilistic networks, we hope that we will have helped others to extend their application to currently unimagined horizons.

Appendix A
Conjugate Analysis for Discrete Data

A.1 Bernoulli process

Suppose that we observe a sequence of n independent Bernoulli trials, on each of which the probability of success is θ (and thus the probability of failure is $1 - \theta$). The data d will consist of a sequence of successes and failures of length n. Denote by a, b, respectively, the numbers of successes and failures in d (thus $a + b = n$). Then the probability of observing the sequence d is

$$p(d \mid \theta) = \theta^a (1 - \theta)^b. \tag{A.1}$$

If we do not keep the full information in d, but only the numbers a and b of successes and failures, we have instead (assuming that the total sample size was fixed in advance) the binomial formula

$$p(a, b \mid \theta) = \frac{(a+b)!}{a!\,b!}\,\theta^a (1 - \theta)^b. \tag{A.2}$$

Alternatively, the data may have been collected by generating successive trials until exactly a successes had been observed. In this case (A.1) is still valid, but (A.2) is replaced by the negative binomial formula

$$p(a, b \mid \theta) = \frac{(a+b-1)!}{(a-1)!\,b!}\,\theta^a (1 - \theta)^b. \tag{A.3}$$

Other more complex observation schemes can be envisaged. Suppose the observation scheme is *non-informative*, i.e., the reported sequence d incorporates all the data taken into account in deciding when to stop observation. Then (a, b) is a *sufficient statistic*, and the likelihood function for θ in any such case will again satisfy

$$L(\theta) \propto \theta^a (1-\theta)^b. \tag{A.4}$$

Suppose that the probability distribution representing prior uncertainty about the value of θ can taken to be the Beta distribution $\mathrm{Be}(\alpha, \beta)$, with density $b(\theta \mid \alpha, \beta)$ given by

$$b(\theta \mid \alpha, \beta) \propto \theta^{\alpha-1}(1-\theta)^{\beta-1}. \tag{A.5}$$

The normalization constant, by which the right-hand side of (A.5) must be divided to give equality, is

$$\int_0^1 x^{\alpha-1}(1-x)^{\beta-1} dx = \frac{\Gamma(\alpha)\Gamma(\beta)}{\Gamma(\alpha+\beta)}, \tag{A.6}$$

where $\Gamma(\cdot)$ denotes the standard gamma function. As a function of θ, the prior density has the same general mathematical form as the likelihood. The non-negative quantities α and β, determining the specific member of the Beta family used, are *hyperparameters*. With this prior the expected probability of success $\mathrm{E}(\theta)$ is given by $\alpha/(\alpha+\beta)$, and this also gives the marginal probability of success on a single trial. The variance of θ is given by

$$\mathrm{var}(\theta) = \frac{\alpha\beta}{(\alpha+\beta)^2(\alpha+\beta+1)}. \tag{A.7}$$

Combining the prior density (A.5) with the likelihood $L(\theta)$ from (A.4), using Bayes' theorem, yields the posterior density

$$p(\theta \mid d) \propto \theta^a(1-\theta)^b \times \theta^{\alpha-1}(1-\theta)^{\beta-1} = \theta^{a+\alpha-1}(1-\theta)^{b+\beta-1}. \tag{A.8}$$

That is, the posterior distribution of θ is $\mathrm{Be}(a+\alpha, b+\beta)$. We see that, if the prior is in the family of Beta distributions, then so too will be the posterior, based on any observations. We say that the Beta family is *conjugate* for sampling from the Bernoulli process.

For prediction, or for Bayesian model comparison (see Section 11.3), we need the marginal probability of the data, given by

$$p(d \mid \alpha, \beta) = \int p(d \mid \theta)\, b(\theta \mid \alpha, \beta)\, d\theta.$$

This is easily shown to yield

$$p(d \mid \alpha, \beta) = \frac{\Gamma(\alpha+\beta)}{\Gamma(\alpha)\Gamma(\beta)} \frac{\Gamma(\alpha+a)\Gamma(\beta+b)}{\Gamma(\alpha+\beta+a+b)}. \tag{A.9}$$

If, instead of the full data-sequence d, we want the marginal probability of observing the sufficient statistic values (a, b), we must multiply (A.9) by $(a+b)!/a!\,b!$ for binomial sampling, or by $(a+b-1)!/(a-1)!\,b!$ for negative binomial sampling. However, so long as we use the same data in all cases, such additional factors will not affect marginal likelihood comparisons among competing models.

A.2 Multinomial process

Now consider a discrete random variable X that can take on one of a set of k mutually exclusive and exhaustive values $\{x_1, \dots, x_k\}$ with probabilities $\theta \equiv (\theta_1, \dots, \theta_k)$, where $\theta_i > 0$, $\sum_i \theta_i = 1$. That is, $P(X = x_i \mid \theta) = \theta_i$, for $i = 1, \dots, k$. For a sequence d of n independent observations distributed as X, of which n_1 are of type x_1, n_2 of type x_2, etc., the likelihood is given by

$$p(d \mid \theta) = \prod_{i=1}^{k} \theta_i^{n_i}, \tag{A.10}$$

and again a formula proportional to (A.10) will give the likelihood based on the sufficient statistic $(n_1, \dots, n_k)$, under an arbitrary (non-informative) observation process. For example, if n is fixed in advance, we have the multinomial formula

$$p(n_1, \dots, n_k \mid \theta) = \frac{n!}{\prod n_j!} \prod_j \theta_j^{n_j}. \tag{A.11}$$

Suppose that the probability distribution representing prior uncertainty about θ is the Dirichlet distribution $\mathcal{D}(\alpha_1, \dots, \alpha_k)$ having hyperparameter $\alpha \equiv (\alpha_1, \dots, \alpha_k)$ with each $\alpha_i > 0$. That is, its density is

$$p(\theta) = \frac{\Gamma(\alpha_+)}{\prod_{i=1}^{k} \Gamma(\alpha_i)} \prod_{i=1}^{k} \theta_i^{\alpha_i - 1}, \tag{A.12}$$

where we have introduced $\alpha_+ := \sum_i \alpha_i$. The marginal probability of observing $X = x_j$ is $\mathrm{E}(\theta_j) = \alpha_j/\alpha_+$.

The Dirichlet density again has a similar mathematical form to the likelihood, and thus the family of all Dirichlet distributions is conjugate for the multinomial sampling process. Posterior to observing data d (or the sufficient statistic $(n_1, \dots, n_k)$), the revised distribution for θ will have density

$$p(\theta \mid d) \propto \prod_{i=1}^{k} \theta_i^{n_i + \alpha_i - 1}, \tag{A.13}$$

i.e., the posterior distribution is Dirichlet $\mathcal{D}(\alpha_1 + n_1, \ldots, \alpha_k + n_k)$. Again, the posterior is obtained by simple updating of each of the component hyperparameters, by adding on the number of cases observed to yield the associated outcome.

The marginal probability of the data (needed for example for purposes of prediction or model comparison) can be shown to be

$$p(d \mid \alpha_1, \ldots, \alpha_k) = \frac{\Gamma(\alpha_+)}{\prod_j \Gamma(\alpha_j)} \frac{\prod_j \Gamma(\alpha_j + n_j)}{\Gamma(\alpha_+ + n)}. \tag{A.14}$$

Under multinomial sampling with fixed total sample size n, the marginal probability of the sufficient statistic, i.e., of the counts $(n_1, \ldots, n_k)$, is obtained by multiplying (A.14) by $n!/\prod_j n_j!$.

Appendix B
Gibbs Sampling

The probabilistic networks considered in Chapters 6 and 7 allow exact local computation for the special cases of discrete domains, Gaussian domains, and mixed discrete-Gaussian domains. While useful for many applications, this imposes limitations of two different natures in applications. While quite large and complex networks of discrete random variables have been handled successfully, there can come a point at which the cardinality of the clique and separator potentials become too large to manipulate with current computers. An example of this arises in the analysis of time-series using dynamic graphs (Kjærulff 1995). A second limitation is that the distributions handled are too restrictive for many applications, and that more general distributions are desirable.

These sets of problems are approximately solved by the stochastic simulation technique known as Markov chain Monte Carlo (MCMC), which simulates samples from the required posterior distribution and uses these samples to estimate expectations of desired quantities. A particular form of MCMC is called *Gibbs sampling*, and can be shown to be especially appropriate for use on graphical models.

B.1 Gibbs sampling

Suppose that we have a discrete set of random variables $X = \{X_v : v \in V\}$, whose joint distribution P has *everywhere positive* density $p(x) > 0$.

Suppose further that we can find the *full conditional* distribution,

$$\pounds(X_v \mid X_{V\setminus\{v\}}), \tag{B.1}$$

for every variable. Then one can generate a sample from P as follows. First one initializes each of the n variables in X to some permissible configuration $(x_1^0, x_2^0, \dots, x_n^0)$. Next, using (B.1) one samples in turn each individual variable X_v from its distribution conditional upon the current instantiation of the other variables, replacing the current instantiation of the variable with its sampled state. This process is repeated. Thus, at the (i,j)th stage we have a current instantiation $(x_1^j, x_2^j, \dots, x_{i-1}^j, x_i^{j-1}, \dots, x_n^{j-1})$. One then samples the ith variable from $\pounds(X_i \mid x_1^j, x_2^j, \dots, x_{i-1}^j, x_{i+1}^{j-1}, \dots, x_n^{j-1})$, to obtain $X_i = x_i^j$, say. It can be shown that, under broad regularity conditions, if this process is repeated a large number I of times one will obtain a configuration, $(X_1^I, \dots, X_n^I)$ say, whose asymptotic distribution for $I \to \infty$ is P. In fact, one can be more general and not require the variables to be updated in sequence, as long as each is selected infinitely often. Each iteration, i.e., updating of all n variables, constitutes a transition of a Markov chain, whose equilibrium distribution is the joint distribution P.

If we could be confident that the Markov chain had converged to its equilibrium distribution by I iterations, then any desired expectation $\mathrm{E}f(X)$ could be estimated by the sample mean of the function f evaluated at a large number m of consecutive simulated values X^i:

$$\mathrm{E}f(X) \approx \frac{1}{m} \sum_{i=I+1}^{I+m} f(X^i), \tag{B.2}$$

and with probability equal to 1 this is an equality in the limit for $m \to \infty$.

How large I should be so that the asymptotic regime has been reached (i.e., the Markov chain has converged to its equilibrium distribution) is a topic of current research — see, for example, Cowles and Carlin (1996) for a review. The problem is that consecutive samples may be highly correlated, and so it may take a large number of iterations for a sample to 'forget' its initial values.

It is generally considered advisable to carry out a number of long runs starting from widely dispersed starting points (Gelman and Rubin 1992), whose samples can be pooled, once one is confident that the simulations have converged to a common distribution. Reducing the correlations between consecutive iterations (improving 'mixing') is also an important area of current research.

Notice the requirement that $p(x) > 0$. The Gibbs sampler will sometimes work if there are zeros in the probability density, but such zeros must not make one part of the state space become isolated from any other part by single variable changes. Thus, the presence of zeros must not generate constraints that act as barriers to prevent the sampler from roaming over the whole of the possible configuration space. Technically, the resulting Markov chain must be *irreducible*.

An example of a reducible Markov chain is as follows. Suppose we have two variables A and B with a joint density $p(a, b)$ having zeros as shown in Figure B.1. Then, if we initialize the Gibbs sampler anywhere in the top left part, it will never be able to make a transition into the bottom right part, and vice versa. If the other entries are close to but not quite zero, then eventually it will (with probability equal to 1) make a transition from the top left to bottom right, but it may take a large number of iterations, of the order of the inverse of the probabilities. So in practice such 'near-irreducibility' can cause severe convergence problems. Irreducibility is a particular problem in models for genetic pedigrees, where one promising way of overcoming this type of problem has been to update large blocks of variables simultaneously in so-called *blocking Gibbs sampling* (Jensen et al. 1995).

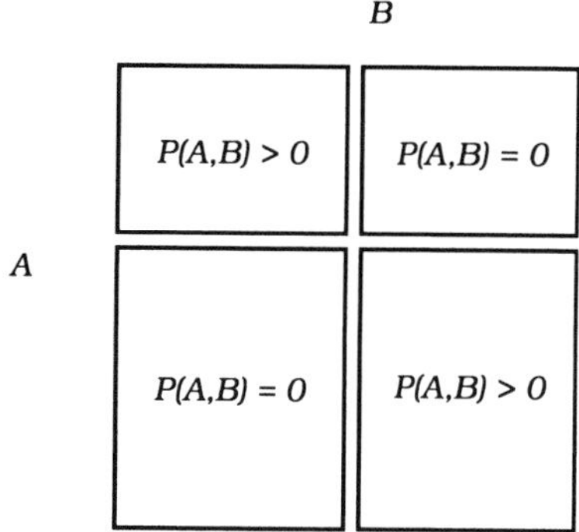

FIGURE B.1. Schematic illustrations of a probability density that can cause problems for a Gibbs sampler. Iterates starting in the top left region remain there and cannot migrate to the bottom right region because of the barrier generated by the zeros of the distribution, and vice versa. Hence, the Markov chain will be reducible.

B.2 Sampling from the moral graph

Having described the basic idea behind the Gibbs sampler, we now describe how it can be applied to probabilistic networks. Consider the simplest case of a probabilistic network consisting of a directed acyclic graph (DAG) of discrete random variables. The recursive factorization of the joint density is given by (5.5):

$$p(x) = \prod_{v \in V} p(x_v \,|\, x_{\mathrm{pa}(v)}). \tag{B.3}$$

To apply the Gibbs sampler, we need to calculate for each node its distribution conditional upon all the other nodes. Now the only terms that involve a node v are the conditional density of X_v given its parents, and the conditional densities associated with each child of v. Taking only these terms yields a function which depends only on x_U, where $U = \{v\} \cup \mathrm{bl}(v)$ and $\mathrm{bl}(v)$ is the Markov blanket of v, defined as a node's parents, children, and co-parents, or equivalently its neighbours in the moral graph (see (5.6)). Hence, we can factorize $p(x)$ into two expressions, one containing all terms that involve x_v and the variables in its Markov blanket $\mathrm{bl}(v)$, and a second expression consisting of the remaining terms, which do not involve x_v and hence are irrelevant to the full conditional distribution. Hence,

$$\begin{aligned} p(x_v \,|\, x_{V\setminus\{v\}}) &= p(x_v \,|\, x_{\mathrm{bl}(v)}) \\ &\propto \text{terms in } p(x) \text{ containing } x_v \\ &= p(x_v \,|\, x_{\mathrm{pa}(v)}) \prod_{w \in \mathrm{ch}(v)} p(x_w \,|\, x_{\mathrm{pa}(w)}). \end{aligned} \tag{B.4}$$

Therefore, to calculate the full conditional distribution for a node v for the Gibbs sampler, one has only to consider the conditional distribution of X_v given its Markov blanket, which is easily calculated by (B.4). The same reasoning holds for undirected and chain graphs, in all cases the Markov blanket of a node being just its neighbours in the relevant moral graph.

Now suppose that one has some evidence $\mathcal{E}$ on some subset of variables. Then Gibbs sampling for the remaining nodes proceeds as above, but keeps the values of the observed nodes fixed throughout the simulation. Two caveats are: (1) the initial configuration must be permissible, and consistent with the evidence, and (2) the evidence must not make the resulting Markov chain reducible.

B.3 General probability densities

In principle, the theory of local computation via message-passing on a junction tree (Chapters 6 and 7) could be applied to probabilistic networks of more general types of random variables, with more complicated dependence of a variable on its parent configuration. The problem is that this generally leads to intractable integrals, so that the local computation cannot be performed in practice. However, this restriction does not apply to the Gibbs sampling procedure. Although the above description was based on discrete variables, it generalizes to arbitrary types of variables and dependencies.

Thus, if we are willing to give up exact local computation for approximate estimation based upon Gibbs sampling, then it is possible to treat much more general types of models. The factorization (B.4) still holds and so again we need only consider terms involving the Markov blanket. Sampling from a distribution proportional to the product of terms in (B.4) is not always straightforward, although a variety of special techniques has been developed.

These ideas form the basis of the BUGS (Bayesian Updating with Gibbs Sampling) computer program, which uses a specially designed high-level language to describe a graphical model (Gilks et al. 1994) and automatically constructs the necessary full conditional sampling procedures. WinBUGS is a prototype version that provides a graphical user interface both for specifying models and viewing results of the simulation. See Appendix C for the Web address.

B.4 Further reading

The use of Gibbs sampling in expert systems was first suggested by Pearl (1987). The PRESS system (Gammerman et al. 1995) incorporates options for both local propagation and Gibbs sampling for mixed discrete-Gaussian models.

Jensen et al. (1995) describe blocking Gibbs sampling, a variant of Gibbs sampling for expert system purposes, when the cliques of a junction tree are too large for practical computation. The basic idea is that variables are removed from the junction tree one at a time in order to reduce the size of cliques; this involves restructuring the junction tree. When the junction tree is of a computationally tractable size for exact local propagation, it is used to generate samples by the method described in Section 6.4.3, by a local exact computation. Thus, one samples clusters or *blocks* of variables simultaneously, so speeding up the convergence of the Gibbs sampler.

Dawid et al. (1995) have examined hybrid propagation in junction trees in which the natures of the potentials in the cliques and separators are heterogeneous and possibly change after receiving messages. Their scheme permits exact computation between some cliques and sampling in others.

The literature on the theory and application of Gibbs sampling is growing rapidly. For an exposition of recent developments see Gilks et al. (1996) and references therein; see also the Web addresses given in Appendix C. Recent surveys concerning convergence rates of Markov chains are Diaconis and Saloff-Coste (1995) and Jerrum and Sinclair (1996). Propp and Wilson (1996) showed that it is possible to sample exactly from the steady state distribution of certain types of Markov chain, using coupled Markov chains. Their work has stimulated much interest in algorithms of this type, called *exact sampling* or *perfect sampling.* Appendix C gives the address of a Web page devoted to such techniques.

Appendix C

Information and Software on the World Wide Web

The field of probabilistic networks is developing rapidly, and most of the dissemination of non-academic information and software is via the World Wide Web. Below we provide a selection of sites that currently provide useful material and lists of links, so some exploration should rapidly reveal the current status of these sites and new ones that have been developed. A good starting point is the *Association for Uncertainty in Artificial Intelligence* site. Clearly, no recommendation or guarantee can be given as to the accuracy or suitability of any of the software mentioned.

The sites below were all functioning in April 1999. They are placed in alphabetical order of their URLs.

C.1 Information about probabilistic networks

`http://bayes.stat.washington.edu/almond/belief.html`

Russell Almond's page on software for manipulating belief networks:

Contains an extensive, but not necessarily up-to-date, list of both non-commercial and commercial software, with a glossary and reference list.

`http://http.cs.berkeley.edu/~murphyk/Bayes/bayes.html`

Kevin Murphy's introduction to Bayesian networks:

Contains a review, a good list of recommended reading with links to downloadable versions, and a page of free Bayesian network software (see below).

`http://www.afit.af.mil/Schools/EN/AI/`

US Air Force Institute of Technology Artificial Intelligence Laboratory:

Describes its PESKI software and has a lot of information and links on Bayesian networks. Included are a tutorial and summary of current research, while of particular interest are references to papers describing applications and reports from people in industry who have developed working systems.

`http://www.auai.org/`

Association for Uncertainty in Artificial Intelligence:

Contains links to proceedings of past Uncertainty in Artificial Intelligence conferences, as well as testimonials on the use of Bayesian networks, downloadable tutorials, and an excellent list of links to related sites of organizations, companies, and individuals.

`http://www.cs.auc.dk/research/DSS/`

Decision Support Systems Group at Aalborg University:

Describes their work on methodological development and practical application of Bayesian networks and influence diagrams.

`http://www.maths.nott.ac.uk/hsss/`

Highly Structured Stochastic Systems homepage:

Describes the background and aims of this initiative of the European Science Foundation, which brings together researchers who use stochastic models that exploit conditional independence. It puts research on probabilistic expert systems into a broader context, including Bayesian computation, genetics, environmental modelling, and image analysis.

`http://www.research.microsoft.com/research/dtg/`

Microsoft Research Decision Theory and Adaptive Systems group:

Gives brief description of projects being undertaken in information retrieval, diagnostics and troubleshooting, intelligent user interfaces, and so on, and how this work is being incorporated into Microsoft products. Some publications, particularly on learning models from data, are downloadable. Their free software, Microsoft Belief Networks (MSBN), is available (see next section).

C.2 Software for probabilistic networks

Just a few links are given here — see the above pages for a fuller and more up-to-date guide.

`http://hss.cmu.edu/html/departments/philosophy/TETRAD/tetrad.html`

The TETRAD project homepage:

Provides an overview and download information for the TETRAD software for building causal models from statistical data, and provides a publication list of project participants.

`http://http.cs.Berkeley.edu/~murphyk/Bayes/bnsoft.html`

Kevin Murphy's page of free Bayesian network software:

Gives links to a wide range of free packages, including his own Bayes Net Toolbox, and at the time of writing it is up-to-date.

`http://kmi.open.ac.uk/projects/bkd`

Bayesian Knowledge Discovery Project homepage:

Includes the free Bayesian Knowledge Discover program for automatically constructing discrete Bayesian networks from databases.

`http://www.city.ac.uk/~rgc`

Robert Cowell's home page:

Has a link to the freeware program XBAIES for building chain graph probabilistic and decision networks, available for several platforms.

`http://www.cs.cmu.edu/~javabayes/Home/`

Fabio Cozman's free JavaBayes program:

Includes download information and program details, with further links on Bayesian networks.

`http://www.math.auc.dk/~jhb/CoCo/cocoinfo.html`

Jens Henrik Badsberg's page for his CoCo software:

Describes the software for graphical modelling of discrete variables, but is also a base for links and distribution of various other graphical modelling programs, including MIM, DIGRAM, BIFROST , and GAMES.

`http://www.mrc-bsu.cam.ac.uk/bugs/Welcome.html`

Home page of the BUGS project:

Includes the free WINBUGS program to build Bayesian graphical models. All inferences are by Markov chain Monte Carlo methods.

`http://www2.sis.pitt.edu/~genie/`

The GeNIe homepage:

A freeware program, developed by the Decisions System Laboratory at the University of Pittsburgh, for inference in Bayesian networks by a variety of methods. It can read and save files in most of the formats used by other similar programs. The Decisions System Laboratory also distributes a set of C++ classes, called SMILE, which programmers can use to build their own Bayesian network programs.

Commercial Bayesian network software includes:

- Hugin `http://www.hugin.dk/`
- DXpress `http://www.kic.com/`
- netica `http://www.norsys.com/netica.html`.

These pages contain numerous links to sites of interest.

C.3 Markov chain Monte Carlo methods

`http://dimacs.rutgers.edu/~dbwilson/exact.html`

This page contains a bibliography and pointers to papers concerned with perfect sampling from Markov chains.

`http://www.stats.bris.ac.uk/MCMC/`

Home page of a preprint service for papers concerned with Markov chain Monte Carlo methods.

Bibliography

Aczél, J. (1966). *Lectures on Functional Equations and their Applications.* Academic Press, New York.

Adams, I. D., Chan, M., Clifford, P. C., Cooke, W. M., Dallos, V., de Dombal, F. T., Edwards, M. H., Hancock, D. M., Hewett, D. J., McIntyre, N., Somerville, P. G., Spiegelhalter, D. J., Wellwood, J., and Wilson, D. H. (1986). Computer-aided diagnosis of acute abdominal pain: a multi-centre study. *British Medical Journal*, **292**, 800–4.

Aleliunas, R. (1990). A summary of a new normative theory of probabilistic logic. In *Uncertainty in Artificial Intelligence 4*, (ed. R. D. Shachter, T. S. Levitt, L. N. Kanal, and J. F. Lemmer), pp. 199–206. North-Holland, Amsterdam, The Netherlands.

Almond, R. (1995). *Graphical Belief Modeling.* Chapman and Hall, London, United Kingdon.

Andersen, K. A. and Hooker, J. N. (1994). Bayesian logic. *Decision Support Systems*, **11**, 191–210.

Andersen, L. R., Krebs, J. H., and Andersen, J. D. (1991). STENO: An expert system for medical diagnosis based on graphical models and model search. *Journal of Applied Statistics*, **18**, 139–53.

Andersen, S. K., Olesen, K. G., Jensen, F. V., and Jensen, F. (1989). HUGIN – A shell for building Bayesian belief universes for expert systems. In *Proceedings of the 11th International Joint Conference on Artificial Intelligence*, (ed. N. S. Sridharan), pp. 1080–5. Morgan Kaufmann, San Mateo, California. Also reprinted in Shafer and Pearl (1990).

Andersson, S. A., Madigan, D., and Perlman, M. D. (1996a). An alternative Markov property for chain graphs. In *Proceedings of the 12th Annual Conference on Uncertainty in Artificial Intelligence*, (ed. E. Horvitz and F. V. Jensen), pp. 40–8. Morgan Kaufmann, San Francisco, California.

Andersson, S. A., Madigan, D., and Perlman, M. D. (1996b). A characterization of Markov equivalence classes for acyclic digraphs. *Annals of Statistics*, **25**, 505–41.

Andersson, S. A., Madigan, D., Perlman, M. D., and Richardson, T. (1998). Graphical Markov models in multivariate analysis. In *Multivariate Analysis, Design of Experiments and Survey Sampling*, (ed. S. Ghosh). Marcel Dekker Inc., New York.

Andreassen, S., Jensen, F. V., Andersen, S. K., Falck, B., Kjærulff, U., Woldbye, M., Sørensen, A., Rosenfalck, A., and Jensen, F. (1989). MUNIN — an expert EMG assistant. In *Computer-Aided Electromyography and Expert Systems*, (ed. J. E. Desmedt), pp. 255–77. Elsevier Science Publishers B.V. (North-Holland), Amsterdam, The Netherlands.

Arnborg, S., Corneil, D. G., and Proskurowski, A. (1987). Complexity of finding embeddings in a k-tree. *SIAM Journal on Algebraic and Discrete Methods*, **8**, 277–84.

Bahl, L., Cocke, J., Jelinek, F., and Raviv, J. (1974). Optimal decoding of linear codes for minimizing symbol error rate. *IEEE Transactions on Information Theory*, **20**, 284–7.

Barndorff-Nielsen, O. E. (1978). *Information and Exponential Families in Statistical Theory.* John Wiley and Sons, New York.

Bartlett, M. S. (1935). Contingency table interactions. *Journal of the Royal Statistical Society, Supplement*, **2**, 248–52.

Bauer, E., Koller, D., and Singer, Y. (1997). Update rules for parameter estimation in Bayesian networks. In *Proceedings of the 13th Annual Conference on Uncertainty in Artificial Intelligence*, (ed. D. Geiger and P. Shenoy), pp. 3–13. Morgan Kaufmann, San Francisco, California.

Baum, L. E. (1972). An equality and associated maximization technique in statistical estimation for probabilistic functions of Markov processes. *Inequalities*, **3**, 1–8.

Becker, A. and Geiger, D. (1996). A sufficiently fast algorithm for finding close to optimal junction trees. In *Proceedings of the 12th Annual Conference on Uncertainty in Artificial Intelligence*, (ed. E. Horvitz and F. V. Jensen), pp. 81–9. Morgan Kaufmann, San Francisco, California.

Beeri, C., Fagin, R., Maier, D., Mendelzon, A., Ullman, J., and Yannakakis, M. (1981). Properties of acyclic database schemes. In *Proceedings of the*

13th Annual ACM Symposium on the Theory of Computing, pp. 355–62. Association of Computing Machinery, New York.

Beeri, C., Fagin, R., Maier, D., and Yannakakis, M. (1983). On the desirability of acyclic database schemes. *Journal of the Association of Computing Machinery*, **30**, 479–513.

Berge, C. (1973). *Graphs and Hypergraphs.* North-Holland, Amsterdam, The Netherlands. Translated from French by E. Minieka.

Bernardinelli, L., Pascutto, C., Best, N. G., and Gilks, W. R. (1997). Disease mapping with errors in covariates. *Statistics in Medicine*, **16**, 741–52.

Bernardo, J. M. (1979). Expected information as expected utility. *Annals of Statistics*, **7**, 686–90.

Bernardo, J. M. and Girón, F. J. (1988). A Bayesian analysis of simple mixture problems. In *Bayesian Statistics*, (ed. J. M. Bernardo, M. H. DeGroot, D. V. Lindley, and A. F. M. Smith), pp. 67–78 (with discussion). Clarendon Press, Oxford, United Kingdom.

Bernardo, J. M. and Smith, A. F. M. (1994). *Bayesian Theory.* John Wiley and Sons, Chichester, United Kingdom.

Besag, J. and Green, P. J. (1993). Spatial statistics and Bayesian computation (with discussion). *Journal of the Royal Statistical Society, Series B*, **55**, 25–38.

Besag, J., Green, P. J., Higdon, D., and Mengersen, K. (1995). Bayesian computation and stochastic systems (with discussion). *Statistical Science*, **10**, 3–66.

Best, N. G., Spiegelhalter, D. J., Thomas, A., and Brayne, C. E. G. (1996). Bayesian analysis of realistically complex models. *Journal of the Royal Statistical Society, Series A*, **159**, 323–42.

Bielza, C. and Shenoy, P. P. (1997). A comparison of decision trees, influence diagrams and valuation networks for asymmetric decision problems. In *Preliminary Papers of the 6th International Workshop on Artificial Intelligence and Statistics*, pp. 39–46. Fort Lauderdale, Florida.

Bloemeke, M. and Valtorta, M. (1998). A hybrid algorithm to compute marginal and joint beliefs in Bayesian networks and its complexity. In *Proceedings of the 14th Annual Conference on Uncertainty in Artificial Intelligence*, (ed. G. F. Cooper and S. Moral), pp. 16–23. Morgan Kaufmann, San Francisco California.

Bollen, K. (1988). *Structural Equations with Latent Variables.* John Wiley and Sons, New York.

Box, G. E. P. (1980). Sampling and Bayes inference in scientific modelling and robustness (with discussion). *Journal of the Royal Statistical Society, Series A*, **143**, 383–430.

Box, G. E. P. (1983). An apology for ecumenism in statistics. In *Scientific Inference, Data Analysis and Robustness*, (ed. G. E. P. Box, T. Leonard, and C. F. Wu), pp. 51–84. Academic Press, New York.

Breiman, L., Friedman, J. H., Olshen, R. A., and Stone, C. J. (1984). *Classification and Regression Trees*. Wadsworth, Belmont, California.

Brier, G. W. (1950). Verification of forecasts expressed in terms of probability. *Monthly Weather Review*, **78**, 1–3.

Buntine, W. L. (1994). Operations for learning with graphical models. *Journal of Artificial Intelligence Research*, **2**, 159–225.

Buntine, W. L. (1996). A guide to the literature on learning probabilistic networks from data. *IEEE Transactions on Knowledge and Data Engineering*, **8**, 195–210.

Burnell, L. and Horvitz, E. J. (1995). Structure and chance: melding logic and probability for software debugging. *Communications of the ACM*, **38**, 31–41.

Call, H. J. and Miller, W. A. (1990). A comparison of approaches and implementations for automating decision analysis. *Reliability Engineering and System Safety*, **30**, 115–62.

Cannings, C., Thompson, E. A., and Skolnick, M. H. (1976). Recursive derivation of likelihoods on pedigrees of arbitrary complexity. *Advances in Applied Probability*, **8**, 622–5.

Cannings, C., Thompson, E. A., and Skolnick, M. H. (1978). Probability functions on complex pedigrees. *Advances in Applied Probability*, **10**, 26–61.

Carnap, R. (1950). *Logical Foundations of Probability*. University of Chicago Press, Chicago, Illinois.

Castillo, E., Gutiérrez, J. M., and Hadi, A. S. (1997). *Expert Systems and Probabilistic Network Models*. Springer–Verlag, New York.

Charnes, J. M. and Shenoy, P. P. (1997). A forward Monte Carlo method for solving influence diagrams using local computation. In *Preliminary Papers of the 6th International Workshop on Artificial Intelligence and Statistics*, pp. 75–82. Fort Lauderdale, Florida.

Charniak, E. (1991). Bayesian networks without tears. *AI Magazine*, **12**, 50–63.

Cheeseman, P. and Stutz, J. (1997). Bayesian classification (Autoclass): theory and results. In *Advances in Knowledge Discovery and Data Mining*, (ed. U. Fayyad, G. Pietesky-Shapiro, P. Smyth, and R. Uthurusamy), pp. 153–80. AAAI Press, Menlo Park, California.

Chen, T. and Fienberg, S. E. (1974). Two-dimensional contingency tables with both complete and partially cross-classified data. *Biometrics*, **30**, 629–42.

Chen, T. and Fienberg, S. E. (1976). The analysis of contingency tables with incompletely classified data. *Biometrics*, **32**, 133–44.

Chickering, D. M. (1996). Learning Bayesian networks is NP-complete. In *Learning from Data: Artificial Intelligence and Statistics V*, (ed. D. Fisher and H.-J. Lenz), pp. 121–30. Springer–Verlag, New York.

Chickering, D. M. and Heckerman, D. (1997). Efficient approximations for the marginal likelihood of incomplete data given a Bayesian network. *Machine Learning*, **29**, 181–212.

Chickering, D. M., Heckerman, D., and Meek, C. (1997). A Bayesian approach to learning Bayesian networks with local structure. In *Proceedings of the 13th Annual Conference on Uncertainty in Artificial Intelligence*, (ed. D. Geiger and P. Shenoy), pp. 80–9. Morgan Kaufmann, San Francisco, California.

Chow, C. K. and Liu, C. N. (1968). Approximating discrete probability distributions with dependence trees. *IEEE Transactions On Information Theory*, **14**, 462–7.

Christensen, R. (1990). *Log-Linear Models*. Springer–Verlag, New York.

Consonni, G. and Giudici, P. (1993). Learning in probabilistic expert systems. In *S.I.S. Workshop on Probabilistic Expert Systems, Roma, Oct. 14–15*, (ed. R. Scozzafava), pp. 57–78.

Cooper, G. F. (1990). The computational complexity of probabilistic inference using belief networks. *Artificial Intelligence*, **42**, 393–405.

Cooper, G. F. and Herskovits, E. (1992). A Bayesian method for the induction of probabilistic networks from data. *Machine Learning*, **9**, 309–47.

Covaliu, Z. and Oliver, R. M. (1995). Representation and solution of decision problems using sequential decision diagrams. *Management Science*, **41**, 1860–81.

Cowell, R. G. (1992). Calculating moments of decomposable functions in Bayesian networks. *Research Report* **109**, Department of Statistical Science, University College London, London, United Kingdom.

Cowell, R. G. (1994). Decision networks: a new formulation for multistage decision problems. *Research Report* **132**, Department of Statistical Science, University College London, London, United Kingdon.

Cowell, R. G. (1996). On compatible priors for Bayesian networks. *IEEE Transactions on Pattern Analysis and Machine Intelligence*, **18**, 901–11.

Cowell, R. G. (1997). Sampling without replacement in junction trees. Technical Report **15**, Department of Actuarial Science and Statistics, City University London, London, United Kingdon.

Cowell, R. G. (1998). Mixture reduction via predictive scores. *Statistics and Computing*, **8**, 97–103.

Cowell, R. G. and Dawid, A. P. (1992). Fast retraction of evidence in a probabilistic expert system. *Statistics and Computing*, **2**, 37–40.

Cowell, R. G., Dawid, A. P., Hutchinson, T. A., Roden, S. R., and Spiegelhalter, D. J. (1993a). Bayesian networks for the analysis of drug safety. *The Statistician*, **42**, 369–84.

Cowell, R. G., Dawid, A. P., and Sebastiani, P. (1995). A comparison of sequential learning methods for incomplete data. In *Bayesian Statistics 5*, (ed. J. M. Bernardo, J. Berger, A. P. Dawid, and A. F. M. Smith), pp. 533–41. Clarendon Press, Oxford, United Kingdom.

Cowell, R. G., Dawid, A. P., and Spiegelhalter, D. J. (1993b). Sequential model criticism in probabilistic expert systems. *IEEE Transactions on Pattern Analysis and Machine Intelligence*, **15**, 209–19.

Cowles, M. K. and Carlin, B. P. (1996). Markov chain Monte Carlo convergence diagnostics — a comparative review. *Journal of the American Statistical Association*, **91**, 883–904.

Cox, D. R. and Wermuth, N. (1996). *Multivariate Dependencies.* Chapman and Hall, London, United Kingdom.

Cox, R. T. (1946). Probability, frequency and reasonable expectation. *American Journal of Physics*, **14**, 1–13.

Curds, R. M. (1997). *Propagation Techniques in Probabilistic Expert Systems.* Ph.D. Thesis, Department of Statistical Science, University College London.

Darroch, J. N., Lauritzen, S. L., and Speed, T. P. (1980). Markov fields and log-linear interaction models for contingency tables. *Annals of Statistics*, **8**, 522–39.

Dawid, A. P. (1976). Properties of diagnostic data distributions. *Biometrics*, **32**, 647–58.

Dawid, A. P. (1979). Conditional independence in statistical theory (with discussion). *Journal of the Royal Statistical Society, Series B*, **41**, 1–31.

Dawid, A. P. (1980a). A Bayesian look at nuisance parameters. In *Bayesian Statistics*, (ed. J. M. Bernardo, M. DeGroot, D. V. Lindley, and A. F. M. Smith), pp. 167–84. Valencia University Press, Valencia, Spain.

Dawid, A. P. (1980b). Conditional independence for statistical operations. *Annals of Statistics*, **8**, 598–617.

Dawid, A. P. (1982). Intersubjective statistical models. In *Exchangeability in Probability and Statistics*, (ed. G. Koch and F. Spizzichino), pp. 217–32. North-Holland, Amsterdam, The Netherlands.

Dawid, A. P. (1983). Invariant prior distributions. In *Encyclopedia of Statistical Sciences*, (ed. S. Kotz, N. L. Johnson, and C. B. Read), pp. 228–36. Wiley–Interscience, New York.

Dawid, A. P. (1984). Statistical theory. The prequential approach. *Journal of the Royal Statistical Society, Series A*, **147**, 277–305.

Dawid, A. P. (1985). Invariance and independence in multivariate distribution theory. *Journal of Multivariate Analysis*, **17**, 304–15.

Dawid, A. P. (1986). Probability forecasting. In *Encyclopedia of Statistical Sciences*, (ed. S. Kotz, N. L. Johnson, and C. B. Read), pp. 210–8. Wiley–Interscience, New York.

Dawid, A. P. (1992a). Applications of a general propagation algorithm for probabilistic expert systems. *Statistics and Computing*, **2**, 25–36.

Dawid, A. P. (1992b). Prequential data analysis. In *Current Issues in Statistical Inference: Essays in Honor of D. Basu*, IMS Lecture Notes–Monograph Series, **17**, (ed. M. Ghosh and P. K. Pathak), pp. 113–26. Institute of Mathematical Statistics, Hayward, California.

Dawid, A. P. (1998). Conditional independence. In *Encyclopedia of Statistical Sciences, Update Volume 2*, (ed. S. Kotz, C. B. Read, and D. L. Banks), pp. 146–55. Wiley–Interscience, New York.

Dawid, A. P. and Dickey, J. M. (1977). Likelihood and Bayesian inference from selectively reported data. *Journal of the American Statistical Association*, **72**, 845–50.

Dawid, A. P., Kjærulff, U., and Lauritzen, S. L. (1995). Hybrid propagation in junction trees. In *Advances in Intelligent Computing — IPMU 94*, Lecture Notes in Computer Science, **945**, (ed. B. Bouchon-Meunier, R. R. Yager, and L. A. Zadeh), pp. 87–97. Springer–Verlag, New York.

Dawid, A. P. and Lauritzen, S. L. (1993). Hyper Markov laws in the statistical analysis of decomposable graphical models. *Annals of Statistics*, **21**, 1272–317.

Dawid, A. P. and Skene, A. M. (1979). Maximum likelihood estimation of observer error rates using the EM algorithm. *Applied Statistics*, **28**, 20–8.

Dechter, R. (1996). Bucket elimination: A unifying framework for probabilistic inference. In *Proceedings of the 12th Annual Conference on Uncertainty in Artificial Intelligence*, (ed. E. Horvitz and F. V. Jensen), pp. 211–9. Morgan Kaufmann, San Francisco, California.

DeGroot, M. H. (1970). *Optimal Statistical Decisions*. McGraw–Hill, New York.

Dempster, A. P. (1967). Upper and lower probabilities induced by multivalued mapping. *Annals of Mathematical Statistics*, **28**, 325–39.

Dempster, A. P., Laird, N., and Rubin, D. B. (1977). Maximum likelihood from incomplete data via the EM algorithm. *Journal of the Royal Statistical Society, Series B*, **39**, 1–38 (with discussion).

Diaconis, P. and Saloff-Coste, L. (1995). What do we know about the Metropolis algorithm? In *Proceedings of the 27th Annual ACM Sym-*

posium on the Theory of Computing, pp. 112–29. Association of Computing Machinery, New York.

Didelez, V. and Pigeot, I. (1998). Maximum likelihood estimation in graphical models with missing values. *Biometrika*, **85**, 960–6.

Diestel, R. (1987). Simplicial decompositions of graphs – some uniqueness results. *Journal of Combinatorial Theory, Series B*, **42**, 133–45.

Diestel, R. (1990). *Graph Decompositions.* Clarendon Press, Oxford, United Kingdom.

Diggle, P. and Kenward, M. G. (1994). Informative drop-out in longitudinal data analysis (with discussion). *Applied Statistics*, **43**, 49–93.

Dirac, G. A. (1961). On rigid circuit graphs. *Abhandlungen Mathematisches Seminar Hamburg*, **25**, 71–6.

de Dombal, F. T., Leaper, D. J., Staniland, J. R., McCann, A. P., and Horrocks, J. C. (1972). Computer-aided diagnosis of acute abdominal pain. *British Medical Journal*, **2**, 9–13.

Draper, D. (1995). Assessment and propagation of model uncertainty. *Journal of the Royal Statistical Society, Series B*, **57**, 45–97.

Durbin, R., Eddy, S. R., Krogh, A., and Mitchison, G. (1998). *Biological Sequence Analysis.* Cambridge University Press, Cambridge, United Kingdom.

Edwards, D. (1990). Hierarchical interaction models (with discussion). *Journal of the Royal Statistical Society, Series B*, **52**, 3–20 and 51–72.

Edwards, D. (1995). *Introduction to Graphical Modelling.* Springer–Verlag, New York.

Edwards, D. and Havránek, T. (1985). A fast procedure for model search in multidimensional contingency tables. *Biometrika*, **72**, 339–51.

Edwards, D. and Havránek, T. (1987). A fast model selection procedure for large families of models. *Journal of the American Statistical Association*, **82**, 205–11.

de Finetti, B. (1937). Foresight: its logical laws, its subjective sources. *Annales de l'Institut Henri Poincaré. Probabilités et Statistiques*, **7**, 1–68. Translated by H. E. Kyburg in Kyburg and Smokler (1964), 55–118.

de Finetti, B. (1975). *Theory of Probability (Volumes 1 and 2).* John Wiley and Sons, New York. Italian original 1970.

Franklin, R. C. G., Spiegelhalter, D. J., Macartney, F., and Bull, K. (1991). Evaluation of an algorithm for neonates. *British Medical Journal*, **302**, 935–9.

Frey, B. J. (1998). *Graphical Models for Machine Learning and Digital Communication.* MIT Press, Cambridge, Massachusetts.

Friedman, N. (1998). The Bayesian structural EM algorithm. In *Proceedings of the 14th Annual Conference on Uncertainty in Artificial Intelligence*, (ed. G. F. Cooper and S. Moral), pp. 129–38. Morgan Kaufmann, San Francisco, California.

Friedman, N. and Goldszmidt, M. (1998). Learning Bayesian networks with local structure. In *Learning in Graphical Models*, (ed. M. I. Jordan), pp. 421–60. Kluwer Academic Publishers, Dordrecht, The Netherlands.

Frydenberg, M. (1990). The chain graph Markov property. *Scandinavian Journal of Statistics*, **17**, 333–53.

Fuchs, C. (1982). Maximum likelihood estimation and model selection in contingency tables with missing data. *Journal of the American Statistical Association*, **77**, 270–8.

Fung, R. M. and Del Favero, B. (1995). Applying Bayesian networks to information retrieval. *Communications of the ACM*, **38**, 42–8.

Fung, R. M. and Shachter, R. D. (1990). Contingent influence diagrams. Working paper, Department of Engineering–Economic Systems, Stanford University.

Gabriel, K. R. (1969). Simultaneous test procedures — some theory of multiple comparisons. *Annals of Mathematical Statistics*, **40**, 224–50.

Gammerman, A., Luo, Z., Aitkin, C. G. G., and Brewer, M. J. (1995). Exact and approximate algorithms and their implementations in mixed graphical models. In *Probabilistic Reasoning and Bayesian Belief Networks*, (ed. A. Gammerman), pp. 33–53. Alfred Walker, Henley-on-Thames, United Kingdom.

Gavril, T. (1972). Algorithms for minimum coloring, maximum clique, minimum coloring by cliques and maximum independent set of a graph. *SIAM Journal on Computing*, **1**, 180–7.

Geiger, D. and Heckerman, D. (1994). Learning Gaussian networks. Technical Report MST-TR-94-10, Microsoft Corporation, Advanced Technology Division, Microsoft Corporation, Seattle, Washington.

Geiger, D., Heckerman, D., and Meek, C. (1998). Asymptotic model selection for directed networks with hidden variables. In *Learning in Graphical Models*, (ed. M. I. Jordan), pp. 461–78. Kluwer Academic Publishers, Dordrecht, The Netherlands.

Geiger, D. and Meek, C. (1998). Graphical models and exponential families. In *Proceedings of the 14th Annual Conference on Uncertainty in Artificial Intelligence*, (ed. G. F. Cooper and S. Moral), pp. 156–65. Morgan Kaufmann, San Francisco, California.

Geiger, D. and Pearl, J. (1990). On the logic of causal models. In *Uncertainty in Artificial Intelligence 4*, (ed. R. D. Shachter, T. S. Levitt, L. N. Kanal, and J. F. Lemmer), pp. 3–14. North-Holland, Amsterdam, The Netherlands.

Geiger, D. and Pearl, J. (1993). Logical and algorithmic properties of conditional independence and graphical models. *Annals of Statistics*, **21**, 2001–21.

Gelman, A., Carlin, J. B., Stern, H. S., and Rubin, D. B. (1995). *Bayesian Data Analysis*. Chapman and Hall, London, United Kingdom.

Gelman, A. and Rubin, D. B. (1992). A single series from the Gibbs sampler provides a false sense of security. In *Bayesian Statistics 4*, (ed. J. M. Bernardo, J. O. Berger, A. P. Dawid, and A. F. M. Smith), pp. 625–31. Clarendon Press, Oxford, United Kingdom.

Geman, S. and Geman, D. (1984). Stochastic relaxation, Gibbs distributions, and the Bayesian restoration of images. *IEEE Transactions on Pattern Recognition and Machine Intelligence*, **6**, 721–41.

Geng, Z., Asano, C., Ichimura, M., Tao, F., Wan, K., and Kuroda, M. (1996). Partial imputation method in the EM algorithm. In *Compstat 96*, (ed. A. Prat), pp. 259–63. Physica Verlag, Heidelberg, Germany.

Gilks, W. R., Richardson, S., and Spiegelhalter, D. J. (1996). *Markov Chain Monte Carlo Methods in Practice*. Chapman and Hall, New York.

Gilks, W. R., Thomas, A., and Spiegelhalter, D. J. (1994). A language and program for complex Bayesian modelling. *The Statistician*, **43**, 169–78.

Gillies, D. A. (1994). Philosophies of probability. In *Companion Encyclopedia of the History and Philosophy of the Mathematical Sciences*, (ed. I. Gratton-Guinness), pp. 1407–14. Routledge, London, United Kingdom.

Giudici, P. (1996). Learning in graphical Gaussian models. In *Bayesian Statistics 5*, (ed. J. M. Bernardo, J. O. Berger, A. P. Dawid, and A. F. M. Smith), pp. 621–8. Clarendon Press, Oxford, United Kingdom.

Glymour, C., Scheines, R., Spirtes, P., and Kelley, K. (1987). *Discovering Causal Structure*. Academic Press, New York.

Golumbic, M. C. (1980). *Algorithmic Graph Theory and Perfect Graphs*. Academic Press, London, United Kingdom.

Good, I. J. (1952). Rational decisions. *Journal of the Royal Statistical Society, Series B*, **14**, 107–14.

Goutis, C. (1995). A graphical method for solving a decision analysis problem. *IEEE Transactions on Systems, Man and Cybernetics*, **25**, 1181–93.

Green, P. J. (1990). On use of the EM algorithm for penalized likelihood estimation. *Journal of the Royal Statistical Society, Series B*, **52**, 443–52.

Green, P. J. (1995). Reversible jump MCMC computation and Bayesian model determination. *Biometrika*, **82**, 711–32.

Hammersley, J. M. and Clifford, P. E. (1971). Markov fields on finite graphs and lattices. Unpublished manuscript.

Harary, F. (1972). *Graph Theory.* Addison–Wesley, Reading, Massachusetts.

Heckerman, D. (1986). Probabilistic interpretations for MYCIN's certainty factors. In *Uncertainty in Artificial Intelligence,* (ed. L. N. Kanal and J. F. Lemmer), pp. 167–96. North-Holland, Amsterdam, The Netherlands.

Heckerman, D. (1990). Probabilistic similarity networks. *Networks,* **20**, 607–36.

Heckerman, D. (1998). A tutorial on learning with Bayesian networks. In *Learning in Graphical Models,* (ed. M. I. Jordan), pp. 301–54. Kluwer Academic Publishers, Dordrecht, The Netherlands.

Heckerman, D., Breese, J. S., and Rommelse, K. (1995a). Decision-theoretic troubleshooting. *Communications of the ACM,* **38**, 49–57.

Heckerman, D., Geiger, D., and Chickering, D. M. (1995b). Learning Bayesian networks: The combination of knowledge and statistical data. *Machine Learning,* **20**, 197–243.

Heckerman, D., Horvitz, E. J., and Nathwani, B. (1991). Toward normative expert systems I: The pathfinder project. *Methods of Information in Medicine,* **31**, 90–105.

Heckerman, D. and Wellman, M. P. (1995). Bayesian networks. *Communications of the ACM,* **38**, 27–30.

Henrion, M. (1988). Propagation of uncertainty by probabilistic logic sampling in Bayes' networks. In *Uncertainty in Artificial Intelligence 2,* (ed. J. F. Lemmer and L. N. Kanal), pp. 149–63. North-Holland, Amsterdam, The Netherlands.

Henrion, M., Breese, J. S., and Horvitz, E. J. (1991). Decision analysis and expert systems. *AI Magazine,* **12**, 64–91.

Hocking, R. R. and Oxspring, H. H. (1974). The analysis of partially categorized contingency data. *Biometrics,* **30**, 469–83.

Howard, R. A. (1960). *Dynamic Programming and Markov Processes.* MIT Press, Cambridge, Massachusetts.

Howard, R. A. and Matheson, J. E. (1984). Influence diagrams. In *Readings in the Principles and Applications of Decision Analysis,* (ed. R. A. Howard and J. E. Matheson). Strategic Decisions Group, Menlo Park, California.

Højsgaard, S. and Thiesson, B. (1995). BIFROST — *B*lock recursive models *I*nduced *F*rom *R*elevant knowledge, *O*bservations and *S*tatistical *T*echniques. *Computational Statistics and Data Analysis,* **19**, 155–75.

Jeffreys, H. (1939). *Theory of Probability.* Clarendon Press, Oxford, United Kingdom.

Jensen, C. S., Kong, A., and Kjærulff, U. (1995). Blocking-Gibbs sampling in very large probabilistic expert systems. *International Journal of Human-Computer Studies*, **42**, 647–66.

Jensen, F. and Andersen, S. K. (1990). Approximations in Bayesian belief universes for knowledge-based systems. In *Proceedings of the 6th Conference on Uncertainty in Artificial Intelligence*, pp. 162–9. Cambridge, Massachusetts.

Jensen, F., Jensen, F. V., and Dittmer, S. L. (1994). From influence diagrams to junction trees. In *Proceedings of the 10th Conference on Uncertainty in Artificial Intelligence*, (ed. R. L. de Mantaras and D. Poole), pp. 367–73. Morgan Kaufmann, San Francisco, California.

Jensen, F. V. (1995). Cautious propagation in Bayesian networks. In *Proceedings of the 11th Conference on Uncertainty in Artificial Intelligence*, (ed. P. Besnard and S. Hanks), pp. 323–8. Morgan Kaufmann, San Francisco, California.

Jensen, F. V. (1996). *An Introduction to Bayesian Networks.* University College London Press, London, United Kingdom.

Jensen, F. V. and Jensen, F. (1994). Optimal junction trees. In *Proceedings of the 10th Conference on Uncertainty in Artificial Intelligence*, (ed. R. L. de Mantaras and D. Poole), pp. 360–6. Morgan Kaufmann, San Francisco, California.

Jensen, F. V., Lauritzen, S. L., and Olesen, K. G. (1990a). Bayesian updating in causal probabilistic networks by local computation. *Computational Statistics Quarterly*, **4**, 269–82.

Jensen, F. V., Olesen, K. G., and Andersen, S. K. (1990b). An algebra of Bayesian belief universes for knowledge-based systems. *Networks*, **20**, 637–59.

Jerrum, M. and Sinclair, A. (1996). The Markov chain Monte Carlo method: an approach to approximate counting and integration. In *Approximation Algorithms for NP-hard Problems*, (ed. D. S. Hochbaum). PWS Publishing, Boston, Massachusetts.

Jordan, M. I. (ed.) (1998). *Learning in Graphical Models.* Kluwer Academic Publishers, Dordrecht, The Netherlands.

Kalman, R. E. and Bucy, R. (1961). New results in linear filtering and prediction. *Journal of Basic Engineering*, **83 D**, 95–108.

Kass, R. and Raftery, A. E. (1995). Bayes factors and model uncertainty. *Journal of the American Statistical Association*, **90**, 773–95.

Kearns, M. and Saul, L. (1998). Large deviation methods for approximate probabilistic inference. In *Proceedings of the 14th Annual Conference on Uncertainty in Artificial Intelligence*, (ed. G. F. Cooper and S. Moral), pp. 311–9. Morgan Kaufmann, San Francisco, California.

Kellerer, H. G. (1964a). Maßtheoretische Marginalprobleme. *Mathematische Annalen*, **153**, 168–98.

Kellerer, H. G. (1964b). Verteilungsfunktionen mit gegebenen Marginalverteilungen. *Zeitschrift für Wahrscheinlichkeitstheorie und verwandte Gebiete*, **3**, 247–70.

Keynes, J. M. (1921). *A Treatise on Probability.* Macmillan, London, United Kingdom.

Kim, J. H. and Pearl, J. (1983). A computational model for causal and diagnostic reasoning in inference systems. In *Proceedings of the 8th International Joint Conference on Artificial Intelligence*, (ed. A. Bundy), pp. 190–3. William Kaufmann, Los Altos, California.

Kjærulff, U. (1992). Optimal decomposition of probabilistic networks by simulated annealing. *Statistics and Computing*, **2**, 7–17.

Kjærulff, U. (1993). A computational scheme for reasoning in dynamic probabilistic networks. In *Proceedings of the 9th Conference on Uncertainty in Artificial Intelligence*, (ed. D. Heckerman and A. Mamdani), pp. 121–9. Morgan Kaufmann, San Mateo, California.

Kjærulff, U. (1994). Reduction of computational complexity in Bayesian networks through removal of weak dependences. In *Proceedings of the 10th Conference on Uncertainty in Artificial Intelligence*, (ed. R. L. de Mantaras and D. Poole), pp. 374–82. Morgan Kaufmann, San Francisco, California.

Kjærulff, U. (1995). dHUGIN: a computational system for dynamic time-sliced Bayesian networks. *International Journal of Forecasting*, Special Issue on Probability Forecasting, **11**, 89–111.

Kjærulff, U. (1998). Nested junction trees. In *Learning in Graphical Models*, (ed. M. I. Jordan), pp. 51–74. Kluwer Academic Publishers, Dordrecht, The Netherlands.

Koller, D. and Pfeffer, A. (1997). Object-oriented Bayesian networks. In *Proceedings of the 13th Annual Conference on Uncertainty in Artificial Intelligence*, (ed. D. Geiger and P. Shenoy), pp. 302–13. Morgan Kaufmann, San Francisco, California.

Kolmogorov, A. N. (1950). *Foundations of the Theory of Probability.* Chelsea Press, New York. German original 1933.

Kong, A. (1986). *Multivariate Belief Functions and Graphical Models.* Ph.D. Thesis, Department of Statistics, Harvard University, Massachusetts.

Koster, J. T. A. (1996). Markov properties of non-recursive causal models. *Annals of Statistics*, **24**, 2148–77.

Kreiner, S. (1989). User's guide to DiGram — a program for discrete graphical modelling. Technical Report 89–10, Statistical Research Unit, University of Copenhagen.

Kullback, S. and Leibler, R. A. (1951). On information and sufficiency. *Annals of Mathematical Statistics*, **22**, 79–86.

Kyburg, H. E. and Smokler, H. E. (ed.) (1964). *Studies in Subjective Probability*. John Wiley and Sons, New York.

Lam, W. (1998). Bayesian network refinement via machine learning approach. *IEEE Transactions On Pattern Analysis And Machine Intelligence*, **20**, 240–51.

Lam, W. and Bacchus, F. (1994). Learning belief networks: an approach based on the MDL principle. *Computational Intelligence*, **10**.

Larrañaga, P., Kuijpers, C. M. H., Poza, M., and Murga, R. H. (1997). Decomposing Bayesian networks: Triangulation of the moral graph with genetic algorithms. *Statistics and Computing*, **7**, 19–34.

Lauritzen, S. L. (1981). Time series analysis in 1880: A discussion of contributions made by T. N. Thiele. *International Statistical Review*, **49**, 319–31.

Lauritzen, S. L. (1992). Propagation of probabilities, means and variances in mixed graphical association models. *Journal of the American Statistical Association*, **87**, 1098–108.

Lauritzen, S. L. (1995). The EM algorithm for graphical association models with missing data. *Computational Statistics and Data Analysis*, **19**, 191–201.

Lauritzen, S. L. (1996). *Graphical Models*. Clarendon Press, Oxford, United Kingdom.

Lauritzen, S. L., Dawid, A. P., Larsen, B. N., and Leimer, H.-G. (1990). Independence properties of directed Markov fields. *Networks*, **20**, 491–505.

Lauritzen, S. L. and Jensen, F. V. (1997). Local computation with valuations from a commutative semigroup. *Annals of Mathematics and Artificial Intelligence*, **21**, 51–69.

Lauritzen, S. L., Speed, T. P., and Vijayan, K. (1984). Decomposable graphs and hypergraphs. *Journal of the Australian Mathematical Society, Series A*, **36**, 12–29.

Lauritzen, S. L. and Spiegelhalter, D. J. (1988). Local computations with probabilities on graphical structures and their application to expert systems (with discussion). *Journal of the Royal Statistical Society, Series B*, **50**, 157–224.

Lauritzen, S. L., Thiesson, B., and Spiegelhalter, D. J. (1994). Diagnostic systems created by model selection methods: A case study. In *Selecting Models from Data: AI and Statistics IV*, (ed. P. Cheeseman and R. Oldford), pp. 143–52. Springer–Verlag, New York. Lecture Notes in Statistics 89.

Lauritzen, S. L. and Wermuth, N. (1984). Mixed interaction models. Technical Report R 84-8, Institute for Electronic Systems, Aalborg University.

Lauritzen, S. L. and Wermuth, N. (1989). Graphical models for associations between variables, some of which are qualitative and some quantitative. *Annals of Statistics*, **17**, 31–57.

Lazarsfeld, P. F. and Henry, N. W. (1968). *Latent Structure Analysis.* Houghton-Mifflin, Boston, Massachusetts.

Leimer, H.-G. (1989). Triangulated graphs with marked vertices. *Annals of Discrete Mathematics*, **41**, 311–24.

Leimer, H.-G. (1993). Optimal decomposition by clique separators. *Discrete Mathematics*, **113**, 99–123.

Lindley, D. V. (1957). A statistical paradox. *Biometrika*, **44**, 187–92.

Lindley, D. V. (1982). Scoring rules and the inevitability of probability. *International Statistical Review*, **50**, 1–26.

Lindley, D. V. (1985). *Making Decisions.* John Wiley and Sons, Chichester, United Kingdom.

Lucas, P. and van der Gaag, L. (1991). *Principles of Expert Systems.* Addison–Wesley, Reading, Massachusetts.

Madigan, D. and Raftery, A. E. (1994). Model selection and accounting for model uncertainty in graphical models using Occam's window. *Journal of the American Statistical Association*, **89**, 1535–46.

Madigan, D. and York, I. (1994). Bayesian graphical models for discrete data. *International Statistical Review*, **63**, 215–32.

Madsen, A. L. and Jensen, F. V. (1998). Lazy propagation in junction trees. In *Proceedings of the 14th Annual Conference on Uncertainty in Artificial Intelligence*, (ed. G. F. Cooper and S. Moral), pp. 362–9. Morgan Kaufmann, San Francisco, California.

Maier, D. (1983). *The Theory of Relational Databases.* Computer Science Press, Rockville, Maryland.

McEliece, R. J., MacKay, D. J. C., and Cheng, J.-F. (1998). Turbo decoding as an instance of Pearl's belief propagation algorithm. *IEEE Journal on Selected Areas in Communications*, **16**, 140–52.

Meek, C. (1995). Causal inference and causal explanation with background knowledge. In *Proceedings of the 11th Conference on Uncertainty in Artificial Intelligence*, (ed. P. Besnard and S. Hanks), pp. 403–10. Morgan Kaufmann, San Francisco, California.

Meilă, M. and Jordan, M. I. (1997). Triangulation by continuous embedding. In *Advances in Neural Information Processing Systems 9*, (ed. M. C. Mozer, M. I. Jordan, and T. Petsche), pp. 557–63. MIT Press, Cambridge, Massachusetts.

Miller, R. A., Pople, H. E. J., and Myers, J. D. (1982). Internist-1, an experimental computer-based diagnostic consultant for general internal medicine. *New England Journal of Medicine*, **307**, 468–76.

Monti, S. and Cooper, G. F. (1997). Learning Bayesian belief networks with neural network estimators. In *Advances in Neural Information Processing Systems 9*, (ed. M. C. Mozer, M. I. Jordan, and T. Petsche), pp. 579–84. MIT Press, Cambridge, Massachusetts.

Moussouris, J. (1974). Gibbs and Markov random systems with constraints. *Journal of Statistical Physics*, **10**, 11–33.

Murphy, A. H. and Winkler, R. L. (1977). Reliability of subjective probability forecasts of precipitation and temperature. *Applied Statistics*, **26**, 41–7.

Murphy, A. H. and Winkler, R. L. (1984). Probability forecasting in meteorology. *Journal of the American Statistical Association*, **79**, 489–500.

Ndilikilikesha, P. (1994). Potential influence diagrams. *International Journal of Approximate Reasoning*, **10**, 251–85.

Neal, R. (1996). *Bayesian Learning for Neural Networks.* Springer–Verlag, New York.

Neapolitan, E. (1990). *Probabilistic Reasoning in Expert Systems.* John Wiley and Sons, New York.

Neyman, J. and Pearson, E. S. (1967). *Joint Statistical Papers of J. Neyman and E. S. Pearson.* University of California Press, Berkeley, California.

Nicholson, A. E. and Brady, J. M. (1994). Dynamic belief networks for discrete monitoring. *IEEE Transactions on Systems, Man and Cybernetics*, **24**, 1593–610.

Nilsson, D. (1994). An algorithm for finding the most probable configurations of discrete variables that are specified in probabilistic expert systems. M. Sc. Thesis, Department of Mathematical Statistics, University of Copenhagen, Denmark.

Nilsson, D. (1998). An efficient algorithm for finding the M most probable configurations in a probabilistic expert system. *Statistics and Computing*, **8**, 159–73.

Nilsson, N. J. (1986). Probabilistic logic. *Artificial Intelligence*, **28**, 71–87.

Normand, S.-L. and Tritchler, D. (1992). Parameter updating in a Bayes network. *Journal of the American Statistical Association*, **87**, 1109–15.

Olesen, K. G., Lauritzen, S. L., and Jensen, F. V. (1992). aHUGIN: A system creating adaptive causal probabilistic networks. In *Proceedings of the 8th Conference on Uncertainty in Artificial Intelligence*, (ed. D. Dubois, M. P. Wellman, B. D'Ambrosio, and P. Smets), pp. 223–9. Morgan Kaufmann, San Mateo, California.

Oliver, R. M. and Smith, J. Q. (1990). *Influence Diagrams, Belief Nets and Decision Analysis.* John Wiley and Sons, Chichester, United Kingdom.

Olmsted, S. M. (1983). *On Representing and Solving Decision Problems.* Ph.D. Thesis, Department of Engineering–Economic Systems, Stanford University, Stanford, California.

Parter, S. (1961). The use of linear graphs in Gauss elimination. *SIAM Review*, **3**, 119–30.

Pearl, J. (1982). Reverend Bayes on inference engines: a distributed hierarchical approach. In *Proceedings of American Association for Artificial Intelligence National Conference on AI, Pittsburgh, Pennsylvania*, pp. 133–6.

Pearl, J. (1986a). A constraint–propagation approach to probabilistic reasoning. In *Uncertainty in Artificial Intelligence*, (ed. L. N. Kanal and J. F. Lemmer), pp. 357–70. North-Holland, Amsterdam, The Netherlands.

Pearl, J. (1986b). Fusion, propagation and structuring in belief networks. *Artificial Intelligence*, **29**, 241–88.

Pearl, J. (1987). Evidential reasoning using stochastic simulation. *Artificial Intelligence*, **32**, 245–57.

Pearl, J. (1988). *Probabilistic Inference in Intelligent Systems.* Morgan Kaufmann, San Mateo, California.

Pearl, J. (1995). Causal diagrams for empirical research. *Biometrika*, **82**, 669–710.

Pearl, J. and Paz, A. (1987). Graphoids: A graph based logic for reasoning about relevancy relations. In *Advances in Artificial Intelligence – II*, (ed. B. D. Boulay, D. Hogg, and L. Steel), pp. 357–63. North-Holland, Amsterdam, The Netherlands.

Poland, W. B. (1994). *Decision Analysis with Continuous and Discrete Variables: A Mixture Distribution Approach.* Ph.D. Thesis, Department of Engineering–Economic Systems, Stanford University, Stanford, California.

Popper, K. R. (1959). *The Logic of Scientific Discovery.* Hutchinson, London, United Kingdom. German original 1934.

Propp, J. G. and Wilson, D. B. (1996). Exact sampling with coupled Markov chains and applications to statistical mechanics. *Random Structures and Algorithms*, **9**, 223–52.

Puterman, M. L. (1994). *Markov Decision Processes: Discrete Stochastic Dynamic Programming.* John Wiley and Sons, New York.

Rabiner, L. R. and Juang, B. H. (1993). *Fundamentals of Speech Recognition.* Prentice Hall, Englewood Cliffs, New Jersey.

Raiffa, H. (1968). *Decision Analysis.* Addison–Wesley, Reading, Massachusetts.

Raiffa, H. and Schlaifer, R. (1961). *Applied Statistical Decision Theory.* MIT Press, Cambridge, Massachusetts.

Ramoni, M. and Sebastiani, P. (1997a). Learning Bayesian networks from incomplete databases. In *Proceedings of the 13th Annual Conference on Uncertainty in Artificial Intelligence*, (ed. D. Geiger and P. Shenoy), pp. 401–8. Morgan Kaufmann, San Francisco, California.

Ramoni, M. and Sebastiani, P. (1997b). Robust parameter learning in Bayesian networks with missing data. In *Preliminary Papers of the 6th International Workshop on Artificial Intelligence and Statistics*, pp. 339–406. Fort Lauderdale, Florida.

Ramoni, M. and Sebastiani, P. (1997c). The use of exogenous knowledge to learn Bayesian networks from incomplete databases. In *Advances in Intelligent Data Analysis. Reasoning about Data. Proceedings of the 2nd International Symposium on Intelligent Data Analysis Conference, London*, (ed. X. Liu, P. Cohen, and M. Berthold), pp. 537–48. Springer–Verlag, New York.

Ramsey, F. P. (1926). Truth and probability. In *The Foundation of Mathematics and Other Logical Essays.* Routledge and Kegan Paul, London, United Kingdom. Reprinted in Kyburg and Smokler (1964), pp. 61–92.

Richardson, T. (1996). A polynomial time algorithm for deciding Markov equivalence of directed cyclic graphical models. In *Proceedings of the 12th Annual Conference on Uncertainty in Artificial Intelligence*, (ed. E. Horvitz and F. V. Jensen), pp. 462–9. Morgan Kaufmann, San Francisco, California.

Ripley, B. D. (1996). *Pattern Recognition and Neural Networks.* Cambridge University Press, Cambridge, United Kingdom.

Rissanen, J. (1987). Stochastic complexity. *Journal of the Royal Statistical Society, Series B*, **49**, 223–39.

Riva, A. and Bellazzi, R. (1996). Learning temporal probabilistic causal-models from longitudinal data. *Artificial Intelligence in Medicine*, **8**, 217–34.

Robinson, R. W. (1977). Counting unlabelled acyclic digraphs. In *Lecture Notes in Mathematics: Combinatorial Mathematics V*, (ed. C. H. C. Little). Springer–Verlag, New York.

Rose, D. J. (1970). Triangulated graphs and the elimination process. *Journal of Mathematical Analysis and Applications*, **32**, 597–609.

Rose, D. J. (1973). A graph theoretic study of the numerical solution of sparse positive definite systems of linear equation. In *Graph Theory and Computing*, (ed. R. Read), pp. 183–217. Academic Press, New York.

Rose, D. J., Tarjan, R. E., and Lueker, G. S. (1976). Algorithmic aspects of vertex elimination on graphs. *SIAM Journal on Computing*, **5**, 266–83.

Rubin, D. B. (1976). Inference and missing data. *Biometrika*, **63**, 581–92.

Russell, S. J., Binder, J., Koller, D., and Kanazawa, K. (1995). Local learning in probabilistic networks with hidden variables. In *Proceedings of the 14th International Joint Conference on Artificial Intelligence*, (ed. C. S. Mellish), pp. 1146–52. Morgan Kaufmann, San Mateo, California.

Sagiv, Y. and Walecka, S. F. (1982). Subset dependencies and a completeness result for a subclass of embedded multivalued dependencies. *Journal of the Association of Computing Machinery*, **29**, 103–17.

Sanguesa, R. and Cortes, U. (1997). Learning causal networks from data: a survey and a new algorithm for recovering possibilistic causal networks. *AI Communications*, **10**, 31–61.

Saul, L., Jordan, M. I., and Jaakkola, T. (1996). Mean field theory for sigmoid belief networks. *Journal of Artificial Intelligence Research*, **4**, 61–76.

Savage, L. J. (1954). *The Foundations of Statistics.* John Wiley and Sons, New York.

Savage, L. J. (1971). Elicitation of personal probabilities and expectations. *Journal of the American Statistical Association*, **66**, 783–801.

Sebastiani, P. and Ramoni, M. (1998). Decision theoretic foundations of graphical model selection. In *Proceedings of the 14th Annual Conference on Uncertainty in Artificial Intelligence*, (ed. G. F. Cooper and S. Moral), pp. 464–71. Morgan Kaufmann, San Francisco, California.

Seillier-Moiseiwitsch, F. and Dawid, A. P. (1993). On testing the validity of sequential probability forecasts. *Journal of the American Statistical Association*, **88**, 355–9.

Seroussi, B. and Golmard, J. L. (1992). An algorithm for finding the K most probable configurations in Bayesian networks. *International Journal of Approximate Reasoning*, **1**, 205–33.

Settimi, R. and Smith, J. Q. (1998). On the geometry of Bayesian graphical models with hidden variables. In *Proceedings of the 14th Annual Conference on Uncertainty in Artificial Intelligence*, (ed. G. F. Cooper and S. Moral). Morgan Kaufmann, San Francisco, California.

Shachter, R. D. (1986). Evaluating influence diagrams. *Operations Research*, **34**, 871–82.

Shachter, R. D. (1998). Bayes-ball: The rational pastime (for determining irrelevance and requisite information in belief networks and influence diagrams). In *Proceedings of the 14th Annual Conference on Uncertainty in Artificial Intelligence*, (ed. G. F. Cooper and S. Moral), pp. 480–7. Morgan Kaufmann, San Francisco, California.

Shachter, R. D., Andersen, S. K., and Szolovits, P. (1994). Global conditioning for probabilistic inference in belief networks. In *Proceedings of the 10th Conference on Uncertainty in Artificial Intelligence*, (ed. R. L. de Mantaras and D. Poole), pp. 514–21. Morgan Kaufmann, San Francisco, California.

Shachter, R. D. and Kenley, C. (1989). Gaussian influence diagrams. *Management Science*, **35**, 527–50.

Shachter, R. D. and Ndilikilikesha, P. (1993). Using potential influence diagrams for probabilistic inference and decision making. In *Proceedings of the 9th Conference on Uncertainty in Artificial Intelligence*, (ed. D. Heckerman and A. Mamdani), pp. 383–90. Morgan Kaufmann, San Mateo, California.

Shachter, R. D. and Peot, M. A. (1990). Simulation approaches to general probabilistic inference on belief networks. In *Uncertainty in Artificial Intelligence 5*, (ed. M. Henrion, R. D. Shachter, L. N. Kanal, and J. F. Lemmer), pp. 221–31. North-Holland, Amsterdam, The Netherlands.

Shafer, G. R. (1976). *A Mathematical Theory of Evidence.* Princeton University Press, Princeton, New Jersey.

Shafer, G. R. (1996). *Probabilistic Expert Systems.* Society for Industrial and Applied Mathematics, Philadelphia, Pennsylvania.

Shafer, G. R. and Pearl, J. (eds.) (1990). *Readings in Uncertain Reasoning.* Morgan Kaufmann, San Mateo, California.

Shafer, G. R. and Shenoy, P. P. (1988). Local computation in hypertrees. Working Paper 201, School of Business, University of Kansas, Kansas.

Shafer, G. R. and Shenoy, P. P. (1990). Probability propagation. *Annals of Mathematics and Artificial Intelligence*, **2**, 327–52.

Sham, P. (1998). *Statistics in Human Genetics.* Edward Arnold, London, United Kingdon.

Shenoy, P. P. (1991). Valuation-based systems for discrete optimization. In *Uncertainty in Artificial Intelligence 6*, (ed. P. P. Bonissone, M. Henrion, L. N. Kanal, and J. F. Lemmer), pp. 385–400. North-Holland, Amsterdam, The Netherlands.

Shenoy, P. P. (1992). Valuation-based systems for Bayesian decision analysis. *Operations Research*, **40**, 463–84.

Shenoy, P. P. (1994). A comparison of graphical techniques for decision analysis. *European Journal of Operational Research*, **78**, 1–21.

Shenoy, P. P. (1996). Representing and solving asymmetric decision problems using valuation networks. In *Learning from Data: Artificial Intelligence and Statistics V*, (ed. D. Fisher and H.-J. Lenz), pp. 99–108. Springer–Verlag, New York.

Shenoy, P. P. (1997). Binary join trees for computing marginals in the Shenoy–Shafer architecture. *International Journal of Approximate Reasoning*, **17**, 239–63.

Shenoy, P. P. and Shafer, G. R. (1986). Propagating belief functions using local propagation. *IEEE Expert*, **1**, 43–52.

Shenoy, P. P. and Shafer, G. R. (1990). Axioms for probability and belief–function propagation. In *Uncertainty in Artificial Intelligence 4*, (ed.

R. D. Shachter, T. S. Levitt, L. N. Kanal, and J. F. Lemmer), pp. 169–98. North-Holland, Amsterdam, The Netherlands.

Shortliffe, E. H. and Buchanan, B. G. (1975). A model for inexact reasoning in medicine. *Mathematical Biosciences*, **23**, 351–79.

Shwe, M., Middleton, B., Heckerman, D., Henrion, M., Horvitz, E. J., and Lehmann, H. (1991). Probabilistic diagnosis using a reformulation of the INTERNIST-1 / QMR knowledge base I: The probabilistic model and inference algorithms. *Methods of Information in Medicine*, **30**, 241–55.

Smith, A. F. M. and Makov, U. E. (1978). A quasi-Bayes sequential procedure for mixtures. *Journal of the Royal Statistical Society, Series B*, **40**, 106–11.

Smith, J. Q. (1979). A generalisation of the Bayesian steady forecasting model. *Journal of the Royal Statistical Society, Series B*, **41**, 375–87.

Smith, J. Q. (1981). The multiparameter steady model. *Journal of the Royal Statistical Society, Series B*, **43**, 256–60.

Smith, J. Q. (1989a). Influence diagrams for Bayesian decision analysis. *European Journal of Operational Research*, **40**, 363–76.

Smith, J. Q. (1989b). Influence diagrams for statistical modelling. *Annals of Statistics*, **7**, 654–72.

Smyth, P., Heckerman, D., and Jordan, M. I. (1997). Probabilistic independence networks for hidden Markov probability models. *Neural Computation*, **9**, 227–69.

Spiegelhalter, D. J. (1990). Fast algorithms for probabilistic reasoning in influence diagrams, with applications in genetics and expert systems. In *Influence Diagrams, Belief Nets and Decision Analysis*, (ed. R. M. Oliver and J. Q. Smith), pp. 361–84. John Wiley and Sons, Chichester, United Kingdom.

Spiegelhalter, D. J. and Cowell, R. G. (1992). Learning in probabilistic expert systems. In *Bayesian Statistics 4*, (ed. J. M. Bernardo, J. O. Berger, A. P. Dawid, and A. F. M. Smith), pp. 447–65. Clarendon Press, Oxford, United Kingdom.

Spiegelhalter, D. J., Dawid, A. P., Lauritzen, S. L., and Cowell, R. G. (1993). Bayesian analysis in expert systems (with discussion). *Statistical Science*, **8**, 219–83.

Spiegelhalter, D. J., Harris, N. L., Bull, K., and Franklin, R. C. G. (1994). Empirical evaluation of prior beliefs about frequencies: methodology and a case study in congenital heart disease. *Journal of the American Statistical Association*, **89**, 435–43.

Spiegelhalter, D. J. and Lauritzen, S. L. (1990). Sequential updating of conditional probabilities on directed graphical structures. *Networks*, **20**, 579–605.

Spirtes, P., Glymour, C., and Scheines, R. (1993). *Causality, Prediction and Search.* Springer–Verlag, New York.

Spirtes, P., Meek, C., and Richardson, T. (1995). Causal inference in the presence of latent variables and selection bias. In *Proceedings of the 11th Conference on Uncertainty in Artificial Intelligence*, (ed. P. Besnard and S. Hanks), pp. 499–506. Morgan Kaufmann, San Francisco, California.

Spirtes, P. and Richardson, T. (1997). A polynomial-time algorithm for determining DAG equivalence in the presence of latent variables and selection bias. In *Preliminary Papers of the 6th International Workshop on Artificial Intelligence and Statistics*, pp. 489–500. Fort Lauderdale, Florida.

Spohn, W. (1988). Ordinal conditional functions: a dynamic theory of epistemic states. In *Causation in Decision, Belief Change, and Statistics*, (ed. W. L. Harper and B. Skyrms), pp. 105–34. Kluwer Academic Publishers, Dordrecht, The Netherlands.

Stone, M. (1974). Cross-validatory choice and assessment of statistical predictions (with discussion). *Journal of the Royal Statistical Society, Series B*, **36**, 111–47.

Studený, M. (1995). Conditional independence and natural conditional functions. *International Journal of Approximate Reasoning*, **12**, 934–45.

Studený, M. (1997). Comparison of graphical approaches to description of conditional independence structures. In *Proceedings of WUPES'97, January 22–25, 1997, Prague, Czech Republic*, pp. 156–72.

Studený, M. and Bouckaert, R. R. (1998). On chain graph models for description of conditional independence structures. *Annals of Statistics*, **26**, 1434–95.

Sy, B. K. (1993). A recurrence local computation approach towards ordering composite beliefs in Bayesian networks. *International Journal of Approximate Reasoning*, **8**, 17–50.

Tarjan, R. E. (1985). Decomposition by clique separators. *Discrete Mathematics*, **55**, 221–32.

Tarjan, R. E. and Yannakakis, M. (1984). Simple linear-time algorithms to test chordality of graphs, test acyclicity of hypergraphs, and selectively reduce acyclic hypergraphs. *SIAM Journal on Computing*, **13**, 566–79.

Tatman, J. A. and Shachter, R. D. (1990). Dynamic programming and influence diagrams. *IEEE Transactions on Systems, Man, and Cybernetics*, **20**, 365–79.

Thiele, T. N. (1880). Om Anvendelse af mindste Kvadraters Methode i nogle Tilfælde, hvor en Komplikation af visse Slags uensartede tilfældige Fejlkilder giver Fejlene en 'systematisk' Karakter. *Vidensk.*

Selsk. Skr. 5. Rk., naturvid. og mat. Afd., **12**, 381–408. French version: *Sur la Compensation de quelques Erreurs quasi-systématiques par la Méthode des moindres Carrés.* Reitzel, København, 1880.

Thiesson, B. (1991). (G)EM algorithms for maximum likelihood in recursive graphical association models. M. Sc. Thesis, Department of Mathematics and Computer Science, Aalborg University.

Thiesson, B. (1995). Accelerated quantification of Bayesian networks with incomplete data. In *Proceedings of the First International Conference on Knowledge Discovery and Data Mining*, (ed. U. M. Fayyad and R. Uthurusamy), pp. 306–11. AAAI Press, Menlo Park, California.

Thiesson, B. (1997). Score and information for recursive exponential models with incomplete data. In *Proceedings of the 13th Annual Conference on Uncertainty in Artificial Intelligence*, (ed. D. Geiger and P. Shenoy), pp. 453–63. Morgan Kaufmann, San Francisco, California.

Thiesson, B., Meek, C., Chickering, D. M., and Heckerman, D. (1998). Learning mixtures of DAG models. In *Proceedings of the 14th Annual Conference on Uncertainty in Artificial Intelligence*, (ed. G. F. Cooper and S. Moral), pp. 504–14. Morgan Kaufmann, San Francisco, California.

Titterington, D. M. (1976). Updating a diagnostic system using unconfirmed cases. *Applied Statistics*, **25**, 238–47.

Titterington, D. M., Murray, G. D., Murray, L. S., Spiegelhalter, D. J., Skene, A. M., Habbema, J. D. F., and Gelpke, G. J. (1981). Comparison of discrimination techniques applied to a complex data-set of head-injured patients (with discussion). *Journal of the Royal Statistical Society, Series A*, **144**, 145–75.

Titterington, D. M., Smith, A. F. M., and Makov, U. E. (1985). *Statistical Analysis of Finite Mixture Distributions.* John Wiley and Sons, Chichester, United Kingdom.

van der Gaag, L. (1991). Computing probability intervals under independency constraints. In *Uncertainty in Artificial Intelligence 6*, (ed. P. P. Bonissone, M. Henrion, L. N. Kanal, and J. F. Lemmer), pp. 457–66. North-Holland, Amsterdam, The Netherlands.

Venn, J. (1866). *The Logic of Chance.* Macmillan, London, United Kingdom.

Verma, T. and Pearl, J. (1990). Causal networks: Semantics and expressiveness. In *Uncertainty in Artificial Intelligence 4*, (ed. R. D. Shachter, T. S. Levitt, L. N. Kanal, and J. F. Lemmer), pp. 69–76. North-Holland, Amsterdam, The Netherlands.

Verma, T. and Pearl, J. (1991). Equivalence and synthesis of causal models. In *Uncertainty in Artificial Intelligence 6*, (ed. P. P. Bonissone, M. Henrion, L. N. Kanal, and J. F. Lemmer), pp. 255–68. North-Holland, Amsterdam, The Netherlands.

Viterbi, A. J. (1967). Error bounds for convolutional codes and an asymptotically optimum decoding algorithm. *IEEE Transactions on Information Theory*, **13**, 260–9.

von Mises, R. (1939). *Probability, Statistics and Truth.* Hodge, London, United Kingdom. German original 1928.

von Neumann, J. and Morgenstern, O. (1944). *Theory of Games and Economic Behaviour.* Princeton University Press, Princeton, New Jersey.

Vorob'ev, N. N. (1962). Consistent families of measures and their extensions. *Theory of Probability and its Applications*, **7**, 147–63.

Vorob'ev, N. N. (1963). Markov measures and Markov extensions. *Theory of Probability and its Applications*, **8**, 420–9.

Vorob'ev, N. N. (1967). Coalition games. *Theory of Probability and its Applications*, **12**, 250–66.

Wagner, K. (1937). Über eine Eigenschaft der ebenen Komplexe. *Mathematische Annalen*, **114**, 570–90.

Walley, P. (1990). *Statistical Reasoning with Imprecise Probabilities.* Chapman and Hall, London, United Kingdom.

Warner, H. R., Toronto, A. F., Veasey, L. G., and Stephenson, R. (1961). A mathematical approach to medical diagnosis — application to congenital heart disease. *Journal of the American Medical Association*, **177**, 177–84.

Wermuth, N. (1976). Model search among multiplicative models. *Biometrics*, **32**, 253–63.

Wermuth, N. and Lauritzen, S. L. (1990). On substantive research hypotheses, conditional independence graphs and graphical chain models (with discussion). *Journal of the Royal Statistical Society, Series B*, **52**, 21–72.

West, M. (1992). Modelling with mixtures. In *Bayesian Statistics 4*, (ed. J. M. Bernardo, J. O. Berger, A. P. Dawid, and A. F. M. Smith), pp. 503–524 (with discussion). Clarendon Press, Oxford, United Kingdom.

West, M. and Harrison, P. J. (1989). *Bayesian Forecasting and Dynamic Models.* Springer–Verlag, New York.

Whittaker, J. (1990). *Graphical Models in Applied Multivariate Statistics.* John Wiley and Sons, Chichester, United Kingdom.

Winston, P. H. (1984). *Artificial Intelligence*, (2 edn). Addison–Wesley, Reading, Massachusetts.

Wright, S. (1921). Correlation and causation. *Journal of Agricultural Research*, **20**, 557–85.

Wright, S. (1923). The theory of path coefficients: a reply to Niles' criticism. *Genetics*, **8**, 239–55.

Wright, S. (1934). The method of path coefficients. *Annals of Mathematical Statistics*, **5**, 161–215.

Yannakakis, M. (1981). Computing the minimum fill-in is NP-complete. *SIAM Journal on Algebraic and Discrete Methods*, **2**, 77–9.

Zadeh, L. A. (1983). The role of fuzzy logic in the management of uncertainty in expert systems. *Fuzzy Sets and Systems*, **11**, 199–228.

Zhang, N. L. (1996). Irrelevance and parameter learning in Bayesian networks. *Artificial Intelligence*, **88**, 359–73.

Author Index

Subject Index